Elementary Concepts of
Power Electronic Drives

Elementary Concepts of Power Electronic Drives

K Sundareswaran, PhD

CRC Press
Taylor & Francis Group
Boca Raton London New York

CRC Press is an imprint of the
Taylor & Francis Group, an **informa** business

CRC Press
Taylor & Francis Group
6000 Broken Sound Parkway NW, Suite 300
Boca Raton, FL 33487-2742

© 2019 by Taylor & Francis Group, LLC
CRC Press is an imprint of Taylor & Francis Group, an Informa business

No claim to original U.S. Government works

Printed on acid-free paper

International Standard Book Number-13 978-1-138-39049-2 (Hardback)

Library of Congress Cataloging-in-Publication Data

Library of Congress Cataloging-in-Publication Data
Names: Sundareswaran, K. (Kinattingal), 1966- author.
Title: Elementary concepts of power electronic drives / authored by Sundareswaran K.
Description: Boca Raton : Taylor & Francis, 2019.
Identifiers: LCCN 2018045336| ISBN 9781138390492 (hardback : alk. paper) |
ISBN 9780429423284 (e-book)
Subjects: LCSH: Power electronics. | Electric current converters. | Switching power supplies.
Classification: LCC TK7881.15 .S87 2019 | DDC 621.31/042—dc23
LC record available at https://lccn.loc.gov/2018045336

Visit the Taylor & Francis Web site at
http://www.taylorandfrancis.com

and the CRC Press Web site at
http://www.crcpress.com

eResource material is available for this title at https://www.crcpress.com/9781138390492.

This book is submitted to

the lotus feet of Uchi Pillayar, Rockfort Temple, Tiruchirappalli.

Contents

Preface

Power electronics is a specialized area in electrical power engineering that focuses on conversion and control of electrical power. Various applications of power electronics include induction heating, illumination control, regulated power supplies, flexible ac transmission (FACT) systems, variable speed drives, electrochemical process, electric-welding, and many more. This book is concerned with variable speed dc and ac motors using power electronic circuits. This subject matter of this book is a core subject in all universities with different names, such as power electronic drives, solid state drives, industrial drives, etc. The contents of this book are suitably tailored to provide first-hand information on relevant power electronic circuits and operation and their application to respective motor drives. This will enable the student community to learn the variable-speed concept hassle-free. Several numerical examples—both illustrated and unsolved—are provided for better understanding of the content. This book will serve both undergraduate and graduate students studying the field of power electronic drives.

The author expresses sincere thanks to Dr. S. Palani, former Professor and Head of Instrumentation and Control Engineering, Regional Engineering College (now known as the National Institute of Technology), Tiruchirappalli, Tamil Nadu, India, for constant encouragement, useful discussion, and constructive criticism during the preparation of this book.

Author

K Sundareswaran, PhD, was born in Pallassana, Kerala, India, in 1966. He earned a BTech (Hons.) in electrical and electronics engineering and MTech (Hons.) in power electronics at the University of Calicut, Calicut, Kerala, India, in 1988 and 1991, respectively. He earned a PhD in electrical engineering at Bharathidasan University, Tiruchirappalli, Tamil Nadu, India, in 2001. From 2005 to 2006, he was a Professor with the Department of Electrical Engineering, National Institute of Technology, Calicut, Kerala, India. He is currently a Professor with the Department of Electrical and Electronics Engineering, National Institute of Technology, Tiruchirappalli, Tamil Nadu, India. His research interests include power electronics, renewable energy systems, and biologically inspired optimization techniques.

List of Symbols

B	friction constant in N-m/rad/s
ω_r	rotor speed in rad/s
J	moment of inertia in kg-m^2
T_0, T_{L1}, T_{L2}, T_L	load torque in N-m
T_e	electromagnetic torque in N-m
N_s	synchronous speed in rpm
N_r	rotor speed in rpm
N_r'	load side speed in rpm
Vp	speed in m/s in chapter 1
F	force in N
a	acceleration in m/s^2
R_p	radius in meter in chapter 1
S	slip
NT_1, NT_2	number of teeth on wheel
Δt	change in time in sec
$\Delta\omega$	change in speed in rad/s
T_L'	load side torque in N-m
T_{m1}, T_{m2}	motor torque in N-m
ω_r'	load side speed in rad/s
$\omega_{r0}, \omega_{r1}, \omega_{r2}$	speed in rad/s
J_1	load inertia kg-m^2
J_m	moment of inertia of motor in kg-m^2
d_1, d_2	diameter in meters
T_{acc}	acceleration torque in N-m
V_{dc}	source voltage in volts
R	resistance in ohms
L	inductance in H
t	time in seconds
C	capacitance in F
V_m	peak value of sinusoidal voltage in volts
ω_s	angular frequency in rad/s
X_L	inductive reactance in ohms
X_c	capacitive reactance in ohms
Z	impedance in ohms
E_b	back emf/battery voltage in volts
α	SCR firing angle
ϕ	power factor angle
β	extinction angle
V_{AK}	anode to cathode voltage in SCR in volts
I_{AK}	anode to cathode current in SCR in amperes
V_{BO}	break down voltage in volts
I_A	anode current in amperes
V_A	anode voltage in volts
V_{GS}	gate to source voltage in volts

V_{DS}	drain to source voltage in volts
R_{DS}	drain to source resistance in ohms
I_D	drain current in amperes
R_2	resistance of n-layer in ohms
R_1	resistance of p-layer in ohms
i_o	instantaneous output current in amperes
v_o	instantaneous output voltage in volta
I_o	load current in Amperes at $\omega_s t = 0$
$I_{o(rms)}$	rms value of load current in amperes
I_π	current in amperes at $\omega_s t = \pi$
$I_{o(av)}$	average value of load current in amperes
η	angle in radians in chapter 2
$V_{o(av)}$	average output voltage in volts
v_s	source voltage in volts
I_a	average value of armature current in amperes
I_{min}	minimum value of load current in amperes
V_R, V_Y, V_B, V_{ph}	phase voltage in volts
v_R, v_Y, v_B	instantaneous value of phase voltages
V_{RY}, V_{YB}, V_{BR}	line voltages in volts
V_{Lm}	maximum line voltage in volts
i_g	gate current in amperes
i_s	instantaneous Source current in amperes
I_{Srms}	rms value of source current in amperes
V_{rms}	rms value of source voltage in volts
V_f	field voltage in volts
I_f	field current in ampere
Φ_m	field flux in weber
P_a	number of poles
A	parallel path
Z_a	total number of armature conductors
K_b	back emf constant V/rad/s
r_a	armature resistance in ohms
L_a	armature inductance in H
Z_a	armature impedance in ohms
i_a	instantaneous armature current in amperes
v_a	armature voltage in volts
K_p	proportional constant
K_I	integral constant
K_f	series field constant of dc series motor
K_T	torque constant of dc series motor
f_1	source frequency in Hertz
e_b	instantaneously back emf
$G(s)$	dc motor transfer function
$G_c(s)$	controller transfer function
I_1	fundamental rms current in amperes
ϕ_1	phase angle between source voltage and fundamental current
I_h	rms value of harmonic current in amperes
ϕ_n	n^{th} harmonic power factor angle

I_{dc}	average value of source current in amperes
v_s	instantaneous source voltage in volts
DF	displacement factor
THD	total harmonic distortion
a_n, b_n	Fourier coefficients
T	time period in seconds
T_{on}	ON period in seconds
ω_{LC}	angular frequency in rad/s in commutating circuit
D	duty ratio
I_{max}	maximum current in amperes
I_{min}	minimum current in amperes
i_c	instantaneous capacitor current in amperes
v_c	instantaneous capacitor voltage in volts
V_{dc}	dc source voltage in volts
i_L	inductor current in amperes
v_L	inductor voltage in volts
t_x	time in seconds in chapter 6
V_o	constant output voltage in volts
R_b	braking resistance in ohms
T_b	braking torque in N-m
R_D	diverting path resistance in ohms
V_{orms}	rms value of output voltage in volts
v_{ds}	stator direct axis winding voltage in volts
v_{qs}	stator quadrature axis winding voltage in volts
v_{dr}	rotor direct axis winding voltage in volts
v_{qr}	rotor quadrature axis winding voltage in volts
θ_r	angular displacement between stator and rotor axis
ψ	flux linkage
R_s	stator resistance in ohms
R_r	rotor resistance in ohms
L_m	mutual inductance in Hertz
L_r	rotor inductance in Hertz
L_s	stator inductance in Hertz
i_{ds}	stator direct axis current in amperes
i_{qs}	stator quadrature axis current in amperes
i_{dr}	rotor direct axis current in amperes
i_{qs}	rotor quadrature axis current in amperes
V_s	stator voltage in volts
I_s	stator current in amperes
I_r	rotor current in amperes
L_{ls}	leakage inductance of stator in H
L_{ls}	leakage inductance of rotor in H
v_{qs}^s	fixed frame quadrature axis voltage
v_{ds}^s	fixed frame quadrature axis voltage
v_{qs}^a	arbitrary frame quadrature axis voltage
v_{ds}^a	arbitrary frame quadrature axis voltage
ψ_{qs}^a	quadrature axis flux linkage for arbitrary stator reference frame
ψ_{ds}^a	direct axis flux linkage for arbitrary stator reference frame
ψ_{qr}^a	quadrature axis flux linkage for arbitrary rotor reference frame

ψ_{qr}^{a}	direct axis flux linkage for arbitrary rotor reference frame
X_{ls}	leakage reactance of stator in ohms
X_{lr}	leakage reactance of rotor in ohms
X_m	mutual reactance in ohms
m_1	number of phases
P_{ag}	air gap power in watts
P_{cu}	copper loss in watts
P_{mech}	mechanical power in watts
f_1	stator frequency in Hertz
p_1	number of pair of poles
T_s	starting torque in N-m
T_{max}	maximum torque in N-m
S_n	n^{th} harmonic slip
I_n	n^{th} harmonic current in amperes
v_n	n^{th} harmonic voltage in volts
S_{FL}	full load slip
I_{smax}	maximum stator current in amperes
I_{sFL}	full load stator current in ampere
V_L	line voltage in volts
N_T	transformer turns ratio
N	ratio of stator turns to rotor turns
V_{RN}, V_{YN}, V_{BN}	phase voltages in volts
r	resistance in ohms
E_1	stator-induced emf in volts
ω_{sl}	slip speed in rad/s
$V_{s(new)}, V_{s(old)}$	voltage in volts
θ_a	angular displacement between stator reference frame and arbitrary reference frame in degree
ω_a	synchronous speed of arbitrary reference frame
I_r	rotor current in amperes
I_m	magnetizing current in amperes
$I_{qs(syn)}, I_{ds(syn)}$	q and d-axis currents in synchronously rotating reference frame in amperes
V_f	rms value of excitation emf in volts
e_f	instantaneous value of excitation emf in volts
P_i	input power in watts
δ	torque angle in degrees

1

Components of Power Electronic Drive

1.1 Introduction

Mechanical power has been utilized in several areas, including farming, household works, transportation, and various industrial processes. In the ancient days, animals were used in many of these areas, such as farming and transportation, often in conjunction with human labor. Then steam engines were developed, followed by the combustion engine and the use of fuels such as diesel, petrol, etc. This led to larger industrialization as well as new transport methods. With the advent of electricity, most of the mechanical power development is now carried out using electric motors. Motor pumps in the agricultural fields, wet grinders in the kitchen, washing machines, and various industrial drives are but a few examples. The automobile industry is now moving toward electric vehicles, as are other industries, such as transportation systems, pumps and fans, paper mills, textile mills, robotic applications, lifts and elevators, and so on.

All of these applications require mechanical power, and suitable speeds of this mechanical power are mandatory for different processes. For example, the household ceiling fan needs to be able to run at low speed during winter, but at high speed in summer. Washing of clothes occurs at low speed, but spinning the clothes to drain water is done at high speed. Thus, mechanical power needs to be controlled as and when required in various operations. In this context, a piece of machine equipment that converts electrical energy to mechanical energy and provides complete control over the process may be defined as an electric drive.

It is evident that an electric motor is an essential unit in an electric drive. It is also obvious that a power source feeds the motor. Because power control is required, a power electronic converter appears between the source and motor. In the case of closed-loop control, a dedicated controller is an integral part of any electric drive. The basic block diagram of a power electronic converter fed motor drive system is shown in Fig. 1.1, and the major components of this block diagram are discussed below.

1.1.1 Power Source

In most cases, the power source is the existing single-phase or three-phase alternating voltage at line frequency. The source is a dc supply in the case of dc traction, battery-powered vehicles, and solar systems. Renewable sources such as wind and solar power, which are utilized in several applications, are climate dependent and hence stochastic in nature.

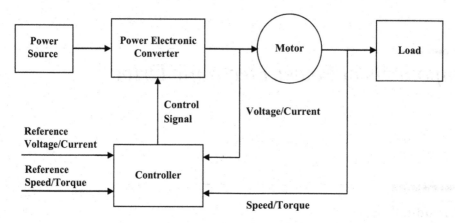

FIGURE 1.1
Schematic of power electronic converter–fed motor drive system.

1.1.2 Power Electronic Converter

There are four basic types of power electronic converters:

a. **ac-dc converters:** These circuits are also known as controlled rectifiers and have numerous topologies, such as half-wave, full-wave, single-phase, three-phase, etc. The input supply is from either single-phase or three-phase, and the output is a controlled dc voltage. Such dc motors are connected at the output for variable-speed operation. These converters also serve as the front-end supply to inverters and dc-dc converters.

b. **dc-dc converters:** Popularly known as chopper circuits, these types of converters are supplied from a constant dc source and provide regulated dc voltage at the output. These converters allow for dc motor speed control.

c. **ac-ac converters:** In this topology, there are two subdivisions:

   i. **ac voltage regulators:** Here, as alternating voltage of fixed frequency, the line voltage is converted to variable-amplitude voltage without a change in frequency. Starting large induction motors is generally carried out with these converters, as is the speed control of single-phase and three-phase induction motors with fan and pump loads.

   ii. **Cyclo converters:** These circuits directly convert existing ac voltage (constant voltage and frequency) into variable voltage, variable frequency at the output. Generally, output frequency is limited to 33–50% of line frequency to avoid excess harmonic distortion. These converters are used for low-speed operation of large-capacity induction motors.

d. **dc-ac converters:** These are the most commonly used converters in the present era. These converters change the input dc voltage into variable frequency, variable voltage/current at the output. There are two types, namely voltage source inverters and current source inverters, and they are used to drive variable-speed induction motors and synchronous motors.

1.1.3 Electric Motor

There are three types of electric motors:

a. dc motors, such as shunt motors, series motors, compound motors, and separately excited dc motors

b. ac motors, such as induction (asynchronous) motors, wound-rotor motors, and synchronous motors

c. Special motors, such as switched reluctance motors

1.2 Motor Load System

1.2.1 The Mechanical System

In power electronic drives, load is sometimes referred as the mechanical system, where mechanical power is developed and utilized. The load is characterized by its speed and torque. The torque component of mechanical system consists of the following:

a. Friction: This occurs between moving and fixed parts of system and can be approximated as $B\omega_r$, where B is the friction constant and has the unit of N-m/rad/s, and ω_r is the rotor speed in rad/s.

b. Windage: This is used to agitate or pump the air surrounding the moving parts of the system. This torque is proportional to the square of speed; however, in all practical cases, this torque can be neglected.

c. Acceleration: This torque is used in transient conditions to overcome the mechanical inertia of the system. Acceleration torque is $J\dfrac{d\omega_r}{dt}$, where J is the moment of inertia in kg-m^2.

d. Mechanical work: This torque performs the required work. This torque depends upon the nature of application and is represented by T_L.

Thus, if T_e represents electromagnetic torque, then

$$T_e = B\omega_r + J\frac{d\omega_r}{dt} + T_L \tag{1.1}$$

EXAMPLE 1.1

An electric motor that develops a starting torque of 16 N-m starts with a load torque of 8 N-m on its shaft. If the acceleration at start is 2 rad/s^2, compute the moment of inertia of the system neglecting viscous and coulomb friction.

SOLUTION:

$$T_e - T_L = J\frac{d\omega_r}{dt}$$

$$16 - 8 = J\frac{\Delta\omega_r}{\Delta t}$$

$$\frac{\Delta\omega_r}{\Delta t} = 2 \text{ rad/s}^2 \text{ (given)}$$

$$8 = 2J$$

$$J = 4 \text{ kg-m}^2$$

EXAMPLE 1.2

In a speed-controlled dc motor drive, the load torque is 40 N-m. At time t = 0, the operation is under steady state and speed is 400 rpm. Under this condition at t = 0, the generated torque instantly increases to 100 N-m. The inertia of the drive is 0.01 N-ms²/rad. The friction is negligible.

   a. Write the differential equation governing the speed of the drive for t > 0.

   b. Evaluate the time taken for the speed to reach 900 rpm.

SOLUTION:

a. $T_e - T_L = J\dfrac{d\omega_r}{dt}$

b. $\quad \Delta\omega_r = 900 - 400$

$\qquad\qquad = 500 \text{ rpm}$

$\qquad\qquad = 52.35 \text{ rad/s}$

$\quad T_e - T_L = 100 - 40$

$\qquad\qquad = 60 \text{ N-m}$

$\quad J = 0.01$

$\quad \Delta t = \dfrac{J\Delta\omega_r}{T_e - T_L}$

$\qquad = \dfrac{0.01 \times 52.35}{60}$

$\quad \Delta t = 8.726 \text{ ms}$

1.2.2 Different Types of Loads

While the load is characterized by its speed and torque requirement, it is possible to categorize a few commonly seen loads.

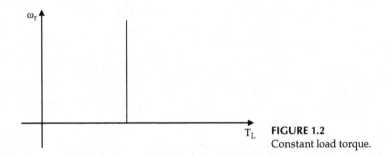

FIGURE 1.2
Constant load torque.

1.2.2.1 Constant Torque Loads

In this case, the load torque demand remains the same for a given period, but the speed requirement may change. For example, in the fabrication of machine tools, cutting and finishing are done at a constant torque but at different suitable speeds. Thus,

$$T_L = \text{constant} \tag{1.2}$$

This is given in Fig. 1.2.

1.2.2.2 Fan and Pump Loads

In these loads, load torque varies as the square of the speed and is expressed as

$$T_L \, \alpha \, \omega_r^2 \tag{1.3}$$

The torque–speed curve is given in Fig. 1.3.

1.2.2.3 Constant Power Loads

Consider Fig. 1.4, where rolling paper on a reel is shown. The tension on the paper strip has to remain the same for satisfactory formation of the paper roll. Let F be the force in N-m acting tangential to strip, which is held constant. The strip emerges from mill rolls at constant speed of V_p m/s and R_p is radius in meters (m) of the paper roll to be formed. Thus, the roll under formation has an angular velocity of ω_r given by

$$\omega_r = \frac{V_p}{R_p} \text{rad/s} \tag{1.4}$$

The power exerted by the motor driving the mandrel,

$$\left(FV_p\right) = \text{constant} \tag{1.5}$$

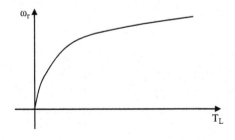

FIGURE 1.3
Fan and pump loads torque-speed characteristics.

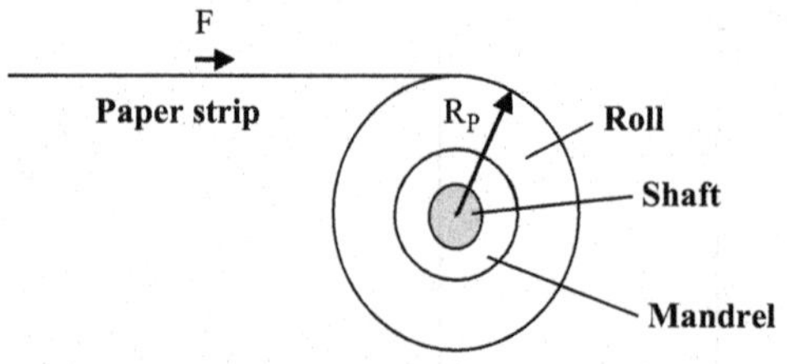

FIGURE 1.4
Constant power load.

In other words, as time advances, R_p increases, and because V_p is constant, ω_r falls. Thus, the curve of T_L against ω_r is a rectangular hyperbola as shown in Fig. 1.5.

EXAMPLE 1.3

In a paper mill, an ac motor is used to form a paper roll on a mandrel having a diameter of 30 cm. The paper strip emerges at a speed of 10 m/s with a tension of 6.8 N. The maximum thickness of the paper roll is 25 cm over the mandrel. Compute the power drawn by the motor assuming its efficiency at 80%. Then compute the minimum motor rpm.

SOLUTION:

Tension on the paper: F = 6.8 N
Velocity with which the paper emerges: V = 10 m/s
Power required = F × V_p = 6.8 × 10 = 68 N-m/s = 68 W
Therefore, the power rating of the motor is 68 W.

$$\text{Input power drawn} = \frac{\text{Power rating of the motor}}{\text{Efficiency}}$$

$$= \frac{68}{0.8}$$

$$= 85 \text{ W}$$

Maximum radius of paper roll:

R_p = Radius of the mandrel + Maximum thickness of paper roll

    = 15 cm + 25 cm = 40 cm

$$\text{Minimum motor rpm: } \omega_r = \frac{V_p}{R_p} = \frac{10}{40 \times 10^{-2}} = 25 \text{ rad/s}$$

$$\text{Minimum motor speed: } N_r = \frac{60 \times \omega_r}{2\pi} = \frac{60 \times 25}{2\pi} = 238.73 \text{ rpm}$$

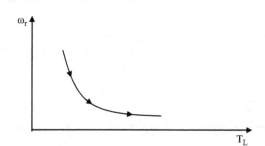

FIGURE 1.5
Constant power load characteristics.

EXAMPLE 1.4

In a paper-takeup system, the radius of the mandrel is 25 cm and the paper moves at a constant speed of 10 cm/s. Find the angular velocity when the thickness of the paper bundle is 50 cm.

SOLUTION:

The total radius at any instant is the sum of the mandrel radius and the thickness of the paper bundle wrapped on it in that instant. So, in this case

$$\text{total radius} = R_p + \text{thickness of the paper bundle}$$

where R_p is mandrel radius
Hence, total radius = 25 cm + 50 cm = 75 cm.
  The velocity of paper movement is

$$V_p = 10 \text{ cm/s}$$

Hence, the angular velocity is

$$\omega_r = \frac{V_p}{\text{total radius}} = \frac{10}{75} = 0.133 \text{ rad/s}$$

EXAMPLE 1.5

In a paper-takeup system, the paper moves at constant speed of 360 cm/s under a constant force of 20 N. If the efficiency of the motor driving the paper roll is 85%, compute the rating of the motor.

SOLUTION:

$$\text{Speed of paper} = 360 \text{ cm/s} = 3.6 \text{ m/s}$$

$$\text{Power required by the system} = 20 \text{ N} \times 3.6 \text{ m/s} = 72 \text{ W}$$

To obtain this power output, the motor must have a greater input considering that its efficiency is not 100%. Hence, the rating of the motor is

$$\text{Rating} = 72 \text{ W}/0.85 = 84.7 \text{ W}$$

1.2.2.4 Transportation Drive Systems

Electric traction and battery-powered vehicles fall into this category. The typical variation of speed and torque of such systems along time axis is given in Fig. 1.6.
  During acceleration, torque is constant and speed rises linearly. In the second stage of acceleration, T_L falls as ω_r increases, resulting in constant power operation. As speed reaches a steady state value, torque falls to a low value. During braking, speed decreases

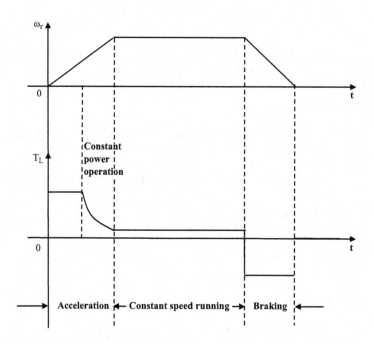

FIGURE 1.6
Time variation of speed and torque of transportation drive.

to zero and T_L is negative. Figure 1.6 can be used to deduce the torque–speed curve of a transportation system and is represented in Fig. 1.7.

EXAMPLE 1.6

A train starts from rest from Station A and attains a speed of 200 km/h in 2 minutes. After the train moves at this speed for a while, the brakes are applied at 1.85 m/s² and the train stops at station B. Compute the acceleration, braking interval, and duration for which the train runs at the constant speed, given that the total time taken for the train to reach station B from station A is 6 minutes. Sketch the speed time curve.

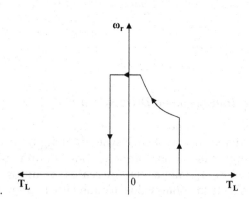

FIGURE 1.7
Torque–speed curve of traction systems.

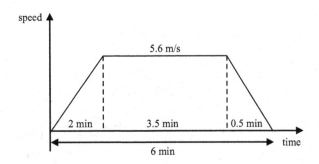

FIGURE 1.8
Speed–time curve.

SOLUTION:

The initial acceleration may be given as

$$a = (\text{Final Speed} - \text{Initial Speed}) / \text{Time taken}$$

$$\text{Final Speed} = 200 \text{ km/h} = 200 \times 1,000/3,600 \text{ m/s} = 55.56 \text{ m/s}$$

$$\text{Time taken} = 2 \text{ m} = 120 \text{ s}$$

$$\text{Acceleration} = 55.56/120 = 0.463 \text{ m/s}^2$$

While braking, the train slows down from 200 km/h to rest with a deceleration of 1.85 m/s². Hence, the braking interval is the change in speed/a′

$$T_{\text{brake}} = 55.56/1.85 = 30.03 \text{ s} = 0.5 \text{ minutes (approx.)}$$

Hence, the time taken for acceleration and deceleration is 2 + 0.5 minutes = 2.5 minutes. Because the total time is 6 minutes, the time for which it runs at constant speed must be 6 − 2.5 = 3.5 minutes.

The speed curve is sketched in Fig. 1.8.

1.3 Load-Gear System

In some applications, the load is attached to the motor shaft through a set of gears. A gear consists primarily of two teethed wheels rotating one over the other, and one of them is connected to a motor shaft while the other is connected to a load. If NT_1 is the number of teeth on one wheel and NT_2 is the number of teeth on the other wheel, then, depending upon $\dfrac{NT_1}{NT_2}$, the speed and torque on either side will vary depending on this ratio.

In this respect, a gear is a mechanical device equivalent to a transformer used to step up or step down the voltage and current. The significance of the gear is that many applications require low-speed operation, whereas it has been observed that motors have lower volume and size at high speed, necessitating the use of gears. The schematic of a motor shaft connected to load through gears is shown in Fig. 1.9.

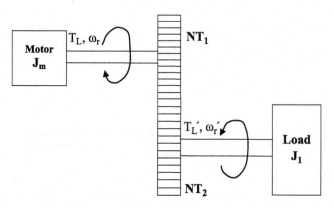

FIGURE 1.9
Motor-load coupling through gears.

Let J_m represents the moment of inertia of motor and J_1 that of load alone. Further assume T_L and ω_r represent torque and speed developed by the motor, and that T_L' and ω_r' are the torque and speed available at the load side. Because the power is same on both sides of the gear, we can write

$$\omega_r\, T_L = \omega_r'\, T_L'$$

$$\text{i.e., } T_L' = T_L\left(\frac{\omega_r}{\omega_r'}\right) \tag{1.6}$$

$$\text{Further, } \frac{\omega_r}{\omega_r'} = \frac{NT_2}{NT_1}$$

$$\text{Thus, } T_L' = T_L\left(\frac{NT_2}{NT_1}\right) \tag{1.7}$$

The load inertia J_1 is influenced by the gear such that, as seen from motor side, this becomes

$$\left(\frac{NT_1}{NT_2}\right)^2 J_1$$

Thus, the net inertia of the system is

$$J_m + \left(\frac{NT_1}{NT_2}\right)^2 J_1 \tag{1.8}$$

In the above analysis, loss in the gear is neglected. If the efficiency of the gear is provided, this should be suitably taken into account.

While a gear system offers robust coupling between motor shaft and load, another method is to use a belt drive as shown in Fig. 1.10. In this arrangement, the pulley with lower diameter of d_1 is connected to motor shaft, while the other one with higher diameter is coupled to load. This scheme is equivalent to a gear drive and Equation (1.7) can

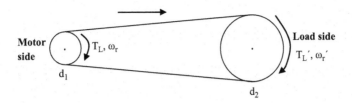

FIGURE 1.10
Belt drive.

be directly applied with NT_1 and NT_2 getting replaced with d_1 and d_2 respectively. The belt drive being simple and cost effective, however has the disadvantages of slippage and oscillations.

EXAMPLE 1.7

A 400 V, 50 Hz, 1,500 rpm, 3ϕ induction motor is running at 3% slip. The motor shaft is coupled to a gear in the ratio of 1:10. Find the load speed.

SOLUTION:

$$\text{Slip: } S = \frac{N_s - N}{N_s}$$

Synchronous speed of the motor: $N_s = 1,500$ rpm

Motor speed: $N = (1-S)\, N_s = (1-0.03) \times 1,500 = 1,455$ rpm

$\therefore$ Angular velocity of motor: $\omega_r = \dfrac{2\pi N}{60} = \dfrac{2\pi \times 1455}{60} = 152.37$ rad/s

Gear ratio $= 1{:}10$

We know that $\dfrac{\omega_r'}{\omega_r} = \dfrac{NT_1}{NT_2} = \dfrac{1}{10}$

$\therefore$ Angular velocity of load: $\omega_r' = 15.237$ rad/s

$\therefore$ Load speed $= \dfrac{60 \times \omega_r'}{2\pi} = \dfrac{60 \times 15.237}{2\pi} = 145.5$ rpm

---

EXAMPLE 1.8

A 400 V, 50 Hz, 750 rpm squirrel cage induction motor is driving a lift in a multistoried housing apartment. The lift is coupled to the motor shaft through a toothed chain drive with an efficiency of 89%. The lift raises 10 people with an average weight of 60 kg to a height of 10 m in 1 min. The motor efficiency is 85%. Compute the power drawn by the motor.

SOLUTION:

Energy required by the lift = $m \times g \times h = 10 \times 60 \times 9.81 \times 10 = 58,860$ J

$$\text{Power required} = \frac{\text{Energy required by the lift}}{\text{Time taken}} = \frac{58860\,(\text{J})}{60\,(\text{sec})} = 981 \text{ W}$$

Power output, P_0 of the motor is given by

$$P_o = \frac{\text{power required}}{\text{efficiency}} = \frac{981}{0.89} = 1102.25 \text{ W}$$

$$\therefore \text{Power drawn by the motor} = \frac{\text{Power output of motor}}{\text{efficiency}}$$

$$= \frac{1102.25}{0.85}$$

$$= 1,296.76 \text{ W}$$

EXAMPLE 1.9

A grinding machine is coupled to a single-phase motor having a constant speed of 500 rpm. If the maximum speed of the grinding machine is 100 rpm, find the diameter ratio of two pulleys on either side.

SOLUTION:

$$\text{Motor speed: } N_r = 500 \text{ rpm}$$

$$\text{Grinding machine speed limit: } N_r' = 100 \text{ rpm}$$

Let d_1 = diameter of motor pulley and d_2 = diameter of grinding machine.

$$\text{Diameter ratio} = \frac{d_1}{d_2} = \frac{\omega_r'}{\omega_r} = \frac{N_r'}{N_r} = \frac{100}{500}$$

$$d_1 : d_2 = 1:5$$

EXAMPLE 1.10

In a drive system, the accelerating torque is 1.8 N-m. The motor shaft is connected to the load through a gear of ratio 1:10. The moment of inertia of the motor is 0.1 kg-m^2, and that of the load is 0.3 kg-m^2. If the accelerating torque increases the speed of the system by 3 rad/s, compute the time taken for acceleration.

SOLUTION:

$$\text{Accelerating torque: } T_{acc} = 1.8 \text{ N-m}$$

The total moment of inertia is

$$J_{total} = J_m + (\text{Gear ratio})^2 J_l$$
$$J_{total} = 0.1 + (0.1)^2 \times 0.3$$
$$J_{total} = 0.103 \text{ kg-m}^2$$

The relationship between accelerating torque and speed increase is

$$T_{acc} = T_e - T_l = J_{total} \times d\omega_r / dt$$

$$\Delta t = J_{total} \times \Delta\omega_r / T_{acc} = 0.103 \times 3 / 1.8 = 0.172 \text{ s}$$

EXAMPLE 1.11

A grinding machine requires a low speed of 500 rpm (approx.) and is connected to three-phase induction motor having four poles supplied at 50 Hz. What should be the ratio of teeth on the gearbox if employed?

SOLUTION:

$$\text{Synchronous speed} = \frac{120 \times f}{p} = 1,500 \text{ rpm}$$

Let
NT$_1$ = teeth of motor gear and
NT$_2$ = teeth of grinding gear

$$\frac{\omega_m}{\omega_l} = \frac{NT_2}{NT_1}$$
$$\frac{1500}{500} = \frac{NT_2}{NT_1}$$
$$NT_2 : NT_1 = 3 : 1$$

1.4 Closed-Loop Control

At a particular operating point, an electric drive needs to maintain a specific speed and torque. When a power converter–fed electric motor drives a load, there can be disturbances either on the supply side or on the load side, which causes drifting of the operating point. To maintain the operating point at the set value, a closed-loop control is required. In most cases, the speed of the drive needs to be maintained constant for different values of load torque. Hence, a simple block diagram for the closed-loop speed control of the electric drive is given in Fig. 1.11. In this control, the motor speed is continuously sensed and compared with the set value. When the operating point shifts, an error is produced

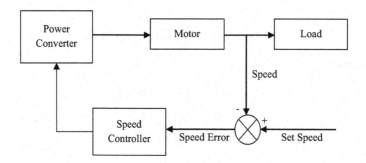

FIGURE 1.11
Block diagram of closed-loop speed control.

and is processed by the speed controller. The output of the speed controller controls the power converter and hence the power converter adjusts the input power to the motor so that the motor speed is brought to the set value. This closed loop control works satisfactorily, but during the control process the motor current may exceed the safe value. Hence, a current-limiting control is generally introduced in the feedback path to ensure that the motor current is a safe value during the control process. This is shown in Fig. 1.12. Closed-loop torque control is also performed in certain applications where the load torque needs to follow the set value. This is shown in Fig. 1.13.

1.5 Advantages of Electric Drives

Salient advantages of an electric drive over its counterparts are listed below:

a. It is the cleanest form of drive, so handling becomes easier.
b. It causes no pollution.
c. It is very compact in size and can be molded into suitable shapes depending on the given application.
d. The losses are minimal and hence the efficiency is higher.
e. These are self-starting equipment.

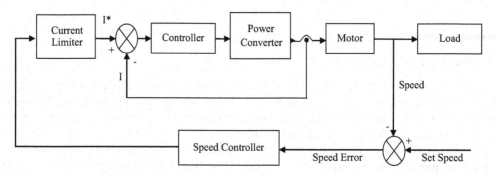

FIGURE 1.12
Block diagram of closed-loop current limit control.

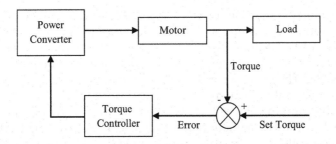

FIGURE 1.13
Block diagram of closed loop torque control.

    f. Very low maintenance and running costs.

    g. The development of power electronic circuits enables efficient speed/torque control, four-quadrant operation, improved steady state and transient responses, regenerative braking, and implementation of optimal control strategies. The digital controllers and power electronic circuits can be suitably integrated with electric drives making it more compact, attractive, and elegant in appearance.

1.6 Four-Quadrant Operation of Drives

To have a better understanding of drive operation, the speed and torque of the drive are taken as variables and are plotted along the X and Y axes. Accordingly, a drive can operate in the first, second, third, or fourth quadrant, depending upon the application. This enables an explanation of the drive operation in the four quadrants of torque–speed axes as indicated in Fig. 1.14.

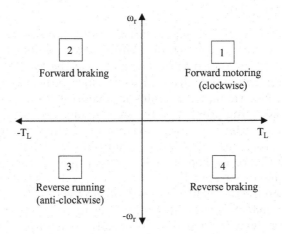

FIGURE 1.14
Four-quadrant operation of drive.

The four-quadrant operation can be well understood with a few practical applications. Consider the operation of a domestic fan that always runs in one direction. In this fan, the torque is also in one direction driving the fan. Thus, both the torque and the speed of the drive are always positive; this corresponds to first quadrant operation and is referred as forward motoring. The operation of a wet grinder and a mixer in the kitchen are other examples of first-quadrant operation.

Now consider the working of a clothes washing machine. The major parts of a washing machine are the drum and an agitator or rotating disc. The clothes are loaded in the drum, which is then filled with water. The agitator is a pole-like attachment with a height up to the middle of the drum and small blades attached along its sides. The agitator moves the clothes and the water back and forth within the drum. When the agitator drives the clothes in the clockwise direction, both speed and torque are positive and the drive is working in the first quadrant. When the agitator rotates the clothes in the opposite direction, both the speed and torque are negative, which corresponds to the third quadrant; third-quadrant operation is also known as reverse motoring.

After washing the clothes, the spin cycle begins. In the spin cycle, the drum rotates at a much higher speed in the forward direction to pull as much water as possible out of the clothes. This process is also a first-quadrant operation. At the end of spin cycle, brakes are applied to stop the drive. During braking, the speed is in the forward direction while the torque acting on the drive is negative; this operation is in the second quadrant.

Consider a passenger lift in a multistory building. When the lift moves up, both the speed and the torque are positive—a first-quadrant operation. When the required floor is reached, the lift drive is stopped by braking, which corresponds to second-quadrant operation. As the lift moves downward, both the speed and the torque are negative, so this falls in the third quadrant. At the end of the downward movement, the lift drive is stopped by braking, is a fourth-quadrant operation.

1.7 Drive Characteristics

As seen earlier, a load has a fixed characteristic along torque–speed axes. An electric motor fed by a power converter can exhibit a family of torque speed curves: the nature of these curves depends on the type of converter as well as the motor. Fig. 1.15 shows a constant load torque and four possible drive characteristics, indicated as 1, 2, 3, and 4. If the power converter–fed electric motor operates in a control setting such that it produces curve 1, then the intersecting point P1 is the steady-state operating point. If the control setting is changed to achieve the drive characteristic 2, the new steady-state operating point is P2. Similarly, control settings can be changed either to curve 3 or curve 4, leading to operating point P3 or P4. For the drive system shown in Fig. 1.15, the control setting 1 produces the maximum speed while setting 4 produces the minimum speed, and the difference between the two values corresponds to the range of controlled speed. It is important to mention that at any operating point the maximum permissible torque available from the drive is always higher than the steady-state torque because some margin of the motor torque is required to accelerate the system and stabilize against transient overloads.

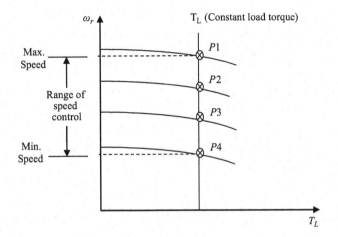

FIGURE 1.15
Adjustable-speed drive characteristics.

1.8 Steady-State Stability of Electric Drives

It is evident from the above discussion that steady-state stability of the drive is guaranteed if and only if the motor torque is higher than the load torque; furthermore, the intersecting point between the load curve and the converter motor curve is the steady-state operating point.

The steady-state operating point will always be under the influence of transient changes due to disturbances either on the load side or the motor side. This is illustrated in Fig. 1.16, where the load is assumed to be a fan or pump type. The steady-state operating point P is characterized by ω_{r0} and T_0. In Fig. 1.16(a), assume a disturbance on the load side causes a reduction of speed to ω_{r1}, shifting the operating point to Q. At this transient operating point, the change in speed is $\Delta\omega_r = \omega_{r0} - \omega_{r1}$, load torque is T_{L1} and motor torque is T_{m1}. Because $T_{m1} > T_{L1}$, the speed of the drive increases and settles again at ω_{r0}. Now consider another case in which the drive speed increases to ω_{r2} due to some disturbance as in Fig. 1.16(b). At this transient point, $\Delta\omega_r = \omega_{r2} - \omega_{r0}$ and the motor and the load torque are T_{m2} and T_{L2}, respectively. Because $T_{L2} < T_{m2}$, the drive decelerates and the operation is restored to P. Thus, the transient changes do not disturb the steady-state operating point, and the drive is said to possess steady-state stability. From the above, it can be concluded that a drive is said to be stable under the steady-state condition if the change in speed $\Delta\omega$ due to a disturbance approaches zero as time tends to infinity.

EXAMPLE 1.12

Figure 1.17 shows a speed–torque characteristic of a typical load marked as L. It is a straight line passing through the origin and (20,100). The load is assumed to be driven by a dc shunt motor, the characteristic of which is plotted as M. The line M passes through (0,154) and (16,80). Compute the steady-state speed and torque of the drive system.

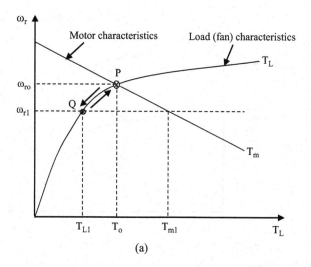

(a)

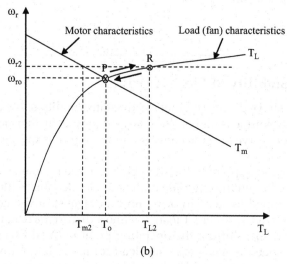

(b)

FIGURE 1.16
Drive operating characteristics under disturbance.

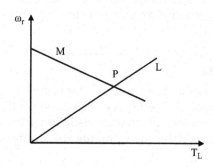

FIGURE 1.17
Figure for Example 1.12.

SOLUTION:

The straight-line equation for line L is obtained by substituting the coordinates (0,0) and (20,100) in

$$\frac{y - y_1}{x - x_1} = \frac{y_2 - y_1}{x_2 - x_1}$$

$$\Rightarrow \frac{y - 0}{x - 0} = \frac{100 - 0}{20 - 0} \tag{1.9}$$

$$\Rightarrow 5x - y = 0$$

The straight-line equation for line M is obtained by substituting the coordinates (0,154) and (16,80) in

$$\frac{y - y_1}{x - x_1} = \frac{y_2 - y_1}{x_2 - x_1}$$

$$\Rightarrow \frac{y - 154}{x - 0} = \frac{80 - 154}{16 - 0} \tag{1.10}$$

$$\Rightarrow 74x + 16y = 2464$$

Let the steady-state operating point P be defined by the coordinates (x', y'). Equations (1.9) and (1.10) have to be satisfied at P:

$$5x' - y' = 0 \tag{1.11}$$

$$\text{and } 74x' + 16y' = 2464 \tag{1.12}$$

Solving (1.11) and (1.12) yields

$$x' = \text{Steady-state torque of the drive} = 16 \text{ N-m}$$

$$y' = \text{Steady-state speed of the drive}$$

$$= \frac{60 \times \omega_r}{2\pi} = \frac{60 \times 80}{2\pi} = 763.94 \text{ rpm}$$

EXAMPLE 1.13

The speed–torque characteristic of a motor is a straight line passing through (0 rpm, 154 N-m) and (80 rpm, 60 N-m). The load curve is a straight line passing through the origin and (30 rpm, 140 N-m). Find the steady state of the system.

SOLUTION:

The steady state is the point of intersection of the two lines.

The equation for the motor characteristic passing through (0 rpm, 154 N-m) and (80 rpm, 60 N-m) is

$$\frac{y - y_1}{x - x_1} = \frac{y_2 - y_1}{x_2 - x_1}$$

Let x = speed (ω_r) and y = torque (T_L).

$$\frac{T_L - 154}{\omega_r - 0} = \frac{60 - 154}{80 - 0}$$

$$T_L = -1.175\,\omega_r + 154 \text{ N-m}$$

The equation for load curve is

$$T_L' = 4.67\,\omega_r' \text{ N-m}$$

Hence, the point of intersection is

$$4.67\,\omega_r = -1.175\,\omega_r + 154$$
$$5.845\,\omega_r = 154$$
$$\omega_r = 26.36 \text{ rpm}$$

The corresponding torque is

$$T_L = 4.67\,\omega_r = 123 \text{ N-m}$$

Therefore, the steady state of the system is located at the coordinates (26.36 rpm, 123 N-m).

EXAMPLE 1.14

Consider the speed–torque characteristics curves of a motor that is connected to a load as depicted in Fig. 1.18, where line AB represent the motor characteristic line and line CD represents the load line. Find the steady-state speed and torque of the combined system.

SOLUTION:

The equation for line AB is obtained by substituting the coordinates (0,0) and (1,400 rpm, 20 N-m).

$$\frac{y - y_1}{x - x_1} = \frac{y_2 - y_1}{x_2 - x_1}$$

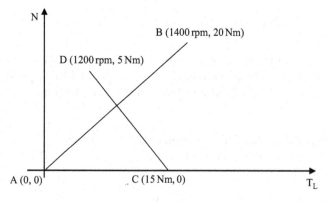

FIGURE 1.18
Figure for Example 1.14.

Let x = speed (ω_r) and y = torque (T).

$$\frac{T-0}{\omega_r - 0} = \frac{20-0}{1400-0}$$

The speed–torque equation of the motor is

$$\frac{T}{20} = \frac{N}{1400} \tag{1.13}$$
$$70T - N = 0$$

The equation for line CD is obtained by substituting the coordinates (15 N-m, 0) and (1,200 rpm, 5 N-m) is
  The speed–torque equation of the load is

$$\frac{T-15}{5-15} = \frac{N}{1200}$$
$$120T - 1,800 = -N \tag{1.14}$$
$$120T + N = 1,800$$

Solving (1.13) and (1.14) yields
Steady-state torque = 9.47 N-m
Speed = 663.15 rpm

Questions

1. Sketch the block diagram of a power electronic drive system, and indicate the name of each block.
2. With a neat block diagram, explain the components of power electronic drive system.
3. Name the power electronic converters used in drive systems.
4. With a specific circuit diagram and related graphs, explain the function of a controller in a power electronic drive system.
5. With necessary graphs and equations, elaborate on the differential types of loads.
6. With neat sketches of speed–torque curves, categorize commonly seen loads.
7. Briefly describe gears and belt drives. Provide neat figures and derivations.
8. Briefly discuss gears in electrical drives.
9. Briefly discuss various types of coupling elements. Provide figures and derive any necessary expressions.
10. Give an example for a drive that operates in (a) the first quadrant, (b) the first and third quadrants, and (c) all four quadrants.

Unsolved Problems

1. A 440 V, 50 Hz, 900 rpm squirrel cage induction motor is driving a lift in multi-story building. The lift is coupled to motor shaft through a toothed chain drive, which has an efficiency of 89%. When the lift carries 15 persons, each with an average weight of 60 kg, to a height of 20 m in 1 minute, the motor takes 23.8 A at rated voltage. If the motor efficiency is 80%, find the power drawn by the motor.

2. The diameter of a coupling belt is 123 cm on one side and 1.8 m on the other side. If a motor shaft running at 100 rad/s is connected to the side of the belt with the larger diameter, find the speed in rpm on the other side.

3. A torque–speed (torque along the X-axis and speed along the Y-axis) characteristic curve of a dc shunt motor passes through (0, 154 rad/s) and (22 N-m, 100 rad/s). If it drives a linear load passing through the origin and (12 N-m, 120 rad/s), find the steady-state operating point.

4. An electric motor takes 2.33 s to start from standstill to a no-load speed of 1,490 rpm. If losses are neglected and the electromagnetic torque required for starting is 23.77 N-m, compute the moment of inertia of the drive.

5. A three-phase, four-pole, 400 V, 50 Hz induction motor is running at 5% slip. The motor shaft is coupled to a load through a gear of ratio 1:20. Find the speed of load.

6. An electric motor developing a starting torque of 15 N-m starts with a load torque of 7.5 N-m on its shaft. If the acceleration at start is 3 rad/s², what is the moment of inertia in the system, neglecting viscous and coulomb friction?

7. Calculate the ratio on the gear box for a wet grinding machine that requires a low-speed operation of 250 rpm and is connected to a three-phase induction motor having 6 poles and 50 Hz.

Answers

1. 4,133.42 W
2. 146.34 rad/s
3. (12.365 N-m, 123.65 rad/s)
4. 0.355 kg-m²
5. 71.25 rpm
6. 2.5 kg-m²
7. 4:1

2

Analysis of Power Electronic Circuits

2.1 Introduction

Essentially, power electronics is concerned with control of power from a source to a load component in a desired manner. The power flow may be through a single power electronic circuit or a suitable combination of several power electronic circuits. A power electronic circuit ideally requires a switch that can be closed and opened at desired instances to facilitate this power flow. A mechanical switch takes a few seconds for closure/opening and therefore is not suitable for effective power control. A semiconductor switching device requires only a few microseconds for closing and opening and can therefore regulate power flow quickly and efficiently. Thus, a power semiconductor working as a switching element is the heart of any power electronic circuit. However, these devices are not ideal switches because they have finite turn-ON and turn-OFF times and conduction losses. These devices largely influence the operation of any power converter circuit, hence it is mandatory to study these devices in detail. While there are a variety of switching devices, only a few are dealt with in this book.

2.2 Diode

A power diode is a P-N junction device and is very similar to a signal diode. The symbol and V-I characteristics are shown in Fig. 2.1. When a diode is forward biased, it conducts with a very small voltage drop of about one volt across it. During reverse bias, the current is the leakage current and can be ignored. Beyond reverse breakdown voltage, avalanche breakdown takes place, which damages the device. In view of this, a diode can be thought of as a closed switch during forward bias and as an open switch during reverse bias. A power diode contains an intrinsic semiconductor layer or first-layer between P and N structure. This first-layer helps to block large reverse voltages.

A power diode can be turned onalmost instantaneously, but turn-OFF requires more time to block the reverse voltage due to the flow of reverse recovery current. The time taken for this current to vanish and for the diode to withstand reverse voltage is known as reverse recovery time. Based on this time, there are two types of power diodes:

- Line frequency diodes: These diodes are connected to line frequency of 50/60 Hz and have higher reverse recovering time.
- Fast recovery diodes: These devices have very low reverse recovery time of the order of few microseconds and are used in high-frequency circuits.

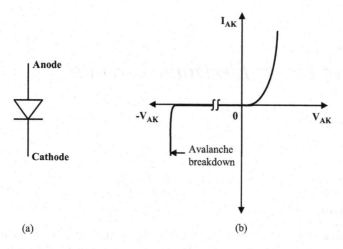

(a)   (b)

FIGURE 2.1
(a) Diode symbol. (b) V-I characteristics.

2.3 Silicon-Controlled Rectifier (SCR)

This device is sometimes called as thyristor, even though this terms stands for a group of devices such as SCR, Triac, GTO, etc. SCR is the traditional major component employed in power converters; however, with the development of gate-commutated devices such as IGBTs, SCRs are mainly employed in ac/dc converters now.

The architecture of the SCR is shown in Fig. 2.2(a). It is a four-layer device forming three junctions, J_1, J_2. and J_3. There are three terminals marked as anode A, cathode K,

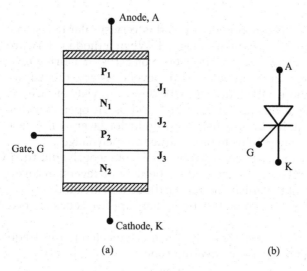

(a)   (b)

FIGURE 2.2
(a) SCR architecture. (b) Symbol.

and gate G. The voltage-current characteristic of the device can be divided into three segments:

i. **Forward blocking state:** When the anode is connected to a positive terminal and the cathode to a negative terminal of a voltage source, the junctions J_1 and J_3 are forward biased and only J_2 is reverse biased. Only a small leakage current will be flowing through the device.

ii. **Reverse blocking state:** The device enters into this state when the anode and cathode are connected to negative and positive terminals of a voltage source, respectively. Junctions J_1 and J_3 are then reverse biased, while J_2 is forward biased. Very small reverse leakage current will be flowing through the SCR.

iii. **ON state (conducting state):** The device comes to conduction if the four layers are filled with charge carriers. The turn-ON process can be achieved through different means, which will be discussed soon. The device can be brought to the conducting state if and only if the device is in forward blocking mode. Once the SCR starts conducting, the voltage drop across the device is 1–2 V and it behaves like a closed mechanical switch.

A better understanding of the device can be gained by examining its V-I characteristic, plotted in Fig. 2.3. Assume that the SCR is in the forward blocking state, and say the voltage across A and K is steadily increasing. Because J_1 and J_3 are in forward biased mode, increased voltage across J_2 causes avalanche breakdown, resulting in large carrier movement through the device and bringing the device to the conduction state. The voltage at which J_2 breaks and allows current flow is termed forward break over voltage, V_{BO}. When the device conducts, the voltage drop across the four layers is 1–2 V, and the current is limited by the load impedance alone.

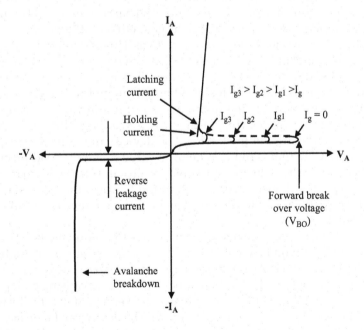

FIGURE 2.3
SCR volt-ampere characteristics.

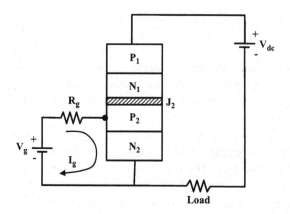

FIGURE 2.4
Gate triggering.

As mentioned, one way to turn on the SCR is to apply a voltage equal to V_{B0} across the device; however, this is infeasible in many cases and further effective power control is impossible. Hence a few methods, namely gate triggering, the $\dfrac{dv}{dt}$ method, and irradiation of the gate-cathode junction were developed. Among these, gate triggering is the most viable method.

When the device is in the forward blocking state, only J_2 is reverse biased. The potential barrier across J_2 can be reduced by injecting holes into layer P_2. Thus, when a current of suitable magnitude is passed from an external dc source, as shown in Fig. 2.4, the current I_g appears as holes in P_2 and electrons in N_2. The excess hole in P_2 lowers the potential hill, and with sufficient holes the potential barrier ceases, allowing free movement of charge carriers. Thus, with sufficient gate current, the SCR can be turned on even with a lower value of anode-to-cathode voltage. This is depicted in Fig. 2.3. Further increase in the amplitude of I_g lowers the V_{AK} at which the device starts conducting, and finally, for $I_g = I_{g3}$, the device behaves like a diode. The magnitude of the gate current is on the order of a few milliamperes. It should be mentioned that successful turn-ON is achieved if the device current is maintained at a minimum value, called latching current, for more than the turn-ON time of the SCR. Once the SCR is turned on, the gate loses control over conduction of the device. The device can be turned off if and only if the anode-to-cathode current is reduced below a value called the holding current. Thus, during the ON state, if the gate current is made zero and the SCR current falls below the holding current, the device reverts back to the forward blocking state (if forward bias continues to exist across the device). The latching current is slightly higher than the holding current.

When a reverse bias is applied across the device, it is then equivalent to two diodes connected in series, with both diodes remaining in reverse bias. Only very small leakage current flows through the device. As reverse voltage increases and reaches a value termed as reverse breakdown voltage, avalanche breakdown takes place, causing damage to the SCR.

Thus, it is evident that the SCR has two stable and reversible states: the forward conduction (ON) state and the reverse blocking (OFF) state. The device can be turned on only during the forward biased state, and the device can be turned off only by lowering the current below holding current. The process of turning off the SCR is called commutation. There are several methods of commutation, such as line commutation, voltage commutation, current commutation, load commutation, etc. In many of these methods, an additional circuit called a commutation circuit is required, which is the major drawback associated with the use of the SCR.

SCR has several specifications, but the most important ones are

i. Forward conduction current

ii. Reverse breakdown voltage

iii. Turn-ON and turn-OFF times

iv. $\dfrac{dv}{dt}$ and $\dfrac{di}{dt}$ values

SCRs today have current ratings of several kA and voltage ratings of several kV. The turn-ON and turn-OFF processes are dynamic in nature. During turn-ON, once the gate current is injected, it has to spread over the entire P_2 region and make the potential barrier across J_2 collapse. As this barrier starts disappearing, anode-to-cathode current starts increasing, initially through a narrow region and finally spreading over the entire layer. The time between the initiation of gate triggering and the moment at which anode-to-cathode current reaches its steady-state value is known as the turn-ON time of the SCR. This is on the order of 2–4 microseconds.

The turn-OFF process of the SCR is known as commutation. As mentioned earlier, this is possible when the device current is reduced below the holding current. This is done with an external source sweeping out electrons from N_2 and holes from P_1. The charge carriers in the inner layers are removed, and the SCR regains forward blocking capability. The turn-OFF time is typically 5–200 microseconds. If the turn-OFF time is low (i.e., below 25 microseconds), such SCRs are named inverter-grade SCRs, which operate at high-frequency switching. Thyristors with large turn-OFF times are known as converter-grade SCRs and are less expensive than inverter-grade SCRs. Converter-grade SCRs are used in line-frequency rectifiers, ac voltage regulators, and cyclo converters.

The magnitude of the SCR gate current is a few tens of milliamperes. A separate source is required for gate-pulse generation in many configurations of power electronic circuits. Generally this power source can be derived from the one that feeds the power converter. A typical gate drive circuit is shown in Fig. 2.5.

If the gate signal is applied as a steady, ripple-free dc current, then heating of the gate–cathode junction will occur. To reduce this heating effect, and to make use of a pulse transformer for electrical isolation, the gate signal is ANDed with a high-frequency pulse train. This is called modulation and is shown in Fig. 2.6.

The gate current can be amplified to the required level using a transistor amplifier operating in the CE mode. The gate current is fed to transistor base and is amplified

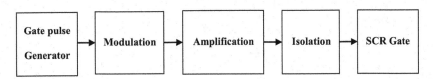

FIGURE 2.5
Block diagram of a typical gate drive circuit for SCR.

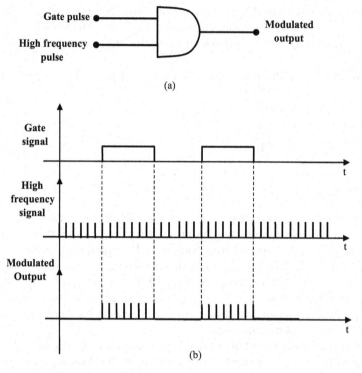

(a)

(b)

FIGURE 2.6
(a) Modulation. (b) Waveforms.

at the collector. The collector lead is connected to the primary of a pulse transformer, which isolates the low-power devices at the gate circuit from the high-voltage source connected to the anode and cathode of the SCR. The amplification and isolation circuit is shown in Fig. 2.7. A diode D_1 is connected across the primary of a pulse transformer to facilitate the primary current to free-wheel during the OFF period of the gating signal.

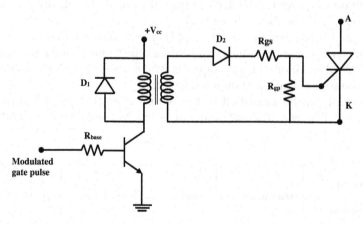

FIGURE 2.7
Amplification and isolation.

Diode D_2 presents reverse current flow across the gate–cathode junction, and resistance Rgs is inserted to limit gate-to-cathode voltage and current. The resistance Rgp, connected as a shunt across the gate–cathode junction, is used to bypass the gate junction leakage current, thereby improving thermal stability.

2.4 Triac

The triac is a low-cost switching device that can replace two SCRs connected in anti-parallel, but it has lower current and voltage ratings. The structure, symbols and V-I characteristic of a triac are delineated in Fig. 2.8. Because the triac does not have a specific

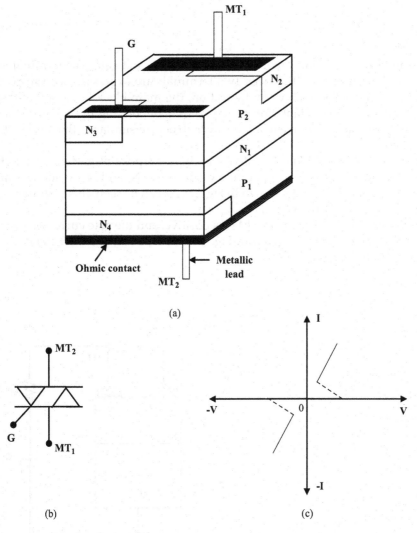

(a)

(b)   (c)

FIGURE 2.8
Triac. (a) Structure. (b) Symbol. (c) V-I characteristics.

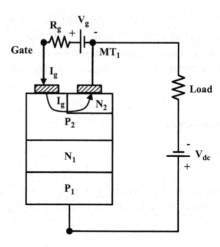

FIGURE 2.9
Triac conduction with positive gate current.

anode/cathode, the terminals are marked as MT_1, MT_2, and gate. The terminal MT_1 is the reference, and the voltage of the other two terminals are expressed with respect to MT_1. The triac has a multiple-junction configuration, the main block being a P-N-P structure. Three N-type layers are suitably integrated so that, depending upon voltage across MT_1 and MT_2 and suitable gate current, a four-layer structure similar to that of the SCR comes into operation.

Consider the case where MT_2 is positive and there is a positive gate current. Referring to Fig. 2.8(a), P_1-N_1-P_2-N_2 comes into operating mode, while N_3 and N_4 do not have any role to play. The equivalent structure is shown in Fig. 2.9. As can be seen, this is similar to the SCR structure, and the device turns ON.

Now consider that MT_1 is negative, MT_2 is positive, and the gate current is negative. The relevant device architecture is given in Fig. 2.10. P_1-N_1-P_2-N_2 is a SCR structure, and the gate is fired through N_3.

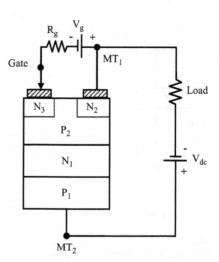

FIGURE 2.10
Triac conduction with negative gate current.

2.5 Power MOSFET

A simple structure of n-channel power MOSFET is given in Fig. 2.11(a). The device has three terminals indicated as drain (D), source (S), and gate (G). Assume that G is open circuited and D is made positive with respect to S. The junction between n and p layer is reverse biased, and no current flows through the device, except for a little leakage current that flows through the device, which is insignificantly small.

Let D be at a positive voltage with respect to S; if a positive voltage is now applied at G with respect to S, the electric field between G and S attracts the electrons from the n^+ region to the p region closer to the gate terminal. If sufficient voltage is applied across G and S, the electrons that migrated in the top region create a channel between the n^- region of D to then n^+ portion of S. This is shown in Fig. 2.11(b). Thus the device turns ON. For practical cases, V_{GS} of amplitude $+12$ V to $+18$ V is sufficient to turn the device on and keep it in conduction mode. When V_{GS} is reduced to zero, the induced channel disappears and the MOSFET turns off. This shows that the power MOSFET is a gate-controlled device. Further, consider a case where D is connected to a negative terminal and S to a positive terminal of a source; the $n^+ - n^-$ and p regions form a diode structure. Thus the functional diagram of the power MOSFET can be represented as a controlled switch in the forward direction with an anti-parallel diode across it. The device symbol and functional diagram are given in Fig. 2.11(c) and 2.11(d).

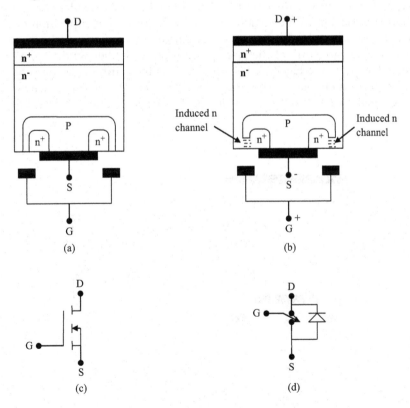

FIGURE 2.11
(a) Structure of MOSFET in the OFF state. (b) Induced n-channel in ON state. (c) Device symbol. (d) Functional diagram.

2.6 Insulated Gate Bipolar Transistor

The construction and principle of operation of the insulated gate bipolar transistor (IGBT) are largely similar to that of the power MOSFET. The ON and OFF processes are gate voltage–controlled operations like the power MOSFET. The junction architecture and symbol of the IGBT are shown in Fig. 2.12. A comparison of the IGBT structure with that of the power MOSFET shows that both are essentially one and the same, except that the IGBT structure has an additional p^+ layer at the top, above the n^+ layer, which is the collector terminal of the device. The collector and emitter are the power terminals with gate as the control terminal in the device. The control voltage is applied across the gate and emitter, turning the device on; withdrawal of the control voltage switches the device off. Though the structure of IGBT is similar to that of power MOSFET, the major difference is that when the device conducts, the top p^+ region injects holes into the adjacent n zone, resulting in reduced device resistance. This effect also enhances the current rating of the IGBT over the power MOSFET. Furthermore, the presence of a p^+ zone shows that there is no anti-parallel diode in the IGBT as was in the case of the power MOSFET.

When the device is in use, the collector terminal polarity is kept positive with respect to the emitter. The device is turned ON by applying a positive gate voltage higher than the threshold value. The conduction mechanism can be explained with the help of Fig. 2.13. Though the structure of the IGBT is similar to the power MOSFET, the turn-ON process and subsequent conduction can be best understood when we consider two parasitic transistors inherently present in the structure. To understand the composition of these transistors, the p and n layers in the structure are rearranged as shown in Fig. 2.13. With this reconfiguration, we can think of the p_1^+-n_1-p_2 transistor together with n_1-p_2-n_2^+, which are interconnected as shown in Fig. 2.13. The MOSFET switch is inherently present, consisting of n^- and n^+ regions with gate G. The resistance R_1 represents the resistance of the p-layer, and R_2 is the resistance of n^- layer. The value of R_2 is variable due to carrier injection from the top p^+ layer.

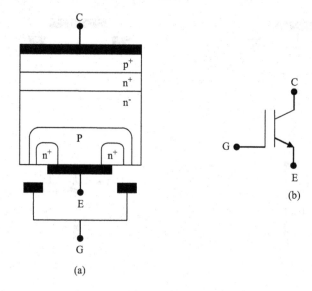

FIGURE 2.12
Insulated gate bipolar transistor. (a) Structure. (b) Symbol.

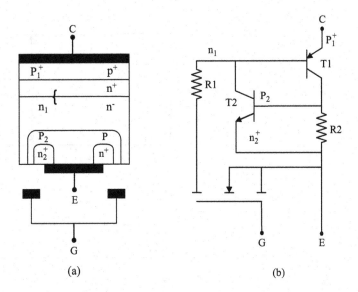

FIGURE 2.13
Insulated gate bipolar transistor structure during the turn-ON process. (a) Structure. (b) Equivalent circuit.

When a positive gate voltage is impressed, a conduction channel is formed from the collector across the top p–n junction, across the middle n zone, and through the channel to the emitter. This channel current serves as the base current for transistor T_1, and its collector current is the amplified version of this base current. The collector current of T_1 appears as the base current of T_2, which in turn amplifies it at its collector end. Collector current of T_2 serves as the base current of T_1. This process results in large-scale injection of carriers across the top p–n layer. The device carries two currents: one is the channel current as started first, and the second one is the current through the two transistors. The resistance values in these two paths are shown in Fig. 2.13 as R_2 and R_1.

The interconnection of two transistors is regenerative in nature; that is, an increase in the base current of T_2 enhances its collector current by transistor action. This current serves as the base current of T_1 and hence collector current T_1 is now boosted, which in turn is the base current of T_2. This cumulative action drives the device into conduction. This implies that, even if the gate voltage is made zero and thus blocks the MOSFET switch in the IGBT section, the IGBT will continue to be in the ON state. Thus gate voltage loses control over the device conduction, which is undesirable. To avoid this situation, values of R_2 and R_1 are established so that most of the power passes through the MOSFET. Then device is conveniently turned off by reducing the gate voltage to zero or a negative amplitude.

2.7 Analysis of Switched Circuits

2.7.1 Introduction

This section explains switched circuits using analytical means. Because switching is an inherent operation in power electronic circuits, this chapter is oriented toward the analysis of switched circuits.

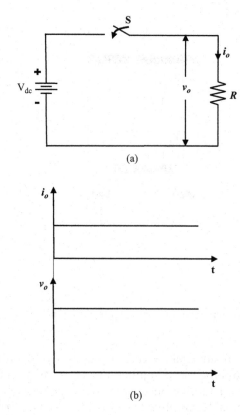

FIGURE 2.14
(a) Switched R-load. (b) Waveforms.

2.7.2 Switched Circuits with Mechanical Switch

The transient nature of electrical variables using a mechanical switch is considered in this section.

2.7.2.1 Ideal Switch with R Load

Consider a constant voltage V_{dc} that is applied to a circuit R as shown in Fig. 2.14(a) at time $t = 0$ by closing the switch S. The transient current i_o can be found by applying Kirchhoff's Voltage Law (KVL).

$$R \times i_o = V_{dc}$$

$$i_o = \frac{V_{dc}}{R}$$

Also, $v_o = V_{dc}$. The waveforms for output voltage and output current are shown in Fig. 2.14(b).

2.7.2.2 Ideal Switch with RL Load

Consider a constant voltage V_{dc} applied to RL circuit as shown in Fig. 2.15(a). At $t = 0$, when switch is closed, the KVL can be written as

$$(R \times i_o) + \left(L \times \frac{di_o}{dt} \right) = V_{dc} \tag{2.1}$$

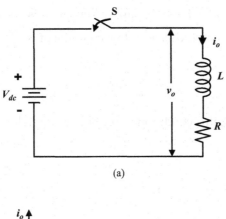

(a)

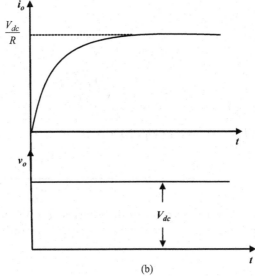

(b)

FIGURE 2.15
(a) Switched RL load. (b) Waveforms.

The solution to this differential equation has two components. One is the steady-state component, and the other is the transient component. The transient component is obtained by forcing V_{dc} in Equation (2.1) to zero.

$$\left(L \times \frac{di_o}{dt}\right) + (R \times i_o) = 0$$

$$\frac{di_o}{dt} + \left(\frac{R}{L} \times i_o\right) = 0$$

$$\left(p + \frac{R}{L}\right) i_o = 0$$

where p denotes $\frac{d}{dt}$ and the solution is given by $i_o = Ke^{\frac{-R}{L}t}$, where K is the constant of integration to be identified from boundary conditions.

Steady state emerges as $t \to \infty$. For a dc supply voltage, L is a short circuit and R alone opposes current flow. Therefore, the steady state component is $\dfrac{V_{dc}}{R}$. Thus,

$$i_o = \frac{V_{dc}}{R} + K e^{\frac{-R}{L}t} \tag{2.2}$$

At $t = 0$, $i_o = 0$. Substituting this condition in the above equation,

$$0 = K + \frac{V_{dc}}{R}$$

$$K = -\frac{V_{dc}}{R}$$

Substituting for K,

$$i_o = \frac{V_{dc}}{R} - \frac{V_{dc}}{R} e^{\frac{-R}{L}t}$$

$$= \frac{V_{dc}}{R}\left(1 - e^{\frac{-R}{L}t}\right) \tag{2.3}$$

The variation of v_o and i_o is shown in Fig. 2.15(b).

EXAMPLE 2.1

A constant voltage of 360 V is switched on to a series combination of resistance and inductance where $R = 230\ \Omega$ and $L = 150$ mH, at $t = 0$. Compute the circuit current at $t = 2$ ms and $t = 100$ ms.

SOLUTION:

$$i_o = \frac{V_{dc}}{R}\left[1 - e^{\frac{-Rt}{L}}\right]$$

$$i_o = \frac{360}{230}\left[1 - e^{\frac{-230t}{150 \times 10^{-3}}}\right]$$

At $t = 2$ ms,
$i_o = 1.492$ A

At $t = 100$ ms,
$i_o = 1.565$ A

EXAMPLE 2.2

A constant dc voltage of 40 V is switched on to a series combination of resistance and inductance, where $R = 150\ \Omega$ at $t = 0$ s. Find the inductance if the time taken is 1 ms for a circuit current to reach (i) 0.15 A and (ii) 0.2 A.

SOLUTION:

$$i_o = \frac{V_{dc}}{R}\left[1 - e^{\frac{-Rt}{L}}\right]$$

i. $i_o = 0.15$ A

$$0.15 = \frac{40}{150}\left[1 - e^{\frac{-0.15}{L}}\right]$$

$$L = 181.44 \text{ mH}$$

ii. $i_o = 0.2$ A

$$0.2 = \frac{40}{150}\left(1 - e^{\frac{-0.15}{L}}\right)$$

$$0.75 = \left(1 - e^{\frac{-0.15}{L}}\right)$$

$$e^{\frac{-0.15}{L}} = 0.25$$

$$\frac{0.15}{L} = -1.386$$

$$L = 108.2 \text{ mH}$$

EXAMPLE 2.3

A constant voltage of 200 V is switched onto a series combination of resistance and inductance where R = 100 Ω and L = 0.2 H, at t = 0. Find the equation for current and voltage and find the time at which $V_R = V_L$.

SOLUTION:

$$i_o = \frac{V_{dc}}{R}\left[1 - e^{\frac{-Rt}{L}}\right]$$

$$= \frac{200}{100}\left[1 - e^{\frac{-100t}{0.2}}\right] A$$

$$= 2\left(1 - e^{-500t}\right) A$$

$$V_R = iR = 200\left(1 - e^{-500t}\right) V$$

$$V_L = L\frac{di}{dt} = 200e^{-5t} V$$

At $V_R = V_L$,

$$200\left(1 - e^{-5t}\right) = 200e^{-5t}$$

$$200 = 400e^{-5t}$$

$$t = 1.386 \text{ ms}$$

2.7.2.3 Ideal Switch with RC Load

Consider a constant voltage V_{dc} is applied to RC circuit shown in Fig. 2.16(a). At t = 0, when the switch is closed, the KVL can be written as

$$R \times i_o + \frac{1}{C}\int i_o \, dt = V_{dc} \tag{2.4}$$

The transient component is obtained by forcing the right-hand part of Equation (2.4) to zero.

$$i_o + \frac{1}{RC}\int i_o \, dt = 0$$

Differentiate and rearrange the equation,

$$\frac{di_o}{dt} + \frac{1}{RC}i_o = 0$$

$$\left(p + \frac{1}{RC}\right)i_o = 0$$

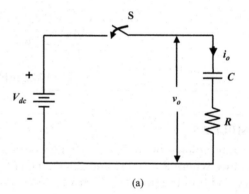

(a)

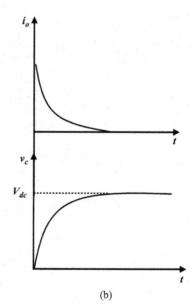

FIGURE 2.16
(a) Switched RC load. (b) Waveforms.   (b)

and the solution is given by

$$i_o = K e^{\frac{-t}{RC}}$$

where K is the constant of integration to be identified from boundary conditions.

Steady state emerges as $t \to \infty$. For a dc supply voltage, C is an open circuit and the current at steady state will be zero. So the steady state component is zero. Thus,

$$i_o = K e^{\frac{-t}{RC}} \tag{2.5}$$

At $t = 0$, when the switch is closed, $v_c = 0$ and hence $i_o = \dfrac{V_{dc}}{R}$. Thus, $\dfrac{V_{dc}}{R} = K$. Substituting for K,

$$i_o = \frac{V_{dc}}{R} e^{\frac{-t}{RC}} \tag{2.6}$$

The voltage across the capacitor is given by

$$v_c = \frac{1}{C} \int i_o \, dt = \frac{1}{C} \int \frac{V_{dc}}{R} e^{\frac{-t}{RC}} \, dt = \frac{V_{dc}}{RC} \left(-RC \, e^{\frac{-t}{RC}} \right) + K$$

where K is the constant of integration to be identified from boundary conditions.

$$v_c = -V_{dc} e^{\frac{-t}{RC}} + K$$

Substituting the initial condition that at $t = 0$, $v_c = 0$, and $0 = -V_{dc} + K$, then $K = V_{dc}$. Substituting for K,

$$v_c = V_{dc} \left(1 - e^{\frac{-t}{RC}} \right) \tag{2.7}$$

The plots of i_o and v_c are shown in Fig. 2.16(b).

EXAMPLE 2.4

A series combination of $R = 25\ \Omega$, $C = 10\ \mu F$ is switched on to $230\ V_{dc}$ at $t = 0$. (i) Find the current in the circuit at $t = 0.5$ ms. (ii) Find the voltage across capacitor at that time. (iii) Find the energy stored in the capacitor.

SOLUTION:

Given $R = 25$, $C = 10\ \mu f$, $V_{dc} = 230\ V$

i. $i_o = \dfrac{V_{dc}}{R} e^{\frac{-t}{RC}}$

$$= \frac{230}{25} \times e^{\frac{-\left(0.5 \times 10^{-3}\right)}{\left(25 \times 10 \times 10^{-6}\right)}}$$

$$= 1.245\ A$$

ii.
$$v_c = V_{dc}\left(1 - e^{\frac{-t}{RC}}\right)$$

$$= 230\left(1 - e^{\left(\frac{-\left(0.5 \times 10^{-3}\right)}{\left(25 \times 10 \times 10^{-6}\right)}\right)}\right)$$

$$= 198.87 \text{ V}$$

iii. energy stored in the capacitor $(w) = \dfrac{1}{2}CV^2$

$$v = v_C(\infty) = 230 \text{ V}$$
$$W = 0.5 \times 10 \times 10^{-6} \times 230^2$$
$$= 0.2645 \text{ J}$$

EXAMPLE 2.5

A series combination of R = 100 Ω, C = 25 μF is switched on to 80 V_{dc} at t = 0. Find the time at which the current reaches to 0.5 A and 0.25 A. Also find the voltage across capacitor at that time.

SOLUTION:

Given R = 100, C = 25 μF, V_{dc} = 80 V

i. For $i_o = 0.5$ A

$$i_o = \frac{V_{dc}}{R}\, e^{\frac{-t}{RC}}$$

Hence, $0.5 = \dfrac{80}{100}e^{\frac{-t}{0.0025}}$

$$t = -0.0025(\ln 0.625)$$
$$t = 1.175 \text{ ms.}$$
$$v_c = V_{dc}\left(1 - \exp^{-t/RC}\right)$$
$$= 80\left(1 - \exp^{-0.47}\right)$$
$$= 30 \text{ V}$$

ii. For $i_o = 0.25$ A

$$0.25 = \frac{80}{100}e^{\frac{-t}{0.0025}}$$
$$t = -0.0025(\ln 0.3125)$$
$$t = 2.908 \text{ ms}$$
$$v_c = V_{dc}\left(1 - e^{\frac{-t}{RC}}\right)$$
$$= 80\left(1 - e^{-1.1632}\right)$$
$$= 55 \text{ V}$$

2.7.2.4 *Ideal Switch with* RLC *Load*

Consider a constant voltage V_{dc} is applied to a RLC circuit as shown in Fig. 2.17(a) by closing the switch S. The transient current i_o can be found by applying KVL.

$$R \times i_o + L \frac{di_o}{dt} + \frac{1}{C} \int i_o \, dt = V_{dc}$$

Differentiating and rearranging the above equation, we get

$$\frac{d^2 i_o}{dt^2} + \frac{R}{L} \frac{di_o}{dt} + \frac{1}{LC} i_o = 0$$

Putting $\frac{d}{dt} = p$, the above equation will become $\left(p^2 + \frac{R}{L} p + \frac{1}{LC} \right) i_o = 0$.

$\therefore p^2 + \frac{R}{L} p + \frac{1}{LC} = 0$, the roots of the equations are given by $\frac{-R}{2L} \pm \sqrt{\left(\frac{R^2}{4L^2} - \frac{1}{LC} \right)}$.

There are three cases depending upon the roots of the equation.

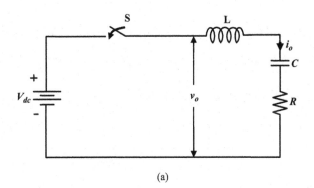

(a)

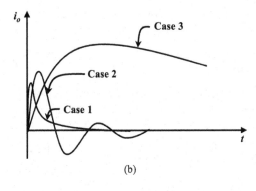

(b)

FIGURE 2.17
(a) Switched RLC load. (b) Waveforms.

Case 1:

For $\dfrac{R^2}{4L^2} > \dfrac{1}{LC}$, there will be two roots and both are real so that

$$i_o = K_1 e_1^{-\alpha t} + K_2 e_2^{-\alpha t} \qquad (2.8)$$

where α_1 and α_2 are the roots of the equation and are given as

$$\alpha_1 = \frac{-R}{2L} + \sqrt{\left(\frac{R^2}{4L^2} - \frac{1}{LC}\right)}$$

$$\alpha_2 = \frac{-R}{2L} - \sqrt{\left(\frac{R^2}{4L^2} - \frac{1}{LC}\right)}$$

Case 2:

For $\dfrac{R^2}{4L^2} < \dfrac{1}{LC}$, there will be two roots with imaginary terms such that the solution is

$$i_o = K_1 e^{-\alpha t}\left(A \cos \beta t + B \sin \beta t\right) \qquad (2.9)$$

where A and B are constants and α and β are the roots of the equation:

$$\alpha = \frac{-R}{2L} \text{ and } \beta = \sqrt{\left(\frac{R^2}{4L^2} - \frac{1}{LC}\right)}$$

Case 3:

For $\dfrac{R^2}{4L^2} = \dfrac{1}{LC}$, there will be two equal roots such that the solution is

$$i_o = (K_1 + K_2 t)e^{-\alpha t} \qquad (2.10)$$

where α is the root of the equation:

$$\alpha = \frac{-R}{2L}$$

The variations of i_o for all the three cases are shown in Fig. 2.17(b).

2.7.3 Circuits with Diodes

With the insertion of a PN junction diode, consistent transients take place in a circuit and are elaborated in this section.

2.7.3.1 Diode with R Load

The circuit diagram is shown in Fig. 2.18(a). The switch is closed at t = 0. The diode conducts for positive half-cycle of the supply while it is an open circuit with negative portion of the supply. Thus,

$$R \times i_o = V_m \sin \omega_s t \qquad 0 \le \omega_s t \le \pi$$

$$= 0 \qquad \pi \le \omega_s t \le 2\pi$$

$$\therefore i_o = \frac{V_m \sin \omega_s t}{R}$$

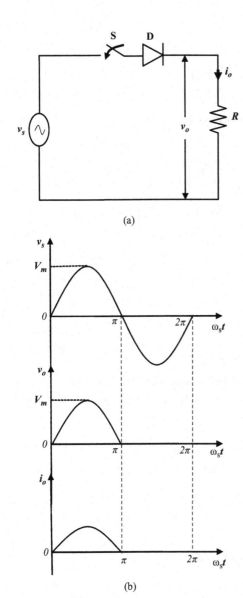

(a)

(b)

FIGURE 2.18
(a) Switched R-load with diode. (b) Waveforms.

The variation of v_s, v_o, and i_o is shown in Fig. 2.18(b).
The average value of i_o is given by

$$I_{o(avg)} = \frac{1}{2\pi} \int_0^\pi \frac{V_m \sin \omega_s t}{R} d\omega_s t$$

$$= \frac{V_m}{2\pi R} [-\cos \omega_s t]_0^\pi$$

$$= \frac{V_m}{\pi R}$$

The rms value of the current is given by

$$I_{o(rms)}^2 = \frac{1}{2\pi} \int_0^\pi \frac{V_m^2 \sin^2 \omega_s t}{R^2} d\omega_s t$$

$$= \frac{V_m^2}{2\pi R^2} \int_0^\pi \left[\frac{1 - \cos 2\omega_s t}{2} \right] d\omega_s t$$

$$= \frac{V_m^2}{4\pi R^2} \left[\omega_s t - \frac{\sin 2\omega_s t}{2} \right]_0^\pi$$

$$= \frac{V_m^2}{4\pi R^2} \left[\pi - \frac{\sin 2\pi}{2} - \left(0 - \frac{\sin 0}{2} \right) \right]$$

$$= \frac{V_m^2}{4R^2}$$

$$I_{o(rms)} = \frac{V_m}{2R}$$

EXAMPLE 2.6

An electric heater rated 2 kW, 220 V, and 50 Hz is supplied from a 230-V, 50-Hz ac main with a semiconductor diode in series. Calculate the average rms currents and the power delivered to heater.

Heater coil resistance, $R_h = \dfrac{(\text{Rated Voltage})^2}{\text{Rated Power}} = \dfrac{220^2}{2000} = 24.2\ \Omega$

Average current, $I_{o(avg)} = \dfrac{V_m}{\pi R} = \dfrac{230\sqrt{2}}{\pi \times 24.2} = 4.27\ A$

rms value of current, $I_{o(rms)} = \dfrac{V_m}{2R} = \dfrac{230\sqrt{2}}{2 \times 24.2} = 6.720\ A$

Power delivered to heater is given by $I_{o(rms)}^2 \times R_h = 6.720^2 \times 24.2 = 1.092\ kW$

2.7.3.2 Diode with RC load

The circuit diagram is shown in Fig. 2.19(a). The switch is closed at the time t = 0. Here,

$$Ri_o + \frac{1}{C} \int i_o\, dt = V_m \sin \omega_s t \quad 0 < \omega_s t < \beta \tag{2.11}$$

The transient component is given by solving

$$R \times i_o + \frac{1}{C} \int i_o\, dt = 0$$

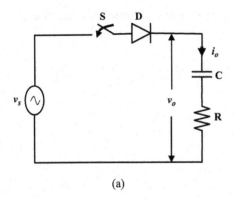

(a)

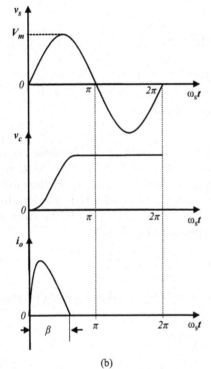

(b)

FIGURE 2.19
(a) Switched RC load with diode. (b) Waveforms.

Differentiating and rearranging the equation,

$$\frac{di_o}{dt} + \frac{1}{RC} i_o = 0$$

$$\left(p + \frac{1}{RC} \right) i_o = 0$$

$\therefore p = -\dfrac{1}{RC}$ and the solution is given by

$$i_o = Ke^{\frac{-t}{RC}}$$

where K is the constant of integration to be identified from boundary conditions. The steady state component is given by

$$\frac{V_m}{Z}\sin(\omega_s t + \phi),$$

where $Z = \sqrt{(R^2 + X_C^2)}$ and $\phi = \tan^{-1}\left(\frac{X_c}{R}\right)$.

Thus,

$$i_o = Ke^{\frac{-t}{RC}} + \frac{V_m}{Z}\sin(\omega_s t + \phi) \tag{2.12}$$

Substituting the initial condition that $i_o = 0$ at $t = 0$,

$$K = -\frac{V_m}{Z}\sin\phi$$

Substituting the value of Kin Equation (2.12) gives,

$$i_o = \frac{V_m}{Z}\left[\sin(\omega_s t + \phi) - \sin\phi \, e^{\frac{-t}{RC}}\right] \tag{2.13}$$

In the RC circuit, current leads the voltage. Hence, the current will become zero at an angle β which will be less than π. At $\omega_s t = \beta$, the device turns off and $v_c < V_m$. At π, the voltage goes to zero and the capacitor will be charged to a value less than V_m. In the next positive cycle, the diode will start conducting after $V_m \sin \omega_s t > v_c$, and it will stop conducting at an angle $\beta_1 < \beta$. This process will continue until the capacitor charges to the value of V_m. The variation of v_s, v_c, and i_o for the first cycle is shown in Fig. 2.19(b).

2.7.3.3 Diode with RL Load

The circuit diagram is shown in Fig. 2.20(a). The switch is closed at $t = 0$. Applying KVL during positive half cycle,

$$R \times i_o + L\frac{di_o}{dt} = V_m \sin \omega_s t \quad 0 < \omega_s t < \beta \tag{2.14}$$

The transient component is obtained by forcing the right-hand part of Equation (2.14) to zero.

$$L\frac{di_o}{dt} + R \times i_o = 0,$$

$$\frac{di_o}{dt} + \frac{R}{L}\times i_o = 0,$$

$$\left(p + \frac{R}{L}\right)i_o = 0$$

and the solution is given by

$$i_o = Ke^{\frac{-R}{L}t}$$

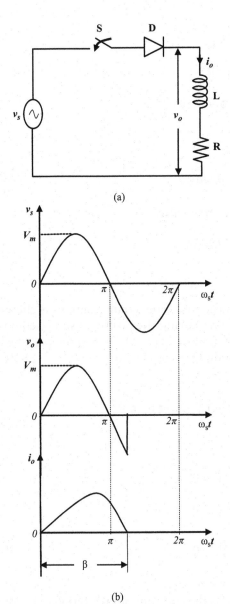

FIGURE 2.20
(a) Switched RL load with diode. (b) Waveforms.

where K is the constant of integration to be identified from boundary conditions.

The steady state component is given by $\dfrac{V_m}{Z}\sin(\omega_s t - \phi)$ where

$$Z = \sqrt{\left(R^2 + X_L^2\right)} \text{ and } \phi = \tan^{-1}\left(\frac{X_L}{R}\right)$$

(2.15)

$$i_o = Ke^{\frac{-R}{L}t} + \frac{V_m}{Z}\sin(\omega_s t - \phi)$$

Substituting the initial condition, the value of K can be determined.

At $t = 0$, $i_o = 0$, therefore $K = \dfrac{V_m}{Z} \sin \phi$.

Substituting the value of K in Equation (2.15) gives

$$i_o = \frac{V_m}{Z}\left(\sin(\omega_s t - \phi) + \sin\phi\, e^{\frac{-R}{L}t}\right)$$ (2.16)

At $\omega_s t = \beta$, $i_o = 0$, or

$$0 = \frac{V_m}{Z}\left(\sin(\beta - \phi) + \sin\phi\, e^{\frac{-R\beta}{L\omega_s}}\right)$$ (2.17)

β can be obtained by solving Equation (2.17) numerically.

The variation of v_o, v_s, and i_o is shown in Fig. 2.20(b).

2.7.3.4 R-L *Circuit with a Freewheeling Diode*

The load with a freewheeling diode is now considered. The circuit diagram is shown in Fig. 2.21(a). There are two modes of operation, one during the positive half cycle and the other during the negative half cycle. The switch is closed at $t = 0$. During the positive half cycle, diode D_1 will be forward biased and diode D_2 will be in reverse bias. So KVL for mode 1 can be written as

$$R \times i_o + L\frac{di_o}{dt} = V_m \sin\omega_s t \quad 0 < \omega_s t < \pi$$ (2.18)

The transient component is given by

$$L\frac{di_o}{dt} + R \times i_o = 0$$

$$\frac{di_o}{dt} + \frac{R}{L}i_o = 0$$

$$\left(p + \frac{R}{L}\right)i_o = 0$$

and the solution is given by $i_o = Ke^{\frac{-R}{L}t}$, where K is the constant of integration to be identified from boundary conditions.

The steady state component is given by $\dfrac{V_m}{Z}\sin(\omega_s t - \phi)$. Thus,

$$i_o = Ke^{\frac{-R}{L}t} + \frac{V_m}{Z}\sin(\omega_s t - \phi).$$

Substituting the initial condition that at $t = 0$, $i_o = I_o$,

$$I_o = K + \frac{V_m}{Z}\sin(-\phi)$$

The value of K is given by $K = I_o + \dfrac{V_m}{Z}\sin\phi$.

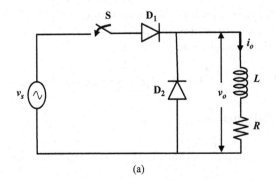

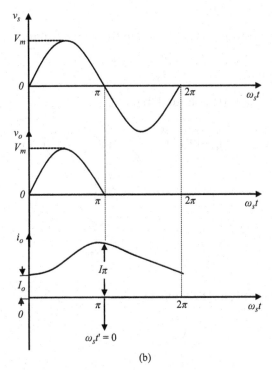

FIGURE 2.21
(a) Switched R-L load with freewheeling diode. (b) Waveforms.

Substituting the value of K in the above equation yields

$$i_o = \left(I_o + \frac{V_m}{Z} \sin\phi \right) e^{\frac{-R}{L}t} + \frac{V_m}{Z} \sin(\omega_s t - \phi)$$

At $\omega_s t = \pi \left(\text{i.e., } t = \dfrac{\pi}{\omega_s} \right)$, $i_o = I_\pi$ and hence

$$I_\pi = \left(I_o + \frac{V_m}{Z} \sin\phi \right) e^{\frac{-R\pi}{L\omega_s}} + \frac{V_m}{Z} \sin(\pi - \phi)$$

$$= \left(I_o + \frac{V_m}{Z} \sin\phi \right) e^{\frac{-R\pi}{L\omega_s}} + \frac{V_m}{Z} \sin(\phi)$$

(2.19)

During negative half cycle, which is the second mode (also known as the freewheeling mode), diode D_1 will be in reverse bias and diode D_2 will be in forward bias. The load current freewheels through diode D_2. Thus,

$$R \times i_o + L \frac{di_o}{dt'} = 0, \ 0 \le \omega_s t' \le \pi \text{ where } \omega_s t' = \omega_s t - \pi$$

Solving the differential equation,

$$i_o = K_2 e^{\frac{-R}{L}t'}. \text{ At } \omega_s t' = 0, \ i_o = I_\pi \text{ and hence } K_2 = I_\pi.$$

$$\text{Thus } i_o = I_\pi \, e^{\frac{-R}{L}t'}$$

at $\omega_s t' = \pi$, $i_o = I_o$. Substituting this condition in the above equation yields

$$I_o = I_\pi e^{\frac{-\pi R}{L\omega_s}} \tag{2.20}$$

Substituting the value of I_π from Equation (2.19),

$$I_o = \left[I_o e^{\frac{-\pi R}{L\omega}} + \frac{V_m}{Z} \sin\phi \, e^{\frac{-\pi R}{L\omega}} + \frac{V_m}{Z} \sin\phi \right] e^{\frac{-\pi R}{L\omega_s}}$$

$$I_o \left(e^{\frac{\pi R}{L\omega_s}} - e^{\frac{-\pi R}{L\omega}} \right) = \frac{V_m}{Z} \sin\phi \left(1 + e^{\frac{-\pi R}{L\omega_s}} \right) \tag{2.21}$$

$$I_o = \frac{V_m}{Z} \sin\phi \left(\frac{1 + e^{\frac{-\pi R}{L\omega_s}}}{e^{\frac{\pi R}{L\omega_s}} - e^{\frac{-\pi R}{L\omega_s}}} \right)$$

Substituting the value of I_o in (2.20) yields

$$I_\pi = \frac{V_m}{Z} \sin\phi \left(\frac{1 + e^{\frac{\pi R}{L\omega_s}}}{e^{\frac{\pi R}{L\omega_s}} - e^{\frac{-\pi R}{L\omega_s}}} \right) \tag{2.22}$$

The variation of v_s, v_o, and i_o is shown in Fig. 2.21(b).

EXAMPLE 2.7

An RL load with a freewheeling diode connected in parallel is supplied from a 220-V, 50-Hz ac main through a single diode. Calculate the current at $\omega t = 0$ and $\omega t = \pi$ if $R = 10 \ \Omega$ and $L = 100$ mH.

Angular frequency, $\omega_s = 2\pi f = 2\pi \times 50 = 314.159$ rad/s

Inductive reactance, $X_L = L\omega_s = 100 \times 10^{-3} \times 314.159 = 31.42 \ \Omega$

Load impedance, $Z = \sqrt{R^2 + X_L^2} = \sqrt{10^2 + 31.42^2} = 32.97 \ \Omega$

Power factor angle, $\phi = \tan^{-1} \frac{L\omega_s}{R} = \tan^{-1} \frac{31.42}{10} = 72.34^0$

Load current at $\omega_s t = 0$ is given by,

$$I_o = \frac{V_m}{Z} \sin \phi \left(\frac{1 + e^{\frac{-\pi R}{L\omega_s}}}{e^{\frac{\pi R}{L\omega_s}} - e^{\frac{-\pi R}{L\omega_s}}} \right)$$

$$= \frac{220\sqrt{2}}{32.97} \sin 72.34 \left(\frac{1 + e^{\frac{-\pi \times 10}{31.42}}}{e^{\frac{\pi \times 10}{31.42}} - e^{\frac{-\pi \times 10}{31.42}}} \right) = 0.58 \text{ A}$$

Load current at $\omega_s t = \pi$ is given by,

$$I_\pi = \frac{V_m}{Z} \sin \phi \left(\frac{1 + e^{\frac{\pi R}{L\omega_s}}}{e^{\frac{\pi R}{L\omega_s}} - e^{\frac{-\pi R}{L\omega_s}}} \right)$$

$$= \frac{220\sqrt{2}}{32.97} \sin 72.34 \left(\frac{1 + e^{\frac{\pi \times 10}{31.42}}}{e^{\frac{\pi \times 10}{31.42}} - e^{\frac{-\pi \times 10}{31.42}}} \right) = 1.58 \text{ A}$$

2.7.3.5 Diode with R-L-E_b Load

Consider a voltage $V_m \sin \omega_s t$ applied to a R-L-E_b circuit shown in Fig. 2.22(a) by closing the switch S. Referring to the figure, it is evident that diode D is reverse biased until $v_s > E_b$. Thus, during the positive half cycle, when $v_s = E_b$, diode starts conducting. Let $\omega_s t = \eta$ for $v_s = E_b$. From the figure, $V_m \sin \omega_s t |_{\omega_s t = \eta} = E_b$.

$$V_m \sin \eta = E_b$$

$$\therefore \eta = \sin^{-1} \left(\frac{E_b}{V_m} \right)$$

The transient current i_o can be found by applying KVL.

$$Ri_o + L\frac{di_o}{dt} + E_b = V_m \sin \omega_s t \quad \eta < \omega_s t < \beta \qquad (2.23)$$

The transient component is obtained as follows.

$$L\frac{di_o}{dt} + R \times i_o = 0$$

$$\frac{di_o}{dt} + \frac{R}{L}i_o = 0$$

$$\left(p + \frac{R}{L} \right) i_o = 0$$

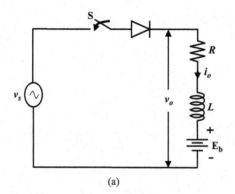

(a)

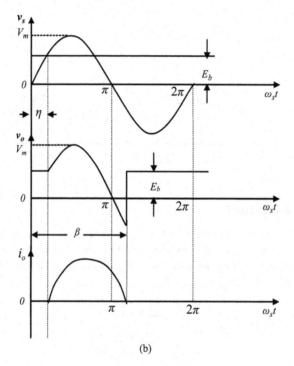

(b)

FIGURE 2.22
(a) Diode with R-L-E_b load. (b) Waveforms.

and the solution is given by

$$i_o = Ke^{\frac{-R}{L}t}$$

where K is the constant of integration to be identified from boundary conditions. The steady state current has two parts, one is from the source voltage and other from back emf, which will be in opposite direction, i.e., $\dfrac{V_m}{Z}\sin(\omega_s t - \phi) - \dfrac{E_b}{R}$, where $Z = \sqrt{\left(R^2 + X_L^2\right)}$, $\phi = \tan^{-1}\left(\dfrac{X_L}{R}\right)$. Thus,

$$i_o = Ke^{\frac{-R}{L}t} + \frac{V_m}{Z}\sin\left(\omega_s t - \phi\right) - \frac{E_b}{R} \tag{2.24}$$

Substituting the initial condition, i.e., at $\omega_s t = \eta$, $i_o = 0$, then

$$K = \left[\frac{E_b}{R} - \frac{V_m}{Z}\sin(\eta - \phi)\right]e^{\frac{\eta R}{L\omega_s}}$$

Substituting the value of K in Equation (2.24), we get

$$i_o = \left[\frac{E_b}{R} - \frac{V_m}{Z}\sin(\eta - \phi)\right]e^{\frac{-R}{L}\left(t - \frac{\eta}{\omega_s}\right)} + \frac{V_m}{Z}\sin(\omega_s t - \phi) - \frac{E_b}{R} \tag{2.25}$$

The variation of v_o, v_s, and i_o is shown in Fig. 2.22(b).

2.7.3.6 *Diode with R-E$_b$ Load*

Consider a voltage $V_m \sin \omega_s t$ is connected to a R-E$_b$ circuit shown in Fig. 2.23(a) by closing the switch S. The current i_o can be found by applying KVL.

$$R \times i_o + E_b = V_m \sin \omega_s t, \quad \eta < \omega_s t < \beta \tag{2.26}$$

$$i_o = \frac{V_m \sin \omega_s t - E_b}{R} \tag{2.27}$$

The average value of current is given by

$$I_{o(av)} = \frac{1}{2\pi R}\left[\int_{\eta}^{\pi - \eta}(V_m \sin \omega_s t - E_b)d\omega_s t\right]$$

$$= \frac{1}{2\pi R}\left[\left[-V_m \cos \omega_s t\right]_{\eta}^{\pi - \eta} - E_b\left[\omega_s t\right]_{\eta}^{\pi - \eta}\right]$$

$$= \frac{1}{2\pi R}\left[\left[-V_m \cos(\pi - \eta) + V_m \cos \eta\right] - E_b\left[\pi - \eta - \eta\right]\right]$$

$$= \frac{1}{2\pi R}\left[\left[V_m \cos \eta + V_m \cos \eta\right] - E_b\left[\pi - 2\eta\right]\right]$$

Thus,

$$I_{o(av)} = \frac{1}{2\pi R}\left[2V_m \cos \eta - E_b\left[\pi - 2\eta\right]\right] \tag{2.28}$$

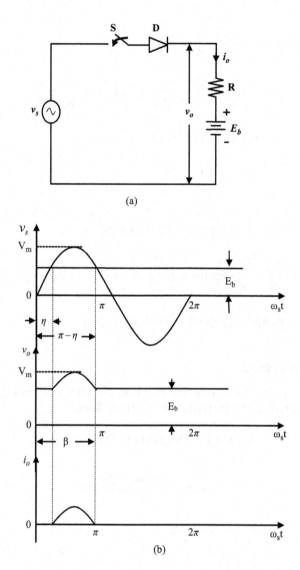

FIGURE 2.23
(a) Diode with R-E_b load. (b) Waveforms.

Similarly, the rms value of the current is given by

$$I_{o(rms)}^2 = \frac{1}{2\pi R^2}\left[\int_{\eta}^{\pi-\eta}\left(V_m\sin\omega_st - E_b\right)^2 d\omega_st\right]$$

$$= \frac{1}{2\pi R^2}\left[\int_{\eta}^{\pi-\eta}\left[V_m^2\sin^2\omega_st + E_b^2 - 2V_mE_b\sin\omega_st\right]d\omega_st\right]$$

$$= \frac{1}{2\pi R^2}\left[\int_{\eta}^{\pi-\eta}\left[\frac{V_m^2}{2} - \frac{V_m^2\cos 2\omega_st}{2} + E_b^2 - 2V_mE_b\sin\omega_st\right]d\omega_st\right]$$

$$= \frac{1}{2\pi R^2}\left[\frac{V_m^2}{2}[\omega_s t]_\eta^{\pi-\eta} - \frac{V_m^2}{4}[\sin 2\omega_s t]_\eta^{\pi-\eta} - 2V_m E_b[-\cos \omega_s t]_\eta^{\pi-\eta} + E_b^2[\omega_s t]_\eta^{\pi-\eta} \right]$$

$$= \frac{1}{2\pi R^2}\left[\frac{V_m^2}{2}[\pi-\eta-\eta] - \frac{V_m^2}{4}[\sin(2\pi-2\eta)-\sin 2\eta] + E_b^2[\pi-\eta-\eta] - 2V_m E_b[-\cos(\pi-\eta)+\cos\eta] \right]$$

$$= \frac{1}{2\pi R^2}\left[\frac{V_m^2}{2}[\pi-2\eta] + \frac{V_m^2}{4}[\sin 2\eta] + \frac{V_m^2}{4}[\sin 2\eta] + E_b^2[\pi-2\eta] - 2V_m E_b[\cos\eta+\cos\eta] \right]$$

$$I_{o(rms)}^2 = \frac{1}{2\pi R^2}\left[\frac{V_m^2}{2}(\pi-2\eta) + \frac{V_m^2}{2}\sin 2\eta + E_b^2(\pi-2\eta) - 4V_m E_b\cos\eta \right]$$

$$I_{o(rms)} = \left[\frac{1}{2\pi R^2}\left[\frac{V_m^2}{2}(\pi-2\eta) + \frac{V_m^2}{2}\sin 2\eta + E_b^2(\pi-2\eta) - 4V_m E_b\cos\eta \right] \right]^{\frac{1}{2}} \qquad (2.29)$$

$$\text{Input power factor} = \frac{\text{o/p power}}{V_{rms}I_{o(rms)}} = \frac{I_{o(rms)}^2 R + I_{o(av)}E_b}{V_{rms}I_{o(rms)}}$$

$$= \frac{I_{o(rms)}^2 R + I_{o(av)}E_b}{\dfrac{V_m}{\sqrt{2}}I_{o(rms)}} \qquad (2.30)$$

The variation of v_o, v_s, and i_o is shown in Fig. 2.23(b).

EXAMPLE 2.8

An ac voltage of rms value 110 V is used to charge a battery with an internal resistance of 1.0 Ω and a voltage of 110 V through the diode. Find the average value, the rms value of the current, and the power factor.

SOLUTION:

Given $V_m = 110\sqrt{2}$ V, $E_b = 110$ V, $r_a = 1.0\,\Omega$,

$$\eta = \sin^{-1}\left(\frac{E_b}{V_m}\right)$$

$$\eta = \sin^{-1}\left(\frac{110}{110\sqrt{2}}\right)$$

$$\eta = \sin^{-1}\left(\frac{1}{\sqrt{2}}\right) = 45° = 0.707 \text{ radian}$$

The rms value of current is computed as

$$I_{0(rms)}^2 = \frac{1}{2\pi R^2}\left[\frac{V_m^2}{2}[\pi-2\eta+\sin 2\eta] + E_b^2[\pi-2\eta] - 4V_m E_b\cos\eta \right]$$

$$= \frac{1}{2\pi(1.0)^2}\left[\frac{\left(110\sqrt{2}\right)^2}{2}[\pi-2(\pi/4)+\sin 2(\pi/4)] + 110^2[\pi-2(\pi/4)] - 4\times 110\sqrt{2}\times 110\cos(\pi/4) \right]$$

$$= \frac{1}{2\pi}\left[110^2[\pi/2+\sin(\pi/2)] + 110^2(\pi/2) - 4\sqrt{2}(110)^2\cos(\pi/4) \right]$$

$$I^2_{o(rms)} = \frac{110^2}{2\pi}[\pi + 1 - 4] = 272.67 \text{ A}$$

$$I_{o(rms)} = 16.51 \text{ A}$$

The average value of current is calculated as

$$I_{o(av)} = \frac{1}{2\pi R}\left[2V_m \cos\eta - E_b(\pi - 2\eta)\right]$$

$$= \frac{1}{2\pi(1.0)}\left[2 \times 110\sqrt{2} \times \cos(\pi/4) - 110 \times (\pi - 2\pi/4)\right]$$

$$I_{o(av)} = 7.51 \text{ A}$$

$$\text{Input power factor} = \left(\frac{I^2_{o(rms)} \times R + I_{o(av)} \times E_b}{V_{rms} \times I_{o(rms)}}\right)$$

$$\text{Input power factor} = \left(\frac{(16.51)^2 \times (1.0) + (7.51) \times (110)}{(110) \times (16.51)}\right)$$

$$= 0.6052\,(\text{lagging})$$

EXAMPLE 2.9

An ac voltage of rms value 230 V is used to charge a battery with an internal resistance of 5.0 Ω and a voltage of 180 V through a diode. Find the average value, the rms value of the current, and the power factor.

SOLUTION:

Given $V_m = 230\sqrt{2}$ V, $E_b = 180$ V, $r_a = 5.0$ Ω,

$$\eta = \sin^{-1}\left(\frac{E_b}{V_m}\right)$$

$$\eta = \sin^{-1}\left(\frac{180}{230\sqrt{2}}\right)$$

$$\eta = \sin^{-1}(0.55) = 0.58 \text{ radian}$$

The rms value of the current is given by

$$I^2_{o(rms)} = \frac{1}{2\pi R^2}\left[\frac{V^2_m}{2}[\pi - 2\eta + \sin 2\eta] + Eb^2[\pi - 2\eta] - 4V_m E_b \cos\eta\right]$$

$$= \frac{1}{2 \times \pi \times 5^2}\left[\frac{(230\sqrt{2})^2}{2}[\pi - 2 \times 0.58 + \sin(2 \times 0.58)] + 180^2[\pi - (2 \times 0.58) - 4 \times (230\sqrt{2}) \times 180 \times \cos(0.58)]\right]$$

$$= \frac{1}{50\pi}[21641.25] = 137.77$$

$$I_{o(rms)} = 11.7377 \text{ A}$$

The average current is computed as

$$I_{o(av)} = \frac{1}{2\pi R}\left[2V_m \cos\eta - E_b(\pi - 2\eta)\right]$$

$$I_{o(av)} = \frac{1}{2\pi(5.0)}\left[2\times 230\sqrt{2}\times\cos(.58) - (180)\times(\pi - 2(.58))\right]$$

$$I_{o(av)} = 5.9667\ A$$

$$\text{Input power factor} = \left(\frac{I_{o(rms)}^2 \times r_a + I_{o(av)} \times E_b}{V_{rms} \times I_{o(rms)}}\right)$$

$$\text{Input power factor} = \left(\frac{(11.737)^2 \times (5.0) + (5.96)\times(180)}{(230)\times(11.73)}\right)$$

$$= 0.6530(\text{lagging})$$

EXAMPLE 2.10

An ac voltage of rms value 330 V is used to charge a battery with an internal resistance of 10.0 Ω and a voltage of 300 V through a diode. Find the average value, the rms value of the current, and the power factor.

SOLUTION:

Given $V_m = 330\sqrt{2}$ V, $E_b = 300$ V, $r_a = 10\Omega$,

The angle at which current starts conducting is given by

$$\eta = \sin^{-1}\left(\frac{E_b}{V_m}\right)$$

$$\eta = \sin^{-1}\left(\frac{300}{330\sqrt{2}}\right)$$

$$\eta = 0.69\ \text{radian}$$

The rms current is

$$I_{o(rms)}^2 = \frac{1}{2\pi R^2}\left[\frac{V_m^2}{2}[\pi - 2\eta + \sin 2\eta] + E_b^2[\pi - 2\eta] - 4V_m E_b \cos\eta\right]$$

$$= \frac{1}{2\pi(10.0)^2}\left[\frac{\left(330\sqrt{2}\right)^2}{2}[\pi - 2\times 0.69 + \sin 2\times 0.69] + 300^2[\pi - 2\times 0.69] - 4\times 330\sqrt{2}\times 300\times\cos 0.69\right]$$

$$= 40.39\ A$$

$$I_{o(rms)} = 6.3559\ A$$

The average value of the current is

$$I_{o(av)} = \frac{1}{2\pi(10.0)}\left[2 \times 330\sqrt{2} \times \cos(.69) - (300) \times (\pi - 2(.69))\right]$$

$$I_{o(av)} = 3.0464 \text{ A}$$

$$\text{Input power factor} = \left(\frac{I_{o(rms)}^2 r_a + I_{o(av)} E_b}{V_{rms} I_{o(rms)}}\right)$$

$$\text{Input power factor} = \left(\frac{(6.3559)^2 \times (10.0) + (3.04) \times (300)}{(330) \times (6.355)}\right)$$

$$= 0.6283\,(\text{lagging})$$

EXAMPLE 2.11

An ac voltage of rms value 200 V is used to charge a battery with an internal resistance of 7.5 Ω and voltage of 40 V through a diode. Find the average value, the rms value of the current, and the power factor.

SOLUTION:

Given $V_m = 200\sqrt{2}$ V, $E_b = 40$ V, $r_a = 7.5\,\Omega$, first we have to evaluate the value of

$$\eta = \sin^{-1}\left(\frac{E_b}{V_m}\right)$$

$$\eta = \sin^{-1}\left(\frac{40}{200\sqrt{2}}\right)$$

$$\eta = \sin^{-1}(0.1414) = 0.14 \text{ radian}$$

To find the rms value of the current,

$$I_{o(rms)}^2 = \frac{1}{2\pi R^2}\left[\frac{V_m^2}{2}[\pi - 2\eta + \sin 2\eta] + E_b^2[\pi - 2\eta] - 4V_m E_b \cos\eta\right]$$

$$= \frac{1}{2\pi(7.5)^2}\left[\frac{\left(200\sqrt{2}\right)^2}{2}[\pi - 2(0.14) + \sin 2(0.14)] + 40^2[\pi - 2(0.14)] - 4\left(200\sqrt{2}\right)(40)\cos(0.14)\right]$$

$$I_{o(rms)}^2 = 241.30 \text{ A}$$
$$I_{o(rms)} = 15.534 \text{ A}$$

To find average value of the current,

$$I_{o(av)} = \frac{1}{2\pi(7.5)}\left[2\left(200\sqrt{2}\right)\cos(0.14) - (300)(\pi - 2(0.14))\right]$$

$$I_{o(av)} = 9.4572 \text{ A}$$

$$\text{Input power factor} = \left(\frac{I_{o(rms)}^2 \times r_a + I_{o(av)} \times E_b}{V_{rms} \times I_{o(rms)}}\right)$$

$$\text{Input power factor} = \left(\frac{(15.534)^2 \times (7.5) + (9.45) \times (40)}{(200) \times (15.53)}\right) = 0.7043\,(\text{lagging})$$

2.7.4 Circuits with Thyristors

Voltage regulation at the output is possible with the introduction of any switching device in the circuit. The analysis of switched circuits with SCR is described below.

2.7.4.1 Thyristor with R Load

The circuit diagram is shown in Fig. 2.24(a). The thyristor is turned on at $\omega_s t = \alpha$, and due to the resistive nature of load, it is turned off at $\omega_s t = \pi$. The KVL can be written as

$$R \times i_o = V_m \sin \omega_s t, \, \alpha < \omega_s t < \pi$$
$$i_o = \frac{V_m \sin \omega_s t}{R}, \, \alpha < \omega_s t < \pi \tag{2.31}$$

The variation of v_o, v_s, and i_o is shown in Fig. 2.24(b). From Fig. 2.24(b), the average output voltage and rms value of output current are derived below.

$$\text{Average output voltage } V_{o(av)} = \frac{1}{2\pi} \int_{\alpha}^{\pi} V_m \sin \omega_s t \, d(\omega_s t)$$

$$= \frac{V_m}{2\pi} (-\cos \omega_s t)_{\alpha}^{\pi}$$

$$= \frac{V_m}{2\pi} (-\cos \pi + \cos \alpha)$$

$$= \frac{V_m}{2\pi} (1 + \cos \alpha) \tag{2.31A}$$

The rms value of the output voltage is

$$i_{o(rms)}^2 = \frac{1}{2\pi} \int_{\alpha}^{\pi} \frac{V_m^2 \sin^2 \omega_s t}{R^2} d(\omega_s t)$$

$$= \frac{V_m^2}{2\pi R^2} \int_{\alpha}^{\pi} \frac{1 - \cos 2\omega_s t}{2} d(\omega_s t)$$

$$= \frac{V_m^2}{4\pi R^2} \int_{\alpha}^{\pi} (1 - \cos 2\omega_s t) d(\omega_s t)$$

$$= \frac{V_m^2}{4\pi R^2} \left[\omega_s t - \frac{\sin 2\omega_s t}{2} \right]_{\alpha}^{\pi}$$

$$= \frac{V_m^2}{4\pi R^2} \left[\pi - \alpha - \left[\frac{\sin 2\pi}{2} - \frac{\sin 2\alpha}{2} \right] \right]$$

$$= \frac{V_m^2}{4\pi R^2} \left[\pi - \alpha + \frac{\sin 2\alpha}{2} \right]$$

$$i_{o(rms)} = \frac{V_m}{2R} \left[\frac{1}{\pi} \left[\pi - \alpha + \frac{\sin 2\alpha}{2} \right] \right]^{\frac{1}{2}} \tag{2.31B}$$

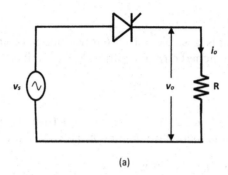

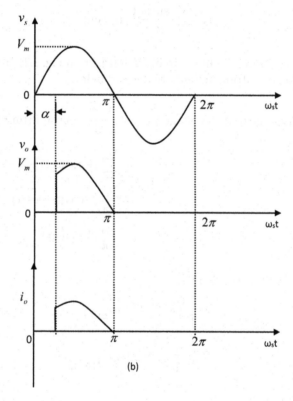

FIGURE 2.24
(a) R load with SCR. (b) Waveforms.

EXAMPLE 2.12

A resistor of 100 Ω is connected to $230\sqrt{2}\sin(314t)$ through a single thyristor fired at $\alpha = 40°$. Compute the supply power factor.

$$\alpha = 40° = 40° \times \frac{\pi}{180°} = 0.698 \text{ rad}$$

$$i_{o(rms)} = \frac{V_m}{2R}\left[\frac{1}{\pi}\left[\pi - \alpha + \frac{\sin 2\alpha}{2}\right]\right]^{\frac{1}{2}}$$

$$i_{o(rms)} = \frac{230\sqrt{2}}{2 \times 100}\left[\frac{1}{\pi}\left[\pi - 0.698 + \frac{\sin(2 \times 0.698)}{2}\right]\right]^{\frac{1}{2}} = 1.575 \text{ A}$$

$$\text{Supply power factor} = \frac{\text{Output power}}{V_{rms} \times i_{rms}}$$

$$\text{Because } i_{rms} = i_{o(rms)}$$

$$\text{Supply power factor} = \frac{i_{o(rms)}^2 \times R}{V_{rms} \times i_{o(rms)}} = \frac{1.575^2 \times 100}{230 \times 1.575} = 0.685 \text{(lagging)}$$

2.7.4.2 Thyristor with RL Load

The circuit diagram is shown in Fig. 2.25(a). At $\omega_s t = \alpha$, when thyristor is turned on, the KVL can be written as

$$R \times i_o + L\frac{di_o}{dt} = V_m \sin \omega_s t, \ \alpha < \omega_s t < \beta \tag{2.32}$$

where β is the angle at which current ceases to zero.

The transient component is obtained by forcing the right-hand part to zero.

$$L\frac{di_o}{dt} + R \times i_o = 0$$

$$\frac{di_o}{dt} + \frac{R}{L}i_o = 0$$

$$\left(p + \frac{R}{L}\right)i_o = 0.$$

and the solution is given by $i_o = Ke^{\frac{-R}{L}t}$, where K is the constant of integration to be identified from boundary conditions.

The steady state component is given by $\frac{V_m \sin(\omega_s t - \phi)}{Z}$, where

$$Z = \sqrt{(R^2 + X_L^2)}, \phi = \tan^{-1}\left(\frac{X_L}{R}\right)$$

$$i_o = Ke^{\frac{-R}{L}t} + \frac{V_m \sin(\omega_s t - \phi)}{Z} \tag{2.33}$$

Substituting the initial condition, the value of K can be found.
At $\omega_s t = \alpha$ and $i_o = 0$,

$$0 = Ke^{\frac{-\alpha R}{L\omega_s}} + \frac{V_m \sin(\alpha - \phi)}{Z}$$

$$K = -\frac{V_m \sin(\alpha - \phi)}{Z}e^{\frac{\alpha R}{L\omega_s}}$$

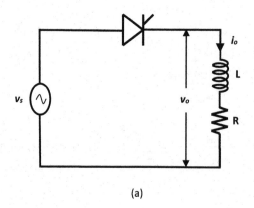

(a)

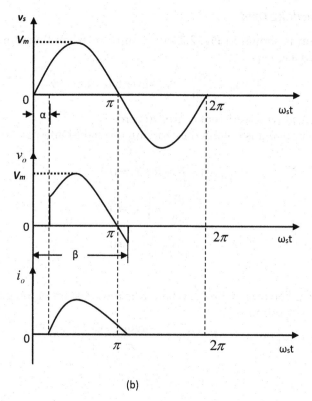

(b)

FIGURE 2.25
(a) R-L load with SCR. (b) Waveforms.

Substituting in the above equation, we get

$$i_o = \frac{V_m \sin(\omega_s t - \phi)}{Z} - \frac{V_m \sin(\alpha - \phi)}{Z} e^{\frac{R}{L}\left(\frac{\alpha}{\omega_s} - t\right)} \qquad (2.34)$$

The variation of v_o, v_s, and i_o is shown in Fig. 2.25(b).

EXAMPLE 2.13

An R-L load is supplied from a 100-V dc supply through one SCR with a gate-control circuit. Calculate the minimum duty ratio of the gate pulse at 1 kHz. Given a latching current of SCR = 1,000 mA, a holding current of 500 mA, R = 10 Ω, L = 50 mH.

SOLUTION:

Gate pulse has to be maintained until the anode current rises to latching current.

For an RL circuit,

$$V = L\frac{di(t)}{dt} + Ri(t)$$

$$i(t) = \frac{V}{R}\left(1 - e^{-\frac{R}{L}t}\right)$$

$$1,000 \times 10^{-3}\,A = \frac{100}{10}\left(1 - e^{-\frac{10}{50\times10^{-3}}t}\right)$$

$$1 = 10\left(1 - e^{-200t}\right)$$

$$-200t = \ln(0.9)$$

$$t = 0.5\,\text{ms}$$

$$\text{Gate pulse width} = \frac{1}{1\,\text{kHz}} = 10^{-3}\text{s} = 1\,\text{ms}$$

Hence, minimum pulse duration is 0.5 ms, which gives a duty ratio of 50% because the pulse duration is 1 ms.

2.7.4.3 Thyristor with R-L-E$_b$ Load

Consider the voltage $V_m \sin\omega_s t$ is applied to a R-L-E$_b$ circuit shown in Fig. 2.26(a). At $\omega t = \alpha$, the thyristor is turned on. The transient current i_o can be found by applying KVL.

$$R \times i_o + L\frac{di_o}{dt} + E_b = V_m \sin\omega_s t, \quad \alpha < \omega_s t < \beta$$

$$R \times i_o + L\frac{di_o}{dt} = V_m \sin\omega_s t - E_b, \quad \alpha < \omega_s t < \beta \qquad (2.35)$$

The solution is given below.

$$L\frac{di_o}{dt} + R \times i_o = 0$$

$$\left(\frac{di_o}{dt} + \frac{R}{L}\right)i_o = 0$$

$$\left(p + \frac{R}{L}\right)i_o = 0$$

and the solution is given by $i_o = Ke^{-\frac{R}{L}t}$, where K is the constant of integration to be identified from boundary conditions.

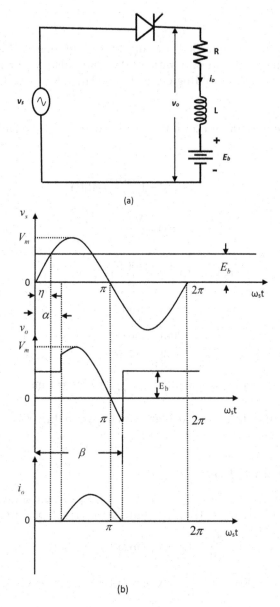

FIGURE 2.26
(a) R-L-E_b load with SCR. (b) Waveforms.

The steady state component has two parts; one is from the supply voltage and other from the back emf, which is in the opposite direction, i.e., $\dfrac{V_m \sin(\omega_s t - \phi)}{Z} - \dfrac{E_b}{R}$, where

$$Z = \sqrt{(R^2 + X_L^2)} \text{ and } \phi = \tan^{-1}\left(\frac{X_L}{R}\right)$$

$$i_o = Ke^{\frac{-R}{L}t} + \frac{V_m \sin(\omega_s t - \phi)}{Z} - \frac{E_b}{R}$$

(2.36)

Up to $\omega_s t = \eta$, the thyristor will be reverse biased by the back emf, so it will be in the OFF state and this state will continue until $\omega_s t = \alpha$. Substituting the initial condition that at $\omega_s t = \alpha$ and $i_o = 0$,

$$0 = Ke^{\frac{-R\alpha}{L\omega}} + \frac{V_m \sin(\alpha - \phi)}{Z} - \frac{E_b}{R}$$

$$K = \left(-\frac{V_m \sin(\alpha - \phi)}{Z} + \frac{E_b}{R} \right) e^{\frac{R\alpha}{L\omega_s}}$$

Substituting in the above equation, we get

$$i_o = \left(-\frac{V_m \sin(\alpha - \phi)}{Z} + \frac{E_b}{R} \right) e^{\frac{R}{L}\left(\frac{\alpha}{\omega_s} - t\right)} + \frac{V_m \sin(\omega_s t - \phi)}{z} - \frac{E_b}{R} \qquad (2.37)$$

The variation of v_o, v_s, and i_o is shown in Fig. 2.26(b).

2.7.4.4 Thyristor with R-E_b Load

Consider a voltage $V_m \sin \omega_s t$ is applied to a R-E_b circuit shown in Fig. 2.27(a). At $\omega t = \alpha$, the thyristor is turned on by giving a triggering pulse. The current i_o can be found by applying KVL.

$$R \times i_o + E_b = V_m \sin \omega_s t, \quad \alpha < \omega_s t < \beta \qquad (2.38)$$

From the figure, at $\omega_s t = \pi - \alpha$ and $v_s = E_b$, the SCR turns off. Hence, $\beta = \pi - \eta$.

$$i_o = \frac{V_m \sin \omega_s t - E_b}{R} \qquad (2.39)$$

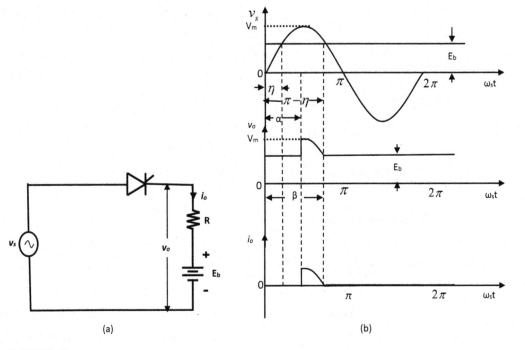

(a)   (b)

FIGURE 2.27
(a) R-E_b load with SCR. (b) Waveforms.

The average value of the current, $I_{o(av)}$, is given by

$$I_{o(av)} = \frac{1}{2\pi R} \int_{\alpha}^{\pi-\eta} \left(V_m \sin \omega_s t - E_b \right) d\omega_s t = \frac{1}{2\pi R} \left\{ \left[-V_m \cos \omega_s t \right]_{\alpha}^{\pi-\eta} - E_b \left[\omega_s t \right]_{\alpha}^{\pi-\eta} \right\}$$

$$= \frac{1}{2\pi R} \left[V_m (-\cos(\pi - \eta) + \cos\alpha) - E_b(\pi - \eta - \alpha) \right] \tag{2.40}$$

$$= \frac{1}{2\pi R} \left[V_m (\cos\eta + \cos\alpha) - E_b(\pi - \eta - \alpha) \right]$$

Further, the rms value of current, $I_{o(rms)}$ is obtained by

$$I_{o(rms)}^2 = \frac{1}{2\pi R^2} \int_{\alpha}^{\pi-\eta} (V_m \sin \omega_s t - E_b)^2 \, d\omega_s t$$

$$= \frac{1}{2\pi R^2} \int_{\alpha}^{\pi-\eta} [V_m^2 \sin^2 \omega_s t + E_b^2 - 2 V_m E_b \sin \omega_s t] d\omega_s t$$

$$= \frac{1}{2\pi R^2} \int_{\alpha}^{\pi-\eta} \left[\frac{V_m^2}{2} - \frac{V_m^2 \cos 2\omega_s t}{2} + E_b^2 - 2 V_m E_b \sin \omega_s t \right] d\omega_s t$$

$$= \frac{1}{2\pi R^2} \left\{ \frac{V_m^2}{2} [\omega_s t]_{\alpha}^{\pi-\eta} - \frac{V_m^2}{4} [\sin 2\omega_s t]_{\alpha}^{\pi-\eta} + E_b^2 [\omega_s t]_{\alpha}^{\pi-\eta} - 2 V_m E_b [-\cos \omega_s t]_{\alpha}^{\pi-\eta} \right\}$$

$$= \frac{1}{2\pi R^2} \left\{ \frac{V_m^2}{2} [\pi - \eta - \alpha] - \frac{V_m^2}{4} [\sin(2\pi - 2\eta) - \sin 2\alpha] + E_b^2[\pi - \eta - \alpha] \right.$$

$$\left. - 2 V_m E_b [-\cos(\pi - \eta) + \cos\alpha] \right\}$$

$$\tag{2.41}$$

$$= \frac{1}{2\pi R^2} \left\{ \frac{V_m^2}{2} [\pi - \eta - \alpha] + \frac{V_m^2}{4} \sin 2\eta + \frac{V_m^2}{4} \sin 2\alpha + E_b^2[\pi - \eta - \alpha] \right.$$

$$\left. - 2 V_m E_b [\cos\eta + \cos\alpha] \right\}$$

$$\therefore I_{o(rms)} = \left[\frac{1}{2\pi R^2} \left\{ \frac{V_m^2}{2} (\pi - \eta - \alpha) + \frac{V_m^2}{4} \sin 2\eta + \frac{V_m^2}{4} \sin 2\alpha + E_b^2(\pi - \eta - \alpha) \right. \right.$$

$$\left. \left. - 2 V_m E_b (\cos\eta + \cos\alpha) \right\} \right]^{\frac{1}{2}}$$

$$\text{Input power factor} = \frac{o/p \text{ power}}{V_{rms} I_{o(rms)}} = \frac{I_{o(rms)}^2 R + I_{o(av)} E_b}{V_{rms} I_{o(rms)}}$$

$$= \frac{I_{o(rms)}^2 R + I_{o(av)} E_b}{\dfrac{V_m}{\sqrt{2}} I_{o(rms)}}$$

The variation of v_o, v_s, and i_o is shown in Fig. 2.27(b).

EXAMPLE 2.14

A battery charger employing a single-phase half wave-controlled rectifier is feeding a battery with terminal voltage = 100 V and internal resistance = 5 Ω. Calculate the rms current and the input power factor for a supply of $230\sqrt{2}\sin(314t)$.

We have

$$230\sqrt{2}\sin(\eta) = 100$$

$$\therefore \eta = \sin^{-1}\left(\frac{100}{230\sqrt{2}}\right) = 17.90° = 0.312 \text{ rad}$$

Output voltage, $V_o = \frac{V_m}{\pi}\cos\alpha$

$$\therefore \alpha = \cos^{-1}\left(\frac{V_o\pi}{V_m}\right) = \cos^{-1}\left(\frac{100 \times \pi}{230\sqrt{2}}\right) = 15.01° = 0.261 \text{ rad}$$

$$\therefore I_{o(rms)} = \left[\frac{1}{2\pi R^2}\left\{\frac{V_m^2}{2}(\pi - \eta - \alpha) + \frac{V_m^2}{4}\sin 2\eta + \frac{V_m^2}{4}\sin 2\alpha + E_b^2(\pi - \eta - \alpha) - 2V_m E_b(\cos\eta + \cos\alpha)\right\}\right]^{\frac{1}{2}}$$

$$= 14.38 \text{ A}$$

$$I_{o(av)} = \frac{1}{2\pi R}[V_m(\cos\eta + \cos\alpha) - E_b(\pi - \eta - \alpha)]$$

$$= \frac{1}{2\pi \times 5}[325.26(0.95 + 0.96) - 100(\pi - 0.132 - 0.26)]$$

$$= 11.59 \text{ A}$$

$$\text{Input power factor} = \frac{I_{o(rms)}^2 R + I_{o(av)}E_b}{\frac{V_m}{\sqrt{2}}I_{o(rms)}} = \frac{14.38^2 \times 5 + 11.59 \times 100}{230 \times 14.38} = 0.66(\text{lagging})$$

EXAMPLE 2.15

A series R-L load with R = 10 Ω and L = 50 mH is connected to a source voltage of $240\sqrt{2}\sin(314t)$ at t = 0 through a diode. If a freewheeling diode is connected across the load, compute (i) amplitudes of the source current at $\omega t = 0$ and $\omega t = \pi$, and (ii) average load voltage and current.

SOLUTION:

$$I_0 = \frac{V_m}{Z}\sin\phi\left[\frac{1 + e^{-\pi/\tan\phi}}{e^{\pi/\tan\phi} + e^{\pi/\tan\phi}}\right]$$

$$I_0 = \frac{333.9}{18.6}\sin(57.5)\left[\frac{1 + e^{-\pi/1.5697}}{e^{\pi/1.5697} + e^{\pi/1.5697}}\right]$$

$$I_0 = \frac{333.9}{18.6}\sin(57.5)\left[\frac{1.1351}{7.2642}\right]$$

$$I_0 = 2.41 \text{ A}$$

$$I_\pi = I_0 e^{\pi/\tan\phi}$$

$$I_\pi = 2.41 e^{\pi/1.5697}$$

$$I_\pi = 17.8325 \text{ A}$$

$$V_{o(avg)} = \frac{V_m}{\pi} = \frac{333.9}{\pi}$$

$$V_{o(avg)} = 108.03 \text{ V}$$

$$I_{dc} = \frac{V_{o(avg)}}{R}$$

$$I_{dc} = 10.8 \text{ A}$$

EXAMPLE 2.16

A dc source V_{dc} is connected across an R-L load. Derive the expression for $i_o(t)$ assuming an initial current of I_o at $t = 0$. If $V_{dc} = 190$ V, $I_o = 60$ A, $R = 2.5\ \Omega$, and $L = 0.5$ H, determine the instant of time at which the current becomes zero.

SOLUTION:

At $t = 0$ $i_o = I_0$

$$i_0 = K_1 e^{\frac{-R}{L}t} + \frac{V_{dc}}{R}$$

$$I_0 = K_1 + \frac{V_{dc}}{R}$$

$$K_1 = \left(I_0 - \frac{V_{dc}}{R} \right)$$

$$i_0(t) = \left(I_0 - \frac{V_{dc}}{R} \right) e^{\frac{-R}{L}t} + \frac{V_{dc}}{R}$$

$$i_0(t) = I_0 e^{\frac{-R}{L}t} + \frac{V_{dc}}{R} \left(1 - e^{\frac{-R}{L}t} \right)$$

Let $t = t_1 i_o(t) = 0$

$$0 = \left(I_0 - \frac{V_{dc}}{R} \right) e^{\frac{-R}{L}t} + \frac{V_{dc}}{R}$$

$$0 = \left(60 + \frac{190}{2.5} \right) e^{\frac{-2.5}{0.5}t} - \frac{190}{2.5}$$

$$\frac{190}{2.5} = \left(60 + \frac{190}{2.5} \right) e^{\frac{-2.5}{0.5}t}$$

$$e^{-5t_1} = 0.5588$$

$$-5t_1 = \ln(0.5588)$$

$$t_1 = 0.116 \text{ s}$$

EXAMPLE 2.17

A dc battery with a resistor R is charged through one SCR. For an ac source of 230 V and 50 Hz, (i) find the value of average charging current for R = 5 Ω and E = 150 V; (ii) find the power supplied to battery and that dissipated in the resistor; and (iii) calculate the supply pf. The SCR is supplied with a continuous gate pulse.

SOLUTION:

i. Find the value of average charging current for R = 5 Ω and E = 150 V:

$$I_{dc} = \frac{1}{2\pi R}\left[V_m\left(\cos\alpha + \cos\eta\right) - E(\pi - \eta - \alpha)\right]$$

$$\eta = \sin^{-1}\left(\frac{E}{V_m}\right) = \sin^{-1}\left(\frac{150}{230\sqrt{2}}\right) = 27.46°$$

$$\beta = \pi - \eta = 152.54°$$

$$2\eta = 54.92° = \frac{54.92 \times \pi}{180}\,rad = 0.9585\,rad$$

if $\alpha = \eta$

$$I_{dc} = \frac{1}{2\pi R}\left[2V_m\cos\eta - E(\pi - 2\eta)\right]$$

$$I_{dc} = \frac{1}{2\pi R}\left[2 \times 230\sqrt{2}\cos(27.46) - E(\pi - 0.9585)\right]$$

$$I_{dc} = \frac{1}{2\pi \times 5}\left[249.78\right] = 7.95\,A$$

ii. Find the power supplied to the battery and that dissipated in the resistor:

$$= E \times I_{dc} = 150 \times 7.95 = 1.192\,kW$$

iii. Calculate the supply pf:

$$\therefore I_{o(rms)} = \left[\frac{1}{2\pi R^2}\left\{\frac{V_m^2}{2}(\pi - \eta - \alpha) + \frac{V_m^2}{4}\sin 2\eta + \frac{V_m^2}{4}\sin 2\alpha + E_b^2(\pi - \eta - \alpha) - 2V_m E_b(\cos\eta + \cos\alpha)\right\}\right]^{\frac{1}{2}}$$

$$\therefore I_{o(rms)} = \left[\frac{1}{2\pi \times 5^2}\left\{\frac{\left(230\sqrt{2}\right)^2}{2}(\pi - 0.4793 - 0.4793) + \frac{\left(230\sqrt{2}\right)^2}{4}\sin(2 \times 0.4793) + \frac{\left(230\sqrt{2}\right)^2}{4}\sin(2 \times 0.4793) + 150^2(\pi - 0.4793 - 0.4793) - 2 \times 230\sqrt{2} \times 150\left(\cos(0.4793) + \cos(0.4793)\right)\right\}\right]^{\frac{1}{2}}$$

$$= 15.53\,A$$

$$\text{Supply power factor} = \frac{\text{Output power}}{V_{rms} \times i_{rms}} = \frac{i_{o(rms)}^2 \times R + E_b \times I_{dc}}{V_{rms} \times i_{rms}}$$

Because $i_{rms} = i_{o(rms)}$

$$= \frac{15.53^2 \times 5 + 150 \times 7.95}{230 \times 15.53} = 0.671(\text{lagging})$$

Questions

1. Explain SCR voltage-current characteristics.
2. How is triac different from SCR?
3. Elaborate the architecture of MOSFET.
4. What is the principle of operation of IGBT?
5. Explain the analysis of Rload, R-L load, R-C load, and R-L-C load with an ideal switch.
6. Describe the behavior of Rload, R-L load with and without a freewheeling diode, R-E_b load, and R-L-E_b load with a diode.
7. Analyze Rload, R-L load, R-E_b load, and R-L-E_b load with a thyristor.

Unsolved Problems

1. A constant voltage of 40 V is switched onto a series combination of resistance and inductance where R = 200 Ω and L = 160 mH, at t = 0. Compute the circuit current at t = 3 ms and t = 105 ms.
2. A constant dc voltage of 45 V is switched onto a series combination of resistance and inductance where R = 200 Ω at t = 0. Find the inductance if the time taken is 2 ms for a circuit current of (i) 0.25 A and (ii) 0.3 A.
3. A constant voltage of 220 V is switched onto a series combination of resistance and inductance where R = 80 Ω and L = 25 H at t = 0. Find the equation for current and voltage and find the time at which $V_R = V_L$.
4. A series combination of R = 20 Ω and C = 8 μF is switched on to 230 V_{dc} at t = 0. (a) Find the current in the circuit at t = 1 ms. (b) Find the voltage across the capacitor at that time. (c) Find the energy lost in the circuit.
5. An electric heater rated 2.2 kW, 230 V, 50 Hz is supplied from a 240-V, 50-Hz ac main with a semiconductor diode in series. Calculate the average current, the rms currents, and the power delivered to the heater.
6. An ac voltage of rms value 120 V is used to charge a battery with an internal resistance of 1.5 Ω and a back emf of 120 V through the diode. Find the average current, the rms current, and the power factor.

Answers

1. 195.3 mA, 0.2 A
2. 246 mH, 182.04 mH
3. 0.2166 s
4. 0.022 A, 229.55 V, 0.2116 J
5. 4.493 A, 7.058 A, 1.197 kW
6. 18.014 A, 5.465 A, 0.528

3

ac/dc Converters

3.1 Introduction

This chapter discusses different circuit topologies and operational characteristics of ac/dc power converters, which are commonly labeled as controlled rectifiers. These power converters convert existing ac supply in to dc voltage of controlled amplitude. Silicon-controlled rectifiers (SCRs) are employed as switching elements, and commutation of these SCRs takes place with the help of input line voltage so that additional commutation circuits are not necessary. These converters are classified as single-phase and three-phase converters based on availability of single-phase or three-phase power supply. The ac/dc converters are also categorized as semi-converters or full converters; semi-converters are single-quadrant converters, whereas full converters operate in two quadrants of the V-I diagram. Line-commutated ac/dc converters are used for variable-speed operation of dc motors, high-voltage dc transmission (HVDC), battery charging, and front-end feeders to inverter circuits.

3.2 Single-Phase Semi-Converter

The power circuit of a single-phase semi-converter is shown in Fig. 3.1. There are two SCRs, marked as T_1 and T_2, and two diodes, D_1 and D_2. The thyristors alone need to be triggered, and the diodes conduct depending upon the polarity of the input supply voltage. The supply voltage is shown in Fig. 3.2(a). From the power circuit, it is evident that thyristor T_1 and diode D_2 are forward-biased during the positive half-cycle of the supply voltage; similarly, thyristor T_2 and diode D_1 are forward-biased during the negative half-cycle of the supply. The firing angle, α, is measured from the zero crossing of the supply voltage, and hence T_1 is triggered at $\omega_s t = \alpha$ and T_2 is triggered at $\omega_s t = \pi + \alpha$. The firing pulses are indicated in Fig. 3.2(b). It should be mentioned that the width of the firing pulse should be more than the SCR turn-ON time. Thus, at $\omega_s t = \alpha$, thyristor T_1 is turned ON and the current starts from the source and passes through T_1-load-D_2-back to the supply, and the load voltage is now the supply voltage. Similarly, at $\omega_s t = \pi + \alpha$, T_2 is turned ON and the load current passes from the source through T_2-load-D_1. The nature of load current depends on the characteristic of load. Three typical loads are discussed now.

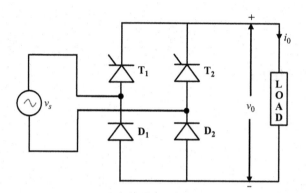

FIGURE 3.1
Power circuit for semi-converter.

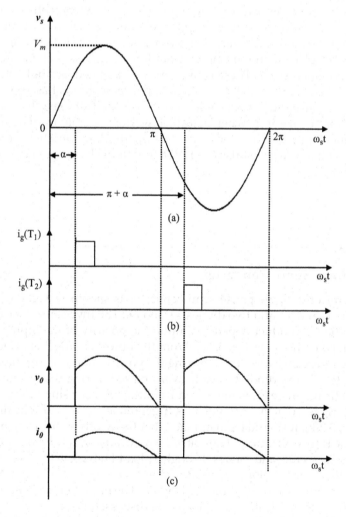

FIGURE 3.2
Waveforms of semi-converter with R load. (a) Supply voltage waveform. (b) Triggering pulses. (c) Output waveforms.

3.2.1 Purely Resistive load

At $\omega_s t = \alpha$, T_1 and D_2 start conducting, and the load voltage and current are indicated in Fig. 3.2(c). With a purely resistive load, the load current i_0 is $\frac{V_o}{R}$ and therefore has a shape identical to that of output voltage. At $\omega_s t = \pi$, v_o goes to zero value and hence i_0 also goes to zero value. Thyristor T_1 and diode D_2 are turned OFF by natural commutation, and the load is disconnected from the supply, leading to zero output voltage. At $\omega_s t = \pi + \alpha$, T_2 is triggered and T_2 and D_1 are conducting, now making output voltage once again positive.

At $\omega_s t = 2\pi$, load voltage and current become zero, turning OFF T_2 and D_1. An expression for the average value of output voltage is derived below:

$$V_{o(av)} = \frac{1}{\pi} \int_{\alpha}^{\pi} V_m \sin(\omega_s t) d(\omega_s t)$$

$$V_{o(av)} = \frac{V_m}{\pi} \left[-\cos(\omega_s t) \right]_{\alpha}^{\pi} \tag{3.1}$$

$$V_{o(av)} = \frac{V_m}{\pi} [1 + \cos \alpha]$$

3.2.2 R-L Load

Now consider the operation of the semi-converter with series resistance-inductance load. It is worth mentioning that firing of T_1 and T_2 remains at α and $\pi + \alpha$ independent of type of load. With R-L load, there are two operating modes.

Continuous conduction mode: At $\omega_s t = \alpha$, when T_1 is triggered, the current starts increasing through the load. Because it is a lagging load, at $\omega_s t = \pi$, while the voltage goes to zero, the load current is not yet zero. At this instant, D_1 is more conducive for conduction than D_2; thus D_2 turns OFF and T_1 and D_1 carry the load current. This causes the output terminals to be short-circuited, resulting in zero output voltage. Thus, load current circulates through T_1-D_1-load. This is termed as freewheeling of load current. If the load inductance to resistance ratio is sufficiently large, then the output current is continuous; otherwise, current will be discontinuous. The relevant waveforms for continuous mode are plotted in Fig. 3.3(c). At $\omega_s t = \pi + \alpha$, T_2 is triggered, a voltage amplitude of $V_m \sin(\pi + \alpha) = -V_m \sin \alpha$ is impressed across T_1, and T_1 is turned OFF by line commutation. Now T_2 and D_1 conduct, making the output voltage and current positive. At $\omega_s t = 2\pi$, load current freewheels through T_2-D_2.

Discontinuous conduction mode: As mentioned earlier, if the ratio of load inductance to resistance is low, the load current is discontinuous. Typical waveforms are given in Fig. 3.4(c). Here, freewheeling of load current takes place through T_1-D_1 from $\omega_s t = \pi$ to β. At $\omega_s t = \beta$, T_1 and D_1 are turned OFF by natural commutation. A similar operation takes place during the negative half-cycle of the supply voltage when T_2 is turned ON at $\omega_s t = \pi + \alpha$. At $\omega_s t = 2\pi$, load current freewheels through T_2 and D_2, making the output voltage zero.

During the freewheeling operation, energy stored in the load inductance is dissipated to the load resistance rather than returning to the source. This means less reactive power is drawn from the source and the freewheeling process thus enhances input power factor. This is qualitatively verified in section 3.4. It may be noted that the freewheeling of load

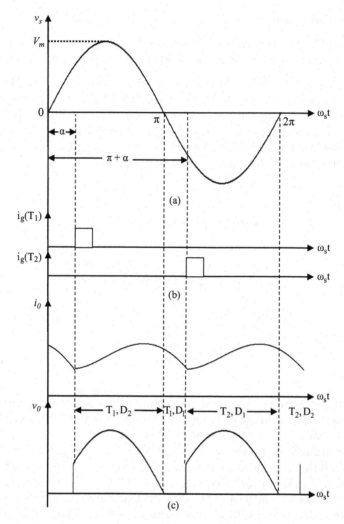

FIGURE 3.3
Waveforms of semi-converter with R-L load (continuous conduction). (a) Supply voltage waveform. (b) Triggering pulses. (c) Output waveform.

current—which is inherent in a semiconductor—is taking place through two devices in series (T_1-D_1 and then T_2-D_2). To cause freewheeling through a single device and also to ensure reliable freewheeling, a single diode marked as FD (often called a freewheeling diode) is connected at the output of the semi-converter as shown in Fig. 3.5.

Thus, because freewheeling is inherent with a single phase semi-converter, the output voltage can never be negative. The expression for the average output voltage is the same as the one derived for a purely resistive load, given by Equation (3.1).

3.2.3 R-L-Back Emf Load

A dc motor load can be thought of R-L-E_b in nature. Here, the load current becomes continuous or otherwise depending upon the value of α and load torque. For a given α, when the motor is running under light load conditions, the average value of current required is

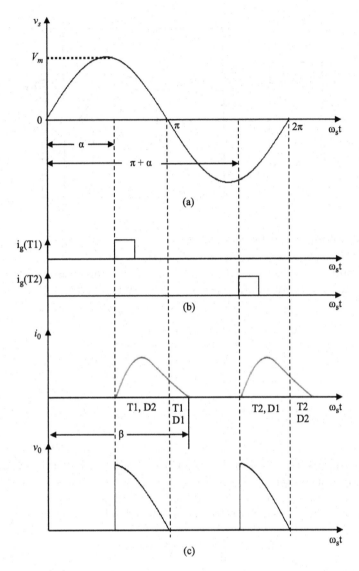

FIGURE 3.4
Waveforms of semi-converter with R-L load (discontinuous conduction). (a) Supply voltage waveform. (b) Triggering pulses. (c) Output waveforms for R-L load.

low. Furthermore, due to increased speed, back emf is large. The supply voltage together with back emf is plotted in Fig. 3.6(a). The angle η in this figure is given as

$$\eta = \sin^{-1}\left(\frac{V_m}{E_b}\right) \tag{3.2}$$

where, V_m is the maximum value of the input supply voltage and E_b is the motor back emf.

This indicates that, under light load conditions, α should be $\geq \eta$ for the SCRs to be turned ON. Typical load voltage and current are shown in Fig. 3.6(c).

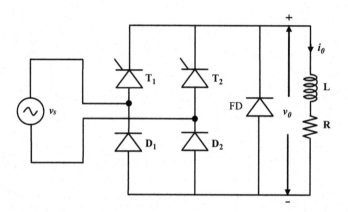

FIGURE 3.5
Power circuit of semi converter with freewheeling diode.

With discontinuous mode, conduction starts at $\omega_s t = \alpha$, and ends at β. The average value of output voltage for discontinuous mode of operation is given as

$$V_{o(av)} = \frac{1}{\pi}\left[\int_{\alpha}^{\pi} V_m \sin(\omega_s t) d\omega t + \int_{\beta}^{\pi+\alpha} E_b d\omega_s t \right]$$

$$V_{o(av)} = \frac{1}{\pi}\left[V_m \left(-\cos(\omega_s t)\right)_{\alpha}^{\pi} + E_b(\pi + \alpha - \beta) \right] \tag{3.3}$$

$$V_{o(av)} = \frac{1}{\pi}\left[V_m (1+\cos\alpha) + E_b(\pi + \alpha - \beta) \right]$$

If the motor is operating under heavy load conditions, the average load current has to be higher, and back emf is low under these conditions. This will result in continuous conduction of load current such that β becomes $\pi + \alpha$; typical characteristics are plotted in Fig. 3.7(c).

At $\omega_s t = \alpha$, thyristor T_1 is turned ON. With T_1 ON, the load is connected to the source through T_1 and D_2. The transient nature of current i_0 can be found by applying KVL.

$$Ri_0 + L\frac{di_0}{dt} + E_b = V_m \sin(\omega_s t)$$

$$Ri_0 + L\frac{di_0}{dt} = V_m \sin(\omega_s t) - E_b$$

The transient component is obtained by forcing left hand side to zero

$$Ri_0 + L\frac{di_0}{dt} = 0$$

$$\frac{R}{L}i_0 + \frac{di_0}{dt} = 0$$

$$\left(\frac{R}{L} + p\right)i_0 = 0$$

and the solution is given by

$$i_0 = -K_1 e^{\frac{-R}{L}t}$$

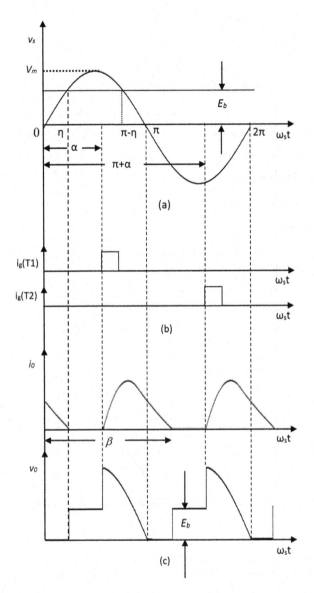

FIGURE 3.6
Waveforms of semi-converter with R-L-E_b load (discontinuous conduction). (a) Supply voltage waveform. (b) Triggering pulses. (c) Output waveforms for R-L-E_b load.

where K_1 is the constant of integration to be identified from boundary conditions.

The steady state component is given by

$$\frac{V_m}{Z}\sin(\omega_s t - \phi) - \frac{E_b}{R}$$

$$Z = \sqrt{\left(R^2 + X_L^2\right)}, \phi = \tan^{-1}\left(\frac{X_L}{R}\right) \tag{3.4}$$

$$i_0 = K_1 e^{\frac{-R}{L}t} + \frac{V_m}{Z}\sin(\omega_s t - \phi) - \frac{E_b}{R}$$

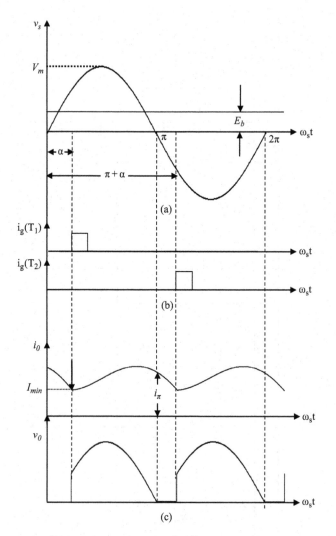

FIGURE 3.7
Waveforms of semi-converter with R-L-E_b load (continuous conduction). (a) Supply voltage waveform. (b) Triggering pulses. (c) Output waveforms for R-L-E_b load.

3.2.3.1 Continuous Mode

Mode 1: In this mode T_1 and D_2 will conduct. At $\omega_s t = \alpha$, $i_0 = I_{min}$ and substituting these conditions in the above equation, the value of K_1 is determined.

$$K_1 = \left[I_{min} - \frac{V_m}{Z} \sin(\alpha - \phi) + \frac{E_b}{R} \right] e^{\frac{R\alpha}{L\omega_s}}$$

Substituting the value of K_1 in Equation (3.4):

$$i_0 = \left[I_{min} - \frac{V_m}{Z} \sin(\alpha - \phi) + \frac{E_b}{R} \right] e^{\frac{R}{L}\left(\frac{\alpha}{\omega_s} - t\right)} + \frac{V_m}{Z} \sin(\omega_s t - \phi) - \frac{E_b}{R}$$

At $\omega_s t = \pi$, $i_0 = I_\pi$

Substituting these conditions, the value of I_π can be determined:

$$I_\pi = \left[I_{min} - \frac{V_m}{Z} \sin(\alpha - \phi) + \frac{E_b}{R} \right] e^{\frac{R}{L\omega_s}(\alpha - \pi)} + \frac{V_m}{Z} \sin\phi - \frac{E_b}{R} \qquad (3.5)$$

Mode 2 (freewheeling period): In this mode T_1 and D_1 will conduct.

$$Ri_0 + L\frac{di_0}{dt'} = -E_b \qquad 0 \le \omega_s t' \le \alpha$$

where $\omega_s t' = \omega_s t - \pi$

The transient component is obtained by forcing left hand side to zero:

$$Ri_0 + L\frac{di_0}{dt'} = 0$$

$$\frac{R}{L} i_0 + \frac{di_0}{dt'} = 0$$

$$\left(\frac{R}{L} + p \right) i_0 = 0$$

$$i_0 = K_2 e^{\frac{-R}{L}t'},$$

where K_2 is the constant of integration to be identified from boundary conditions. The steady state component is given by $-\frac{E_b}{R}$:

$$i_0 = K_2 e^{\frac{-R}{L}t'} - \frac{E_b}{R} \qquad \omega_s t' = 0, \quad i_0 = I_\pi$$

Substituting these conditions, we have

$$I_\pi = K_2 - \frac{E_b}{R}$$

$$K_2 = I_\pi + \frac{E_b}{R}$$

Substituting K_2, we get

$$i_0 = \left(I_\pi + \frac{E_b}{R} \right) e^{\frac{-R}{L}t'} - \frac{E_b}{R} \qquad (3.6)$$

At $\omega_s t' = \alpha$, $i_0 = I_{min}$.

Substituting these conditions in (3.6), we get

$$I_{min} = \left(I_\pi + \frac{E_b}{R} \right) e^{\frac{-R\alpha}{L\omega_s}} - \frac{E_b}{R} \qquad (3.7)$$

Substituting Equation (3.7) in Equation (3.5), the value of I_π can be determined.

$$I_\pi = \left[\left(I_\pi + \frac{E_b}{R}\right)e^{\frac{-R\alpha}{L\omega_s}} - \frac{E_b}{R} - \frac{V_m}{Z}\sin(\alpha-\phi) + \frac{E_b}{R}\right]e^{\frac{R}{L\omega_s}(\alpha-\pi)} + \frac{V_m}{Z}\sin\phi - \frac{E_b}{R}$$

$$I_\pi = \left[I_\pi e^{\frac{R\alpha}{L\omega_s}} + \frac{E_b}{R}e^{-\frac{R\alpha}{L\omega_s}} - \frac{E_b}{R} - \frac{V_m}{Z}\sin(\alpha-\phi) + \frac{E_b}{R}\right]e^{\frac{R}{L\omega_s}(\alpha-\pi)} + \frac{V_m}{Z}\sin\phi - \frac{E_b}{R}$$

$$I_\pi = \left[\frac{\frac{E_b}{R}\left[e^{\frac{-R\pi}{L\omega_s}} - 1\right] + \frac{V_m}{Z}\sin\phi - \frac{V_m}{Z}\sin(\alpha-\phi)e^{\frac{R}{L\omega_s}(\alpha-\pi)}}{1 - e^{\frac{-R\pi}{L\omega_s}}}\right] \tag{3.8}$$

$$I_{min} = \left[\left[\frac{\frac{E_b}{R}\left[e^{\frac{-R\pi}{L\omega_s}} - 1\right] + \frac{V_m}{Z}\sin\phi - \frac{V_m}{Z}\sin(\alpha-\phi)e^{\frac{R}{L\omega_s}(\alpha-\pi)}}{\left[1 - e^{\frac{-R\pi}{L\omega_s}}\right]}\right] + \frac{E_b}{R}e^{\frac{-R\alpha}{L\omega_s}} - \frac{E_b}{R}\right] \tag{3.9}$$

3.2.3.2 Discontinuous Mode

Mode 1: At $\omega_s t = \alpha$, thyristor T_1 is turned ON. With T_1 ON, the load is connected to the source through T_1 and D_2. The transient current i_0 can be found by applying KVL.

$$Ri_0 + L\frac{di_0}{dt} + E_b = V_m\sin(\omega_s t) \qquad \alpha \leq \omega_s t \leq \pi$$

$$Ri_0 + L\frac{di_0}{dt} = V_m\sin(\omega_s t) - E_b$$

The transient component is obtained by forcing left hand side to zero

$$Ri_0 + L\frac{di_0}{dt} = 0$$

$$\frac{R}{L}i_0 + \frac{di_0}{dt} = 0$$

$$\left(\frac{R}{L} + p\right)i_0 = 0$$

and the solution is given by

$$i_0 = K_3 e^{\frac{-R}{L}t},$$

where K_3 is the constant of integration to be identified from boundary conditions. The steady state component is given by

$$\frac{V_m}{Z}\sin(\omega_s t - \phi) - \frac{E_b}{R}$$

$$Z = \sqrt{(R^2 + X_L^2)} \text{ and } \phi = \tan^{-1}\left(\frac{X_L}{R}\right)$$

$$i_0 = K_3 e^{\frac{-R}{L}t} + \frac{V_m}{Z}\sin(\omega_s t - \phi) - \frac{E_b}{R}$$

At $\omega_s t = \alpha$, $i_0 = 0$
Substituting these conditions, we get

$$K_3 = \left[\frac{E_b}{R} - \frac{V_m}{Z}\sin(\alpha - \phi)\right]e^{\frac{R\alpha}{L\omega_s}}$$

Substituting the value of K_3, we get

$$i_0 = \left[\frac{E_b}{R} - \frac{V_m}{Z}\sin(\alpha - \phi)\right]e^{\frac{R}{L}\left(\frac{\alpha}{\omega_s} - t\right)} + \frac{V_m}{Z}\sin(\omega_s t - \phi) - \frac{E_b}{R} \tag{3.10}$$

At $\omega_s t = \pi$, $i = I_\pi$
Substituting these conditions, the value of I_π can be determined

$$I_\pi = \left[\frac{E_b}{R} - \frac{V_m}{Z}\sin(\alpha - \phi)\right]e^{\frac{R}{L\omega_s}(\alpha - \pi)} + \frac{V_m}{Z}\sin\phi - \frac{E_b}{R}$$

Mode 2 (freewheeling): During free-wheeling period T_1 and D_2 conduct and the voltage equation is

$$Ri_0 + L\frac{di_0}{dt'} + E_b = 0 \quad 0 \le \omega_s t' \le \beta$$

where, $\omega_s t' = \omega_s t - \pi$

$$Ri_0 + L\frac{di_0}{dt'} = -E_b$$

$$Ri_0 + L\frac{di_0}{dt'} = 0$$

$$\frac{R}{L}i_0 + \frac{di_0}{dt'} = 0$$

$$\left(\frac{R}{L} + p\right)i_0 = 0$$

and the solution is given by

$$i_0 = K_4 e^{\frac{-R}{L}t'},$$

where K_4 is the constant of integration to be identified from boundary conditions.

The steady state component is given by $-\dfrac{E_b}{R}$

$$i_0 = K_4 e^{\frac{-R}{L}t'} - \frac{E_b}{R}$$

At $\omega_s t' = 0$, $i_0 = I_\pi$

$$K_4 = I_\pi + \frac{E_b}{R}$$

Substituting the value of K_4, we get

$$i_0 = \left(I_\pi + \frac{E_b}{R} \right) e^{\frac{-R}{L}t'} - \frac{E_b}{R}$$

$$i_0 = \left[\left[\frac{E_b}{R} - \frac{V_m}{Z} \sin(\alpha - \phi) \right] e^{\frac{R}{\omega_s L}(\alpha - \pi)} + \frac{V_m}{Z} \sin\phi \right] e^{\frac{-R}{L}t'} - \frac{E_b}{R}$$

At $\omega_s t' = \beta$, $i_0 = 0$

$$\left\{ \left[\frac{E_b}{R} - \frac{V_m}{Z} \sin(\alpha - \phi) \right] e^{\frac{R}{L}\left(\frac{\alpha - \pi}{\omega_s} \right)} + \frac{V_m}{Z} \sin\phi \right\} e^{\frac{-R\beta}{L\omega_s}} - \frac{E_b}{R} = 0$$

and this equation may be solved by numerical methods to find the value of β.

EXAMPLE 3.1

A single-phase semi-converter is triggered at 75°. The converter is supplied from a 230-V, 50-Hz, single-phase supply. The load consists of a pure resistance of 100 Ω. Compute (i) average output load voltage; (ii) average load current; (iii) average load power; (iv) duration of current conduction; and (v) maximum positive and negative voltage across SCR.

SOLUTION:

From Equation (3.1),

i. Average output load voltage $V_{o(av)} = \dfrac{V_m}{\pi}[1 + \cos\alpha]$

$$V_{o(av)} = \frac{325.27}{\pi}[1 + \cos 75°]$$

$$V_{o(av)} = 198.97 \text{ V}$$

ii. Average load current $I_o = \dfrac{V_{o(av)}}{R_L}$

$$I_o = \dfrac{198.97}{100}$$

$$I_o = 1.9897 \text{ A}$$

iii. Average load power $P_L = V_{o(av)} \times I_0 = 198.97 \times 1.9897$

$$P_L = 395.90 \text{ W}$$

iv. Duration of current conduction $= \pi - \alpha = 180° - 75°$
Duration of current conduction $= 105°$

v. Maximum positive voltage across SCR $= V_m \sin(\alpha) = 230\sqrt{2}\sin 75° = 314.14 \text{ V}.$
Maximum negative voltage across SCR $= -V_m = -230\sqrt{2} = -325.22 \text{ V}.$

EXAMPLE 3.2

A 230-V, 50-Hz source supplies a single-phase semi-converter feeding a load with a resistance of 1 Ω, inductance of 10 mH, and back emf of 60 V. The firing angle of the SCR is 30°. Find the minimum steady-state load current.

SOLUTION:

$$X_L = \omega_s L = 2\pi f L = 2 \times \pi \times 50 \times 0.01 = 314 \times 0.01 = 3.14 \ \Omega$$

$$Z = \sqrt{\left(R^2 + X_L^2\right)} = \sqrt{\left(1^2 + 3.14^2\right)}$$

$$Z = 3.29 \ \Omega$$

$$\phi = \tan^{-1}\left(\dfrac{X_L}{R}\right) = \tan^{-1}\left(\dfrac{3.14}{1}\right)$$

$$\phi = 72.33°$$

From Equation (3.7),

$$I_{min} = \left[I_\pi + \dfrac{E_b}{R} \right] e^{-\frac{R\alpha}{L\omega_s}} - \dfrac{E_b}{R}$$

$$I_\pi = \left[\dfrac{\dfrac{E_b}{R}\left[e^{\frac{-R\pi}{L\omega_s}} - 1 \right] + \dfrac{V_m}{Z}\sin\phi - \dfrac{V_m}{Z}\sin(\alpha - \phi)e^{\frac{R}{L\omega_s}(\alpha - \pi)}}{1 - e^{\frac{-R\pi}{L\omega_s}}} \right]$$

$$I_\pi = 134.72 \text{ A}.$$

$$I_{min} = \left[134.72 + \dfrac{60}{1} \right] e^{-\frac{1 \times 30 \times \frac{\pi}{180}}{.01 \times 314}} - \dfrac{60}{1}$$

$$I_{min} = 104.81 \text{ A}$$

Here I_{min} is a positive value. So, it is continuous current.

EXAMPLE 3.3

A 230-V, 50-Hz source supplies a single-phase semi-converter feeding a load with a resistance of 1 Ω, inductance of 10 mH, and back emf of 60 V. The firing angle of the SCR is 60°. Find the minimum steady-state load current.

SOLUTION:

$$X_L = \omega_s L = 2\pi fL = 2 \times \pi \times 50 \times 0.01 = 314 \times 0.01 = 3.14\ \Omega$$

$$Z = \sqrt{\left(R^2 + X_L^2\right)} = \sqrt{\left(1^2 + 3.14^2\right)}$$

$$Z = 3.29\ \Omega$$

$$\phi = \tan^{-1}\left(\frac{X_L}{R}\right) = \tan^{-1}\left(\frac{3.14}{1}\right)$$

$$\phi = 72.33°$$

$$I_{min} = \left[I_\pi + \frac{E_b}{R}\right]e^{-\frac{R\alpha}{L\omega}} - \frac{E_b}{R}$$

$$I_\pi = \left[\frac{\frac{E_b}{R}\left[e^{\frac{-R\pi}{L\omega_s}} - 1\right] + \frac{V_m}{Z}\sin\phi - \frac{V_m}{Z}\sin(\alpha - \phi)e^{\frac{R}{L\omega_s}(\alpha - \pi)}}{1 - e^{\frac{-R\pi}{L\omega_s}}}\right]$$

$$I_\pi = \left[\frac{\frac{60}{1}\left[e^{\frac{-1\times\pi}{0.01\times2\times\pi\times50}} - 1\right] + \frac{230\times\sqrt{2}}{3.29}\sin(72.33°) - \frac{230\times\sqrt{2}}{3.29}\sin(30° - 72.33°)e^{\frac{R}{L\omega}(\alpha - \pi)}}{1 - e^{\frac{-1\times\pi}{0.01\times2\times\pi\times50}}}\right]$$

$$I_\pi = 106.11\ A$$

$$I_{min} = \left[106.11 + \frac{60}{1}\right]e^{-\frac{1\times60\times\frac{\pi}{180}}{.01\times314}} - \frac{60}{1}$$

$$I_{min} = 59.0028\ A$$

Here, I_{min} is a positive value, so it is continuous current.

EXAMPLE 3.4

A 230-V, 50-Hz source supplies a single-phase semi-converter feeding a load with a resistance of 1 Ω, inductance of 10 mH, and back emf of 60 V. The firing angle of the SCR is 120°. Find the minimum steady-state load current.

SOLUTION:

$$X_L = \omega_s L = 2\pi f L = 2 \times \pi \times 50 \times 0.01 = 314 \times 0.01 = 3.14 \, \Omega$$

$$Z = \sqrt{\left(R^2 + X_L^2\right)} = \sqrt{\left(1^2 + 3.14^2\right)}$$

$$Z = 3.29 \, \Omega$$

$$\phi = \tan^{-1}\left(\frac{X_L}{R}\right) = \tan^{-1}\left(\frac{3.14}{1}\right)$$

$$\phi = 72.33°$$

$$I_{min} = \left[I_\pi + \frac{E_b}{R}\right] e^{-\frac{R\alpha}{L\omega_s}} - \frac{E_b}{R}$$

$$I_\pi = \left[\frac{\frac{E_b}{R}\left[e^{\frac{-R\pi}{L\omega_s t}} - 1\right] + \frac{V_m}{Z}\sin\phi - \frac{V_m}{Z}\sin(\alpha - \phi)e^{\frac{R}{L\omega_s}(\alpha - \pi)}}{1 - e^{\frac{-R\pi}{L\omega_s}}}\right]$$

$$I_\pi = \left[\frac{\frac{60}{1}\left[e^{\frac{-1\times\pi}{0.01\times2\times\pi\times50}} - 1\right] + \frac{230\times\sqrt{2}}{3.29}\sin(72.33°) - \frac{230\times\sqrt{2}}{3.29}\sin(120° - 72.33°)e^{\frac{R}{L\omega}(\alpha - \pi)}}{1 - e^{\frac{-1\times\pi}{0.01\times2\times\pi\times50}}}\right]$$

$$I_\pi = 6.0647 \text{ A}.$$

$$I_{min} = \left[6.0647 + \frac{60}{1}\right] e^{-\frac{1\times120\times\frac{\pi}{180}}{.01\times314}} - \frac{60}{1}$$

$$I_{min} = -26.04 \text{ A}$$

Here, I_{min} is a negative value and the current is discontinuous. Hence $I_{min} = 0$.

3.3 Single-Phase Full Converter

The power circuit of a single-phase full converter is shown in Fig. 3.8. Here, there are four SCRs marked as T_1, T_2, T_3, and T_4. The thyristors are triggered in sequence in such a way that, at any given time, any two thyristors—one from the top and the other from bottom—will be conducting. From the power circuit, it is evident that thyristors T_1 and T_3 will be conducting during the positive half-cycle of the supply voltage; similarly, thyristors T_2 and T_4 will be conducting during the negative half-cycle of the supply voltage. The firing angle α is measured from the zero crossing of the supply voltage; hence T_1 and T_3 are triggered at $\omega_s t = \alpha$, and T_2 and T_4 are triggered at $\omega_s t = \pi + \alpha$. The supply voltage and firing pulses are shown in Fig. 3.9(a) and 3.9(b), respectively. Thus, at $\omega_s t = \alpha$, thyristors T_1 and T_3 are turned

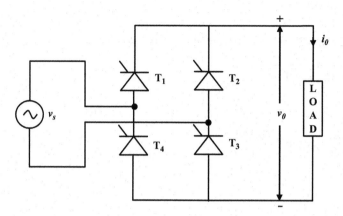

FIGURE 3.8
Power circuit for a full converter.

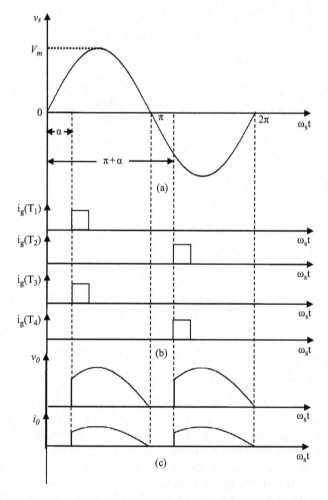

FIGURE 3.9
Waveforms of a full converter with R load. (a) Supply voltage waveform. (b) Triggering pulses. (c) Output waveforms for R load.

ON and the current starts from the source and passes through T_1-load-T_3 back to the supply, and the load voltage is now the supply voltage. Similarly, at $\omega_s t = \pi + \alpha$, T_2 and T_4 are turned ON and the load current flows from the source through T_2-load-T_4. The nature of the load current depends on the characteristics of the load. Three typical loads are discussed below.

3.3.1 Purely Resistive Load

At $\omega_s t = \alpha$, T_1 and T_3 start conducting and the load voltage and current are as indicated in Fig. 3.9(c). With a purely resistive load, the load current i_0 is $\frac{V_o}{R}$, and it therefore has a shape similar to that of the output voltage. At $\omega_s t = \pi$, v_o goes to zero and hence i_0 also goes to zero value. Thyristors T_1 and T_3 are turned OFF by natural commutation, and the load is disconnected from the supply leading to zero output voltage. At $\omega_s t = \pi + \alpha$, T_2 and T_4 are triggered, making the output voltage once again positive. At $\omega_s t = 2\pi$, load voltage and current become zero, thereby turning OFF T_2 and T_4 by natural commutation. An expression for the average value of output voltage is derived below.

$$V_{o(av)} = \frac{1}{\pi} \int_\alpha^\pi V_m \sin \omega_s t \, d(\omega_s t)$$

$$V_{o(av)} = \frac{V_m}{\pi} \left[-\cos(\omega_s t) \right]_\alpha^\pi \tag{3.11}$$

$$V_{o(av)} = \frac{V_m}{\pi} [1 + \cos \alpha]$$

3.3.2 Purely Inductive Load

Consider that the load is a purely inductive one. Presume SCRs T_1 and T_3 are triggered at $\omega_s t = \alpha$ and T_2 and T_4 are triggered at $\omega_s t = \pi + \alpha$. The load current is governed by the following equation:

$$L \frac{di_0}{dt} = V_m \sin(\omega_s t)$$

$$\omega_s L \frac{di_0}{d\omega t} = V_m \sin(\omega_s t)$$

$$i_0 = \frac{V_m}{\omega_s L} \int \sin(\omega_s t) d(\omega_s t) + K_1$$

where K_1 is the constant of integration.

In the above equation, the first limit of integration is α and the last limit of integration is $\omega_s t$, which is the angle at which either SCRs T_2 and T_4 are triggered (i.e., $\omega_s t = \pi + \alpha$ in the case of continuous load current) or the load current goes to zero (in the case of discontinuous load current), whichever is less. Hence,

$$i_0 = \frac{V_m}{\omega_s L} \int_\alpha^{\omega t} \sin(\omega_s t) d(\omega_s t) + K_1$$

$$i_0 = \frac{V_m}{\omega_s L} (\cos \alpha - \cos(\omega_s t)) + K_1$$

At $\omega_s t = \alpha$, $i_0 = 0$ and hence $K_1 = 0$

Hence,

$$i_0 = \frac{V_m}{\omega_s L}(\cos\alpha - \cos(\omega_s t)) \tag{3.12}$$

It is interesting to investigate the effect of α on the continuity of load current.

For $0 \le \alpha < \dfrac{\pi}{2}$, $i_0 \Big|_{\omega_s t = \pi + \alpha}$ is a positive value, indicating continuous conduction. Thus, the current through the load inductance builds up indefinitely; in practical cases, the fuse will blow. The average value of the load voltage is positive in this case.

For $\alpha = \dfrac{\pi}{2}$, $i_0 \Big|_{\omega_s t = \pi + \alpha}$ is a zero, thus the load current is just continuous. Under these conditions, the average value of the load voltage is zero.

For $\alpha > \dfrac{\pi}{2}$, $i_0 \Big|_{\omega_s t = \pi + \alpha}$ i_0 is negative. Because of the circuit topology, the load current cannot be negative; when the load current goes to zero value, the conducting SCRs will be turned OFF.

Hence, $\alpha > \dfrac{\pi}{2}$ corresponds to the discontinuous mode of the load current. Further, because the load current is unidirectional and the average power in an inductor has to be zero, the average voltage across the inductor should be zero.

Thus, $\displaystyle\int_{\alpha}^{\omega_s t} L \frac{di_0}{dt} = 0$. Hence, the load current becomes zero at $\omega_s t = 2\pi - \alpha$. The three different cases explained in this section are illustrated in Fig. 3.10(b)–3.10(d).

3.3.3 R-L Load

Now consider the operation of the full converter with a series resistance-inductance load. Here the major difference of operation compared to a purely resistance type load is the variation of the load current and the output voltage. The load current may be either continuous or discontinuous depending on the value of the ratio of load inductance to resistance and the firing angle α.

Considering a discontinuous mode of operation, at $\omega_s t = \alpha$, when T_1 and T_3 are triggered, current starts increasing through the load. At $\omega_s t = \pi$, the output voltage goes to zero value, but because it is a lagging power factor load, the load current has not yet become zero. This is indicated in Fig. 3.11(c). Hence, T_1 and T_3 continue to conduct until the load current becomes zero at $\omega_s t = \beta$. At this instant, T_1 and T_3 are turned OFF by natural commutation. During the negative half-cycle of the supply voltage, T_2 and T_4 are triggered at $\omega_s t = \pi + \alpha$ and a similar operation takes place. The average output voltage is computed below:

$$V_{o(av)} = \frac{1}{\pi}\left[\int_{\alpha}^{\beta} V_m \sin(\omega_s t)d(\omega_s t)\right]$$

$$V_{o(av)} = \frac{1}{\pi}\left[V_m\left(-\cos(\omega_s t)\right)\Big|_{\alpha}^{\beta}\right] \tag{3.13}$$

$$V_{o(av)} = \frac{1}{\pi}\left[V_m(\cos\alpha - \cos\beta)\right]$$

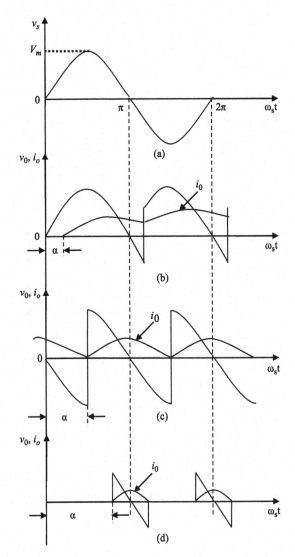

FIGURE 3.10
Waveforms of a full converter for a purely inductive load for different firing angles. (a) Supply voltage waveform.
(b) For $0 \le \alpha < \dfrac{\pi}{2}$. (c) For $\alpha = \dfrac{\pi}{2}$. (d) For $\alpha > \dfrac{\pi}{2}$.

For reduced values of α and increased inductance to resistance ratios, the load current is continuous as shown in Fig. 3.12(c). Here T_1 and T_3 are turned ON at $\omega_s t = \alpha$ and carry the load current. At $\omega_s t = \pi + \alpha$, when T_2 is triggered, the supply voltage appears across T_1 and reverse-biases it. T_1 is turned OFF instantaneously, and the load current is transferred to T_2. Here, T_1 is turned OFF with the use of the line voltage, and such a turning OFF process is called line commutation. Similarly, when T_4 is triggered, T_3 is also turned OFF by line

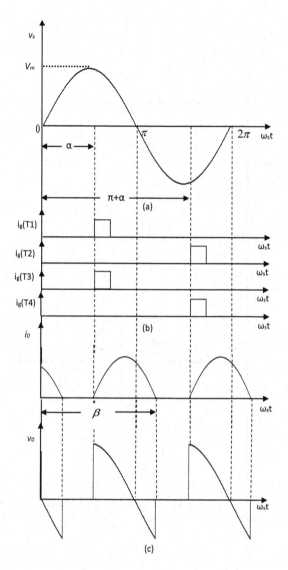

FIGURE 3.11
Waveforms of a full converter with R-L load (discontinuous conduction). (a) Supply voltage waveform. (b) Triggering pulses. (c) Output waveforms.

commutation and the load current is transferred to T_4. The expression for average output voltage is derived below:

$$V_{o(av)} = \frac{1}{\pi} \left[\int_{\alpha}^{\pi+\alpha} V_m \sin(\omega_s t) d\omega_s t \right]$$

$$V_{o(av)} = \frac{1}{\pi} \left[V_m \left(-\cos(\omega_s t) \right)_{\alpha}^{\pi+\alpha} \right]$$

$$V_{o(av)} = \frac{1}{\pi} \left[V_m \left(-\cos(\pi+\alpha) + \cos\alpha \right) \right]$$

$$V_{o(av)} = \frac{2V_m}{\pi} \cos\alpha \tag{3.14}$$

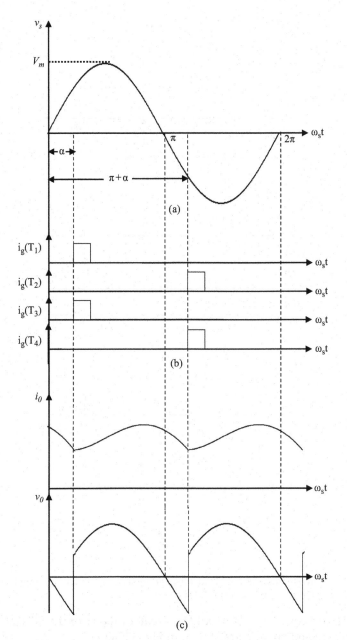

FIGURE 3.12
Waveforms of a full converter with R-L load (continuous conduction). (a) Supply voltage waveform. (b) Triggering pulses. (c) Output waveforms for R-L load.

3.3.4 R-L-Back Emf Load

With a dc motor load, depending upon the magnitude of the load torque and α, the load current will be either continuous or discontinuous. When the motor is running under light load conditions, the average value of current required is low and, due to increased speed, back emf is large. Then motor current is discontinuous as shown in Fig. 3.13. The angle η in this figure is the same as what was explained earlier in Equation (3.2). This indicates

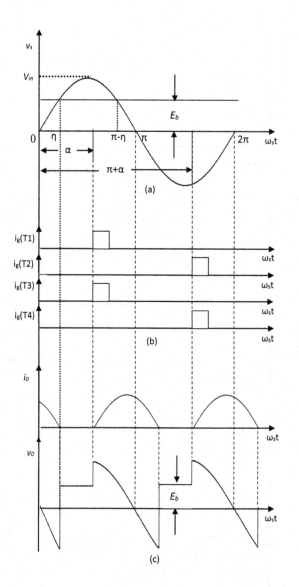

FIGURE 3.13
Waveforms of a full converter with R-L-E_b load (discontinuous conduction). (a) Supply voltage waveform.
(b) Triggering pulses. (c) Output waveforms for R-L-E_b load.

that, under light load conditions, α should be greater than η for the SCRs to be turned ON.
Typical load voltage and current are shown in Fig. 3.13(c).

The expressions for average output voltage is derived below:

$$V_{o(av)} = \frac{1}{\pi}\left[\int_{\alpha}^{\beta} V_m \sin(\omega_s t)d(\omega_s t) + \int_{\beta}^{\pi+\alpha} E_b d(\omega_s t)\right]$$

$$V_{o(av)} = \frac{1}{\pi}\left[V_m\left(-\cos(\omega_s t)\right)_{\alpha}^{\beta} + E_b(\pi+\alpha-\beta)\right]$$

$$V_{o(av)} = \frac{1}{\pi}\left[V_m(\cos\alpha - \cos\beta) + E_b(\pi+\alpha-\beta)\right]$$

(3.15)

When the motor is operating under heavy load conditions, the average load current has to be higher while the back emf is low under these conditions. This will result in continuous conduction of the load current. The typical characteristics are plotted in Fig. 3.14. For this case, the expressions for average output voltage are the same as Equation (3.14).

At $\omega_s t = \alpha$, thyristors T_1 and T_3 are turned ON. With T_1 ON, the load is connected to the source through T_1 and T_3. The transient current i_0 can be found by applying KVL.

$$Ri_0 + L\frac{di_0}{dt} + E_b = V_m \sin \omega_s t$$

$$Ri_0 + L\frac{di_0}{dt} = V_m \sin \omega_s t - E_b$$

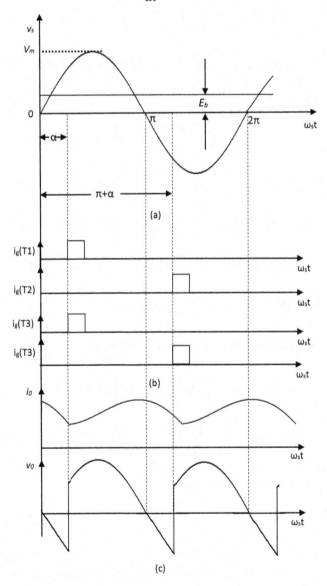

FIGURE 3.14
Waveforms of a full converter with R-L-E_b load (continuous conduction). (a) Supply voltage waveform. (b) Triggering pulses. (c) Output waveforms for R-L-E_b load.

The transient component is obtained by forcing left hand side to zero.

$$Ri_0 + L\frac{di_0}{dt} = 0$$

$$\frac{R}{L}i_0 + \frac{di_0}{dt} = 0$$

$$\left(\frac{R}{L} + p\right)i_0 = 0$$

and the solution is given by

$$i_0 = K_1 e^{\frac{-R}{L}t},$$

where K_1 is the constant of integration to be identified from boundary conditions. The steady state component is given by

$$\frac{V_m}{Z}\sin(\omega_s t - \phi) - \frac{E_b}{R}$$

$$Z = \sqrt{(R^2 + X_L^2)}, \phi = \tan^{-1}\left(\frac{X_L}{R}\right)$$

$$i_0 = K_1 e^{\frac{-R}{L}t} + \frac{V_m}{Z}\sin(\omega_s t - \phi) - \frac{E_b}{R} \qquad (3.16)$$

With a full converter, the current may be continuous or discontinuous.

3.3.4.1 Continuous Mode

At $\omega_s t = \alpha$, current $i_0 = I_{min}$.

Substituting these conditions in Equation (3.16), we have,

$$I_{min} = K_1 e^{\frac{-R\alpha}{\omega_s L}} + \frac{V_m}{Z}\sin(\alpha - \phi) - \frac{E_b}{R}$$

$$K_1 = e^{\frac{R\alpha}{\omega_s L}}\left(\frac{E_b}{R} + I_{min} - \frac{V_m}{Z}\sin(\alpha - \phi)\right)$$

Substituting K_1 in Equation (3.16) we have,

$$i_0 = e^{\frac{R}{L}\left(\frac{\alpha}{\omega_s} - t\right)}\left(I_{min} + \frac{E_b}{R} - \frac{V_m}{Z}\sin(\alpha - \phi)\right) + \frac{V_m}{Z}\sin(\omega_s t - \phi) - \frac{E_b}{R}$$

I_{min}, which depends on the circuit parameters, can be calculated as follows.

At $\omega_s t = \pi + \alpha, i_0 = I_{min}$.

Substituting this in equation above,

$$I_{min} = e^{\frac{R}{\omega_s L}(\alpha - (\pi + \alpha))}\left(I_{min} + \frac{E_b}{R} - \frac{V_m}{Z}\sin(\alpha - \phi)\right) + \frac{V_m}{Z}\sin(\pi + \alpha - \phi) - \frac{E_b}{R}$$

$$I_{min} = e^{\frac{-\pi R}{\omega_s L}}\left(I_{min} + \frac{E_b}{R} - \frac{V_m}{Z}\sin(\alpha - \phi)\right) + \frac{V_m}{Z}\sin(\pi + \alpha - \phi) - \frac{E_b}{R}$$

$$I_{min}(1 - e^{\frac{-\pi R}{\omega_s L}}) = e^{\frac{-\pi R}{\omega_s L}}\left(\frac{E_b}{R} - \frac{V_m}{Z}\sin(\alpha - \phi)\right) + \frac{V_m}{Z}\sin(\pi + \alpha - \phi) - \frac{E_b}{R}$$

$$I_{min}(1 - e^{\frac{-\pi R}{\omega_s L}}) = \frac{E_b}{R}(e^{\frac{-\pi R}{\omega_s L}} - 1) - \frac{V_m}{Z}\sin(\alpha - \phi)(e^{\frac{-\pi R}{\omega_s L}} - 1)$$

$$I_{min} = \left[\frac{\dfrac{E_b}{R}(e^{\frac{-\pi R}{\omega_s L}} - 1) - \dfrac{V_m}{Z}\sin(\alpha - \phi)(e^{\frac{-\pi R}{\omega_s L}} - 1)}{(1 - e^{\frac{-\pi R}{\omega_s L}})}\right]$$

$$I_{min} = \left[\frac{\dfrac{E_b}{R}(e^{\frac{-\pi R}{\omega_s L}} - 1) - \dfrac{V_m}{Z}\sin(\alpha - \phi)(e^{\frac{-\pi R}{\omega_s L}} - 1)}{-(e^{\frac{-\pi R}{\omega_s L}} - 1)}\right]$$

$$I_{min} = \left(\frac{V_m}{Z}\sin(\alpha - \phi) - \frac{E_b}{R}\right)$$

From the above equations, we obtain the minimum current in the circuit at steady state using the circuit parameters. In the discontinuous mode, however, this current is zero.

3.3.4.2 Discontinuous Mode

From Equation (3.16), current is given as,

$$i_0 = K_1 e^{\frac{-tR}{L}} + \frac{V_m}{Z}\sin(\omega_s t - \phi) - \frac{E_b}{R}$$

K_1 can be derived from circuit initial boundary conditions, i.e., at $\omega_s t = \alpha$, $i_0 = 0$. If we substitute this condition in Equation (3.16), we get

$$0 = K_1 e^{-\frac{R\alpha}{\omega_s L}} + \frac{V_m}{Z}\sin(\alpha - \phi) - \frac{E_b}{R}$$

$$K_1 = e^{\frac{R\alpha}{\omega_s L}}\left(\frac{E_b}{R} - \frac{V_m}{Z}\sin(\alpha - \phi)\right)$$

Substituting K_1 in Equation (3.16) above, we have

$$i_0 = e^{\frac{R}{L}\left(\frac{\alpha}{\omega_s} - t\right)}\left(\frac{E_b}{R} - \frac{V_m}{Z}\sin(\alpha - \phi)\right) + \frac{V_m}{Z}\sin(\omega_s t - \phi) - \frac{E_b}{R}$$

This is plotted in Fig. 3.13(c).

EXAMPLE 3.5

A 230-V, 50-Hz source supplies a single-phase full converter feeding a load with a resistance of 1 Ω, inductance of 0.01 H, and emf of 60 V. Find the minimum steady-state load current.

SOLUTION:

For given R and L,

$$Z = \sqrt{R^2 + (\omega_s L)^2}\ ; \omega = 2 \times \pi \times 50 = 2 \times 3.14159 \times 50 \text{ rad/s}$$

$$Z = 3.295\ \Omega$$

$$\phi = \tan^{-1}\left(\frac{\omega_s L}{R}\right);$$

$$\phi = 72°.33^1;$$

The minimum current in the circuit is given by

$$I_{min} = \left(\frac{V_m}{Z} \sin(\alpha - \phi) - \frac{E_b}{R}\right)$$

At $\alpha = 30°$,

$$I_{min} = \left(\frac{230 \times \sqrt{2}}{3.295} \sin(30° - 72.33°) - \frac{60}{1}\right)$$

$$I_{min} = -126.47 \text{ A}$$

At $\alpha = 45°$,

$$I_{min} = \left(\frac{230 \times \sqrt{2}}{3.295} \sin(45° - 72.33°) - \frac{60}{1}\right)$$

$$I_{min} = -105.32 \text{ A}$$

Similarly, at $\alpha = 105°$,

$$I_{min} = \left(\frac{230 \times \sqrt{2}}{3.295} \sin(105° - 72.33°) - \frac{60}{1}\right)$$

$$I_{min} = -6.71 \text{ A}.$$

Because negative current can not flow in a full converter, the current is discontinuous, so the minimum current is taken as zero.

3.4 Comparison of a Single-Phase Semi-Converter and a Full Converter

The striking difference between a single-phase semi-converter and a full converter is that the output voltage is always positive with a semi-converter; therefore, the average output voltage is always positive with a semi-converter. With a full converter, the output voltage

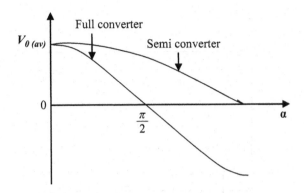

FIGURE 3.15
Variation of average output voltage of a semi-converter and a full converter with respect to α.

waveform contains negative portions of supply voltage, too. The variation of average output voltages of a semi-converter and a full converter with α can be computed using equations 3.1 and 3.14, as shown in Fig. 3.15.

Thus, if the load comprises a back emf of negative polarity, that is $-E_b$, and with $\alpha > \dfrac{\pi}{2}$, the average output voltage of a full converter can be negative. With unidirectional load current, negative average voltage indicates negative average output power. In other words, the load is feeding the power back to mains. This process can take place during the regenerative braking of a dc motor drive.

Consider the operation of a full converter supplying power to a dc motor. During normal operation, the converter's firing angle is maintained below $\dfrac{\pi}{2}$. When the motor has to be decelerated quickly and finally stopped, the motor terminals are interchanged with the converter output terminals to develop $-E_b$. Then α is made larger than $\dfrac{\pi}{2}$ so that the kinetic energy of the rotating motor is converted back to electrical energy and fed back to supply mains. The regenerative braking results in energy saving together with the instantaneous braking of the motor. Thus it is evident that a full converter can be employed for motoring and regenerative braking of a dc motor, whereas a semi-converter can only be used for motoring.

Another major difference between the performance of a semi-converter and a full converter is concerned with the input power factor. In a semi-converter, due to the freewheeling of load current, the input power factor is higher than that of a full converter for the same SCR firing angle. This can be proved as shown below.

Consider a dc motor driving a constant load torque at a fixed speed by taking ripple-free armature current of amplitude I_a. With the semi-converter fired at α and driving the motor, the nature of the source current, together with the supply voltage v_s is shown in Fig. 3.16(b). Due to freewheeling, the supply current is zero from 0 to α and from π to π+α. If the motor is supplied from a full converter, then source current, i_s will be the one shown in Fig. 3.16(c). Assuming constant speed, let E_b represent the motor back emf.

$$\text{Then, the power supplied to motor} = E_b I_a \qquad (3.17)$$

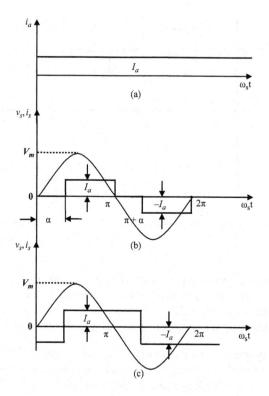

FIGURE 3.16
(a) Motor armature current. (b) Input voltage and current waveforms for a semi-converter. (c) Input voltage and current waveforms for a full converter.

The rms value of source current I_{Srms} with a semi-converter is

$$I_{Sms} = \sqrt{\frac{1}{2\pi}\left(\int_{\alpha}^{\pi} I_a^2\, d\omega_s t + \int_{\pi+\alpha}^{2\pi} I_a^2\, d\omega_s t\right)}$$

$$I_{Sms} = \sqrt{\frac{1}{2\pi}\left(I_a^2\int_{\alpha}^{\pi} d\omega_s t + I_a^2 \int_{\pi+\alpha}^{2\pi} d\omega_s t\right)}$$

$$I_{Sms} = \sqrt{\frac{I_a^2}{2\pi}\left([\omega_s t]_{\alpha}^{\pi} + [\omega_s t]_{\pi+\alpha}^{2\pi}\right)}$$

$$I_{Sms} = \sqrt{\frac{I_a^2}{2\pi}(\pi - \alpha + 2\pi - \pi - \alpha)}$$

$$I_{Sms} = \sqrt{\frac{I_a^2}{2\pi}(2\pi - 2\alpha)}$$

$$I_{Sms} = \sqrt{\frac{I_a^2}{\pi}(\pi - \alpha)}$$

$$I_{Sms} = I_a\sqrt{1 - \left(\frac{\alpha}{\pi}\right)} \tag{3.18}$$

Neglecting losses in the converter,

$$\text{Input power} = \text{Output power}$$

$$\text{Input power} = E_b I_a$$

$$\text{Power factor of semi-converter} = \frac{\text{Input power}}{V_{rms} \cdot I_{rms}} = \frac{E_b I_a}{V_{rms} I_{Srms}} = \frac{E_b I_a}{V_{rms} \cdot I_a \sqrt{1 - \left(\dfrac{\alpha}{\pi}\right)}}$$

$$= \frac{E_b}{V_{rms} \cdot \sqrt{1 - \left(\dfrac{\alpha}{\pi}\right)}} \tag{3.19}$$

The rms value of source current I_{Frms} with a full converter is found as

$$I_{Frms} = I_a$$

$$\text{Input power} = E_b I_a$$

$$\text{Power factor of full converter} = \frac{E_b I_a}{V_{rms} I_{Frms}} = \frac{E_b I_a}{V_{rms} \cdot I_a} = \frac{E_b}{V_{rms}} \tag{3.20}$$

$$\frac{\text{Power factor of Semi converter}}{\text{Power factor of Full converter}} = \frac{1}{\sqrt{1 - \left(\dfrac{\alpha}{\pi}\right)}} = \sqrt{\left(\frac{\pi}{\pi - \alpha}\right)} > 1. \tag{3.21}$$

Thus it is seen that the power factor of a semi-converter is higher than that of a full converter. This aspect is thoroughly investigated in chapter 5.

3.5 Three-Phase Semi-Converter

The power circuit of a three-phase semi-converter is shown in Fig. 3.17. There are three SCRs labeled as T_1, T_2, T_3 and three power diodes named as D_1, D_2 and D_3. Here, SCRs alone need to be triggered, and the diodes will conduct in a sequence depending upon which one is the *most* forward-biased. The phase voltages are shown in Fig. 3.19(a), and here each division corresponds to 30°. Thyristors T_1, T_2, and T_3 are to be triggered in accordance with the relative magnitudes of phase voltages V_R, V_Y, and V_B. Thus, the firing angle α is measured with respect to crossing points of phase voltages. Hence, the reference point for triggering thyristor T_1 is the crossing point of phase voltages V_B and V_R, and is indicated as $\alpha(T_1)$. Similarly, the starting points of firing for T_2 and T_3 are indicated in Fig. 3.19(b) as $\alpha(T_2)$ and $\alpha(T_3)$, respectively. Further, it is decided to trigger the thyristors continuously until their respective phase voltages become negative. The firing pulses thus designed for a typical value of $\alpha = 30°$ are shown in Fig. 3.19(b).

Turning the SCRs OFF takes place in a fixed pattern; when T_2 is triggered, line voltage V_{YR} reverse-biases T_1 and T_1 is instantaneously turned OFF by line commutation. Similarly, turning T_3 ON commutates T_2, and once T_1 is gated, T_3 turns OFF. The cathodes of the diodes are connected to the phase voltages so that, whenever a phase voltage becomes

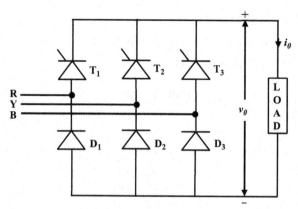

FIGURE 3.17
A 3-phase semi-converter.

more negative than the remaining two phases, the diode connected to that particular phase will be conducting. Thus, the domain of each diode conduction is also well defined and is indicated in the Fig. 3.19(a). At any time, any one thyristor and one diode will be conducting in the semi-converter circuit, thus the output voltage comprises portions of line voltages. Hence it is customary to sketch the line voltages to draw output voltage. Fig. 3.18 shows the relationship between the phase voltages and line voltages.

From Fig. 3.18,

$$\overline{V_R} = |V_R| \angle 0°$$

$$\overline{V_Y} = |V_Y| \angle 120°$$

$$\overline{V_{RY}} = \overline{V_R} - \overline{V_Y}$$

Thus $\overline{V_{RY}}$ is leading $\overline{V_R}$ by 30°. Similarly, V_{YB} and V_{BR} can also be located with respect to V_Y and V_B, respectively. All the line voltages are now sketched with respect to phase voltages; furthermore, the negative portions of line voltages are redrawn as positive because the converter rectifies ac to dc. Thus, Fig. 3.19(c) shows portions of line voltages that will appear at the output.

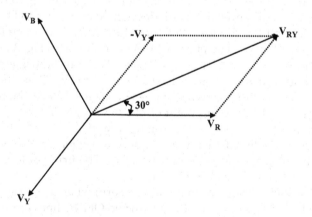

FIGURE 3.18
Relationship between phase and line voltages.

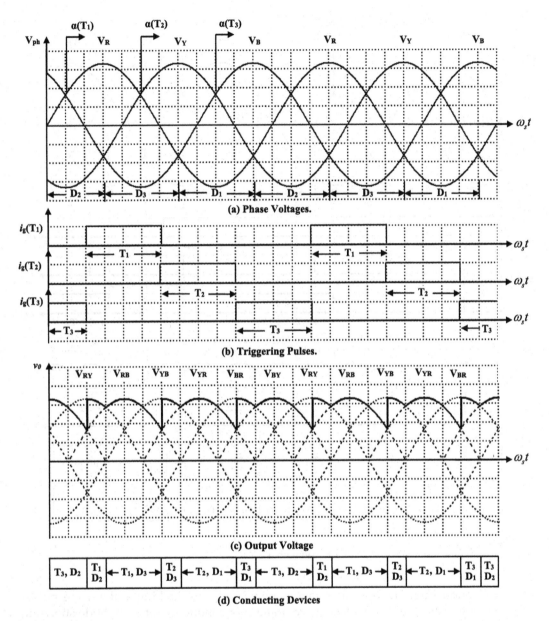

(a) Phase Voltages.

(b) Triggering Pulses.

(c) Output Voltage

| T₃, D₂ | T₁ D₂ | ←T₁, D₃→ | T₂ D₃ | ←T₂, D₁→ | T₃ D₁ | ←T₃, D₂→ | T₁ D₂ | ←T₁, D₃→ | T₂ D₃ | ←T₂, D₁→ | T₃ D₁ | T₃ D₂ |

(d) Conducting Devices

FIGURE 3.19

Three-phase semi-converter waveforms for R-load with $\alpha = 30°$.

To sketch the output voltage, consider the SCR firing angle to be 30° and start from the triggering pulse of T_1. It starts at $\alpha(T_1) = 30°$. From Fig. 3.19(a), at this instant diode D_2 is more forward-biased and hence current starts from R-phase, passes through T_1-LOAD-D_2 and back to Y-phase; thus the output voltage is V_{RY}. After 30°, B-phase becomes more negative, and D_2 is thus turned OFF while D_3 conducts together with T_1. Thus the output voltage is now V_{RB}. When thyristor T_2 is triggered, T_1 is turned OFF by line commutation. Thus, T_2 and D_3 are in series with the load, thereby making the output voltage equal to V_{YB}. When R-phase becomes more negative than B-phase, D_3 is turned OFF and

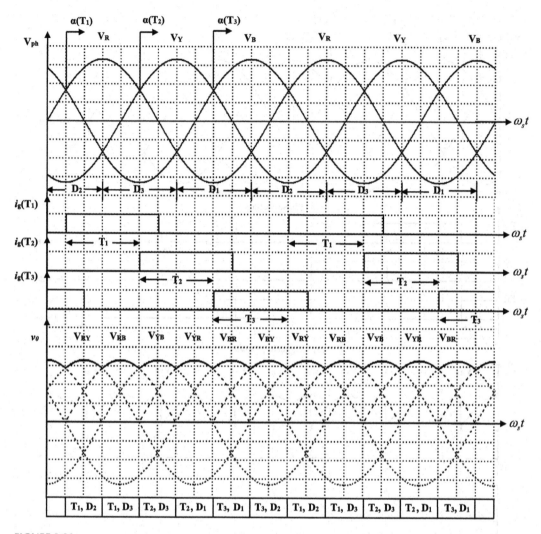

FIGURE 3.20
Three-phase semi-converter waveforms for R-load with $\alpha = 0°$.

D_1 starts conducting. The output voltage is now V_{YR}. Thus, looking at the firing pulses shown in Fig. 3.19(b), and considering the domain of conduction of each diode shown in Fig. 3.19(a), the conduction path at each discrete interval can be identified and the output voltage can be plotted.

For further illustration, Fig. 3.19 through Fig. 3.29 show sketches of the output voltage of a three-phase semi-converter for different firing angles. From these figures, it is evident that the output voltage can be divided into two separate patterns: $< 60°$ and $\geq 60°$.

3.5.1 For $\alpha < 60°$

For $0 \leq \alpha < 60°$, the output voltage contains two line voltages which repeat periodically; further, the output voltage is continuous for all type of loads such as R, R-L and R-L-E$_b$. Also the output voltage ripples are three times that of the supply frequency.

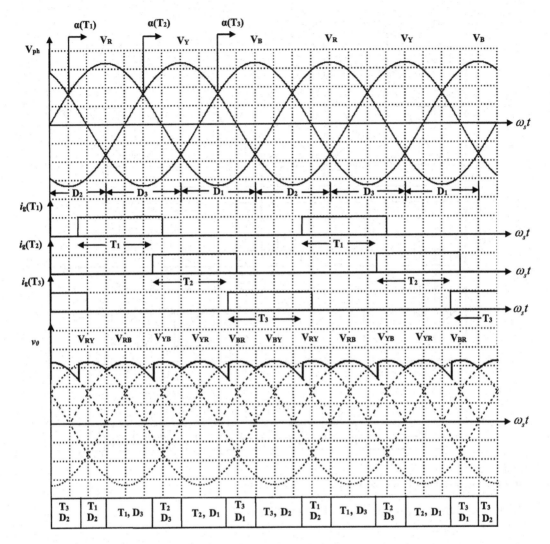

FIGURE 3.21
Three-phase semi-converter waveforms for R-load with $\alpha = 15°$.

An expression for the average value of the output voltage is now derived below.

$$V_{o(av)} = \frac{1}{2\pi/3}\left[\int_{\alpha}^{\pi/3} v_{RY}d\omega_s t + \int_{\pi/3}^{2\pi/3+\alpha} v_{RB}d\omega_s t\right]$$

Taking $\alpha(T_1)$ as the reference point,

$$v_{RY} = V_{Lm}\sin\left(\omega_s t + \pi/3\right) \text{ and}$$

$$v_{RB} = V_{Lm}\sin\omega_s t$$

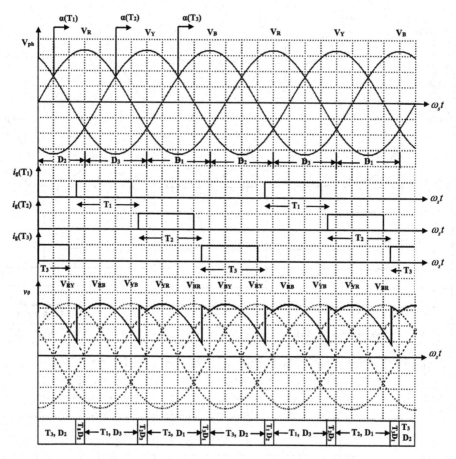

FIGURE 3.22
Three-phase semi-converter waveforms for R-load with $\alpha = 45°$.

where, V_{Lm} is the maximum value of line voltage.

$$V_{o(av)} = \frac{1}{2\pi/3}\left[\int_{\alpha}^{\pi/3} V_{Lm}\sin\left(\omega_s t + \frac{\pi}{3}\right)d\omega_s t + \int_{\pi/3}^{2\pi/3+\alpha} V_{Lm}\sin(\omega_s t)d\omega_s t\right]$$

$$V_{o(av)} = \frac{3V_{Lm}}{2\pi}\left\{\left[-\cos\left(\omega_s t + \frac{\pi}{3}\right)\right]_{\alpha}^{\pi/3} + \left[-\cos\omega_s t\right]_{\pi/3}^{2\pi/3+\alpha}\right\}$$

$$V_{o(av)} = \frac{3V_{Lm}}{2\pi}\left[1 + \cos\left(\alpha + \frac{\pi}{3}\right) - \cos\left(\frac{2\pi}{3} + \alpha\right)\right]$$

$$V_{o(av)} = \frac{3V_{Lm}}{2\pi}\left[1 + \cos\alpha\cdot\cos\left(\frac{\pi}{3}\right) - \sin\alpha\cdot\sin\left(\frac{\pi}{3}\right) - \cos\left(\frac{2\pi}{3}\right)\cos\alpha + \sin\alpha\cdot\sin\left(\frac{2\pi}{3}\right)\right]$$

$$V_{o(av)} = \frac{3V_{Lm}}{2\pi}[1 + 0.5\cos\alpha - (-0.5\cos\alpha)]$$

$$V_{o(av)} = \frac{3V_{Lm}}{2\pi}[1 + \cos\alpha] \tag{3.22}$$

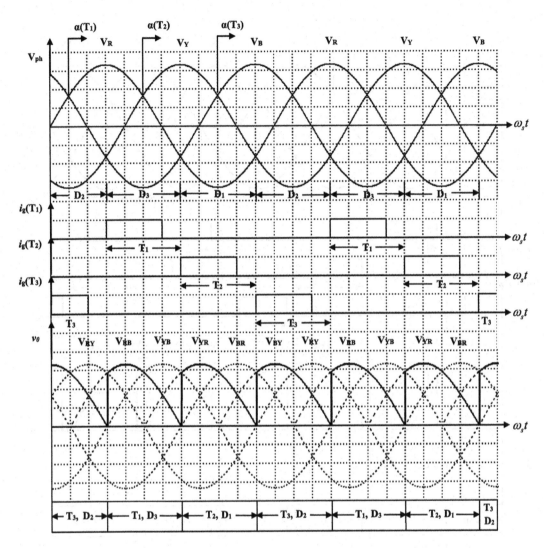

FIGURE 3.23
Three-phase semi-converter waveforms for R-load with $\alpha = 60°$.

3.5.2 For $\alpha \geq 60°$

For any SCR firing angle larger than or equal to 60°, the output voltage contains only one-line voltage, which repeats in sequence. For $\alpha = 60°$, the output voltage is only continuous. For further examination, consider the waveforms plotted for $\alpha = 90°$. Considering the domains of conduction of thyristors and diodes as indicated in Fig. 3.19(a); there is a region where T_1 and D_1 conduct together, thus short-circuiting the output terminals and leading to zero output voltage. Thus for $\alpha > 60°$, inherent freewheeling of the load current exists in a semi-converter. Similarly, freewheeling is existing when there are instants at which T_2 and D_2 conduct together as well as instants at which T_3 and D_3 conduct together. If the load is purely resistive, then the load current will have shape similar to that of the load voltage. If the load is R-L, the load current can be either continuous or discontinuous; in either case, the output voltage will remain the same. If the three-phase semi-converter is

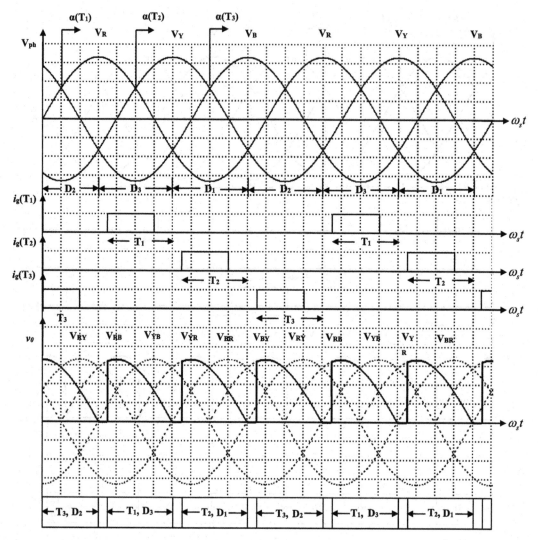

FIGURE 3.24
Three-phase semi-converter waveforms for R-load with $\alpha = 75°$.

driving a dc motor load (which can be treated as R-L-E_b load) with $\alpha > 60°$, the output voltage will be the same as that of R and R-L type, provided the current is continuous. This is given in Fig. 3.24 to Fig. 3.28. If the current is discontinuous with the motor load, then the output voltage will contain motor back emf during discontinuous period as well. A typical case of a three-phase semi converter feeding a dc motor load with discontinuous conduction is shown in Fig. 3.29.

An expression for the average value of output voltage for $\alpha > 60°$ with continuous load current is derived below:

$$V_{o(av)} = \frac{1}{2\pi/3} \int_{\alpha}^{\pi} v_{RY} d\omega_s t$$

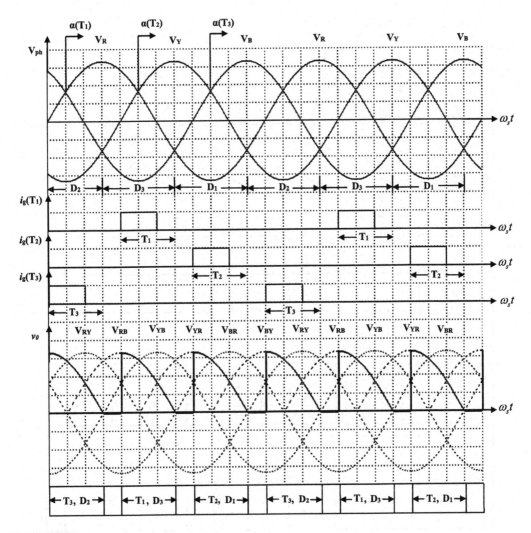

FIGURE 3.25
Three-phase semi-converter waveforms for R-load with $\alpha = 90°$.

Taking $\alpha(T_1)$ as the reference point,

$$v_{RY} = V_{Lm} \sin(\omega_s t)$$

$$V_{o(av)} = \frac{1}{\frac{2\pi}{3}} \int_{\alpha}^{\pi} V_{Lm} \sin(\omega_s t) d\omega_s t$$

$$V_{o(av)} = \frac{3}{2\pi} \int_{\alpha}^{\pi} V_{Lm} \sin(\omega_s t) d\omega_s t$$

$$V_{o(av)} = \frac{3V_{Lm}}{2\pi} \left[-\cos(\omega_s t) \right]_{\alpha}^{\pi}$$

$$V_{o(av)} = \frac{3V_{Lm}}{2\pi} (1 + \cos\alpha) \tag{3.23}$$

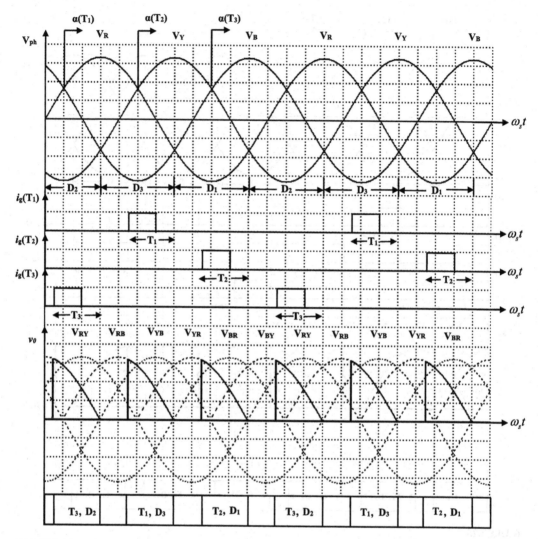

FIGURE 3.26
Three-phase semi-converter waveforms for R-load with $\alpha = 105°$.

3.6 Three-Phase Full-Converter

The power circuit of a three-phase full converter is shown in Fig. 3.30. There are six SCRs, labeled as T_1, T_2, T_3, T_4, T_5, and T_6. The phase voltages are shown in Fig. 3.31(a), with each division corresponding to 30°. Thyristors T_1, T_2, T_3, T_4, T_5, and T_6 are to be triggered in accordance with the relative magnitudes of phase voltages V_R, V_Y, and V_B. Thus, the firing angle α is measured with respect to crossing points of phase voltages. Hence, the reference point for triggering thyristor T_1 is the crossing point of phase voltages V_B and V_R and is indicated as $\alpha(T_1)$. Similarly, the starting points of firing for T_2, T_3, T_4, T_5, and T_6 are also indicated in the Fig. 3.31(a) as $\alpha(T_2)$, $\alpha(T_3)$, $\alpha(T_4)$, $\alpha(T_5)$, and $\alpha(T_6)$ respectively. Furthermore, the

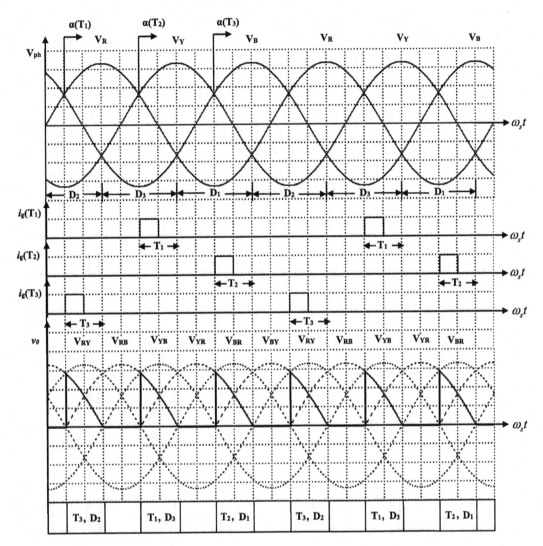

FIGURE 3.27
Three-phase semi-converter waveforms for R-load with α = 120°.

thyristors are triggered continuously until their respective phase voltages become zero. The firing pulses thus designed for a typical value of α = 30° are shown in Fig. 3.31(b).

At any given time, any two thyristors—one from the top and the other from the bottom—will be conducting in the full converter circuit, and hence the output voltage comprises portions of line voltages. It should be noted that commutation of conducting SCRs takes place in a sequential manner; T_1 is turned OFF when T_3 is triggered, and T_3 is commutated when T_5 starts conducting, and T_5 is switched OFF with the firing of T_1. Thyristors T_2, T_4, and T_6 are also commutated in a similar order.

The sketching of the output voltage of the converter has to be done in a systematic manner. Consider that the converter firing angle is 30°. Thus, at α = 30°, the load current starts from R-phase, passes through T_1-load-T_6 and back to Y-phase; thus the output voltage is V_{RY}. When thyristor T_2 is triggered at α(T_2), T_6 is turned OFF by line commutation. The current now starts from R-phase takes the path through T_1-load-T_2-back to B-phase;

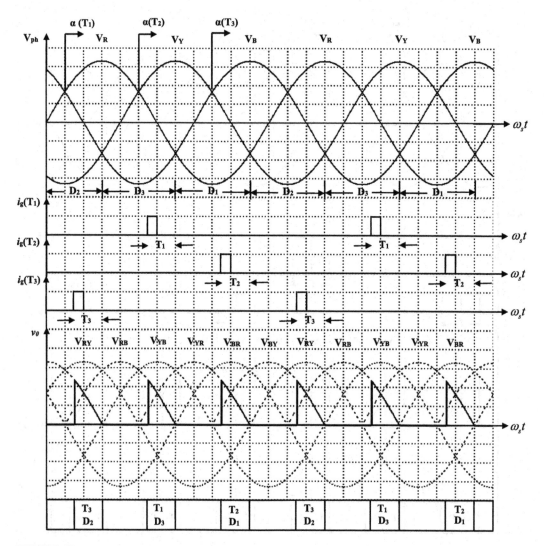

FIGURE 3.28
Three-phase semi-converter waveforms for $\alpha = 135°$.

thus the output voltage is V_{RB}. From Fig. 3.31(c), it is evident that, at any instant, the load current is carried by any pair of thyristors consisting of one from the upper limb and the other from the lower limb. The output voltage has portions of the line voltages, and it repeats in sequence.

For further illustration, sketches of the output voltage of three-phase full converter for different firing angles are shown in Fig. 3.31(d) through Fig. 3.40. From the figures, it is evident that the output voltage can be divided into two distinguished patterns.

3.6.1 For $\alpha < 60°$

For $0 \leq \alpha < 60°$, the output voltage contains two line voltages repeatedly; in addition, the output voltage is continuous, leading to continuous load current for all type of loads, such as R and R-L.

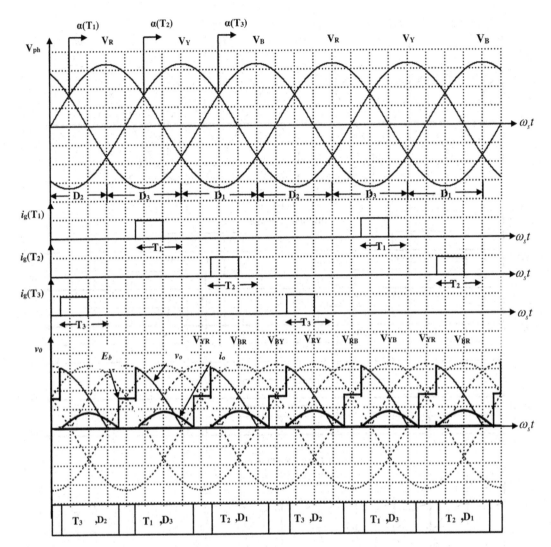

FIGURE 3.29
Three-phase semi-converter waveforms for $\alpha = 105°$ with R-L-E$_b$ load with current discontinuity.

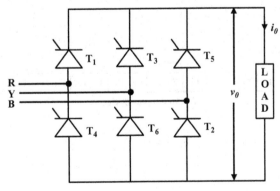

FIGURE 3.30
Three-phase full converter.

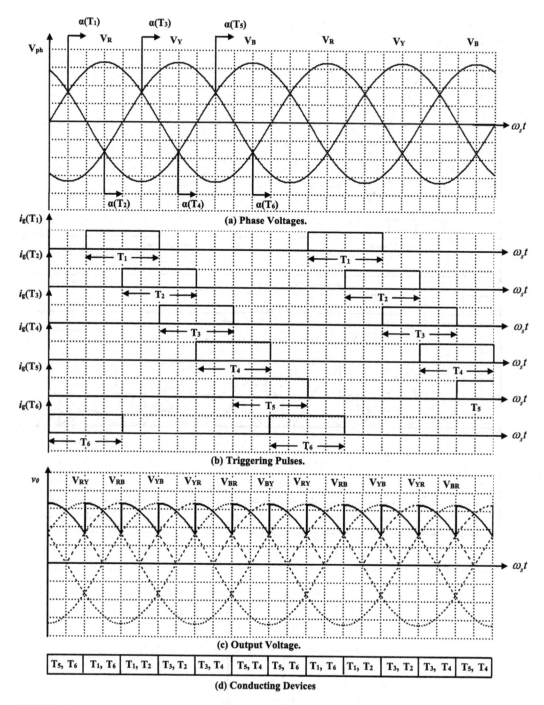

FIGURE 3.31
Three-phase full-converter waveforms for R-load with $\alpha = 30°$.

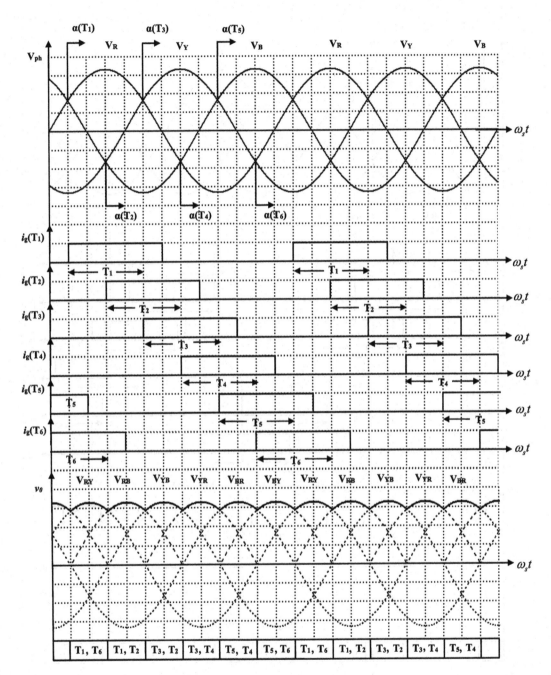

FIGURE 3.32
Three-phase full-converter waveforms for R-load with $\alpha = 0°$.

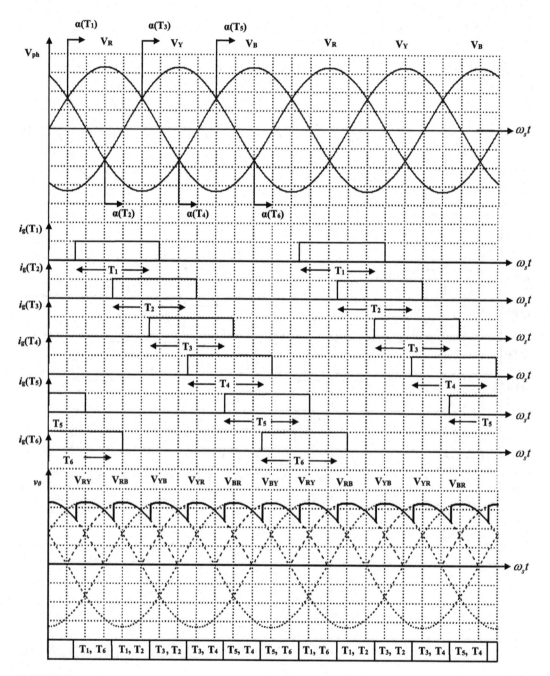

FIGURE 3.33
Three-phase full-converter waveforms for R-load with $\alpha = 15°$.

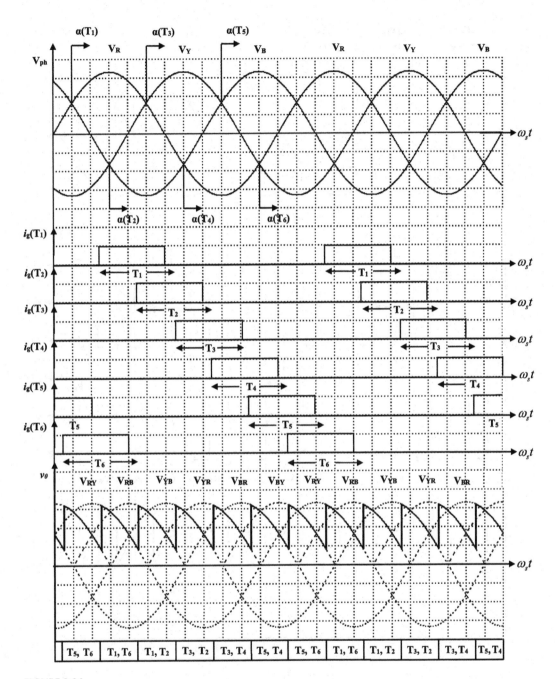

FIGURE 3.34
Three-phase full-converter waveforms for R-load with $\alpha = 45°$.

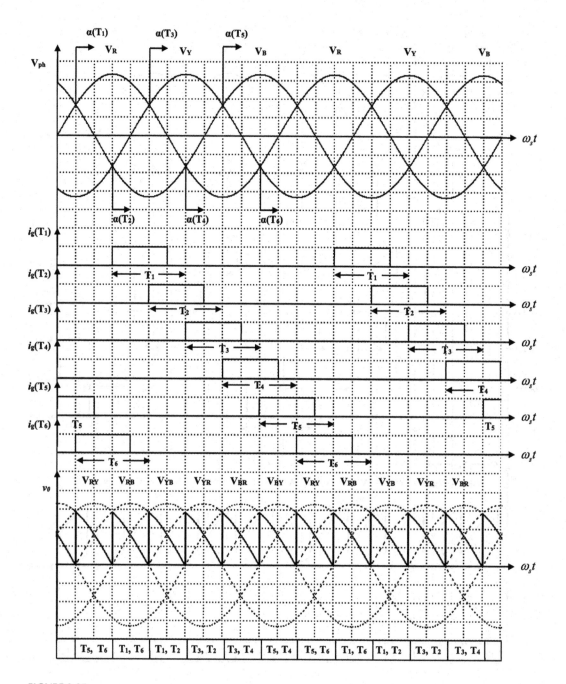

FIGURE 3.35
Three-phase full-converter waveforms for R-load with α = 60°.

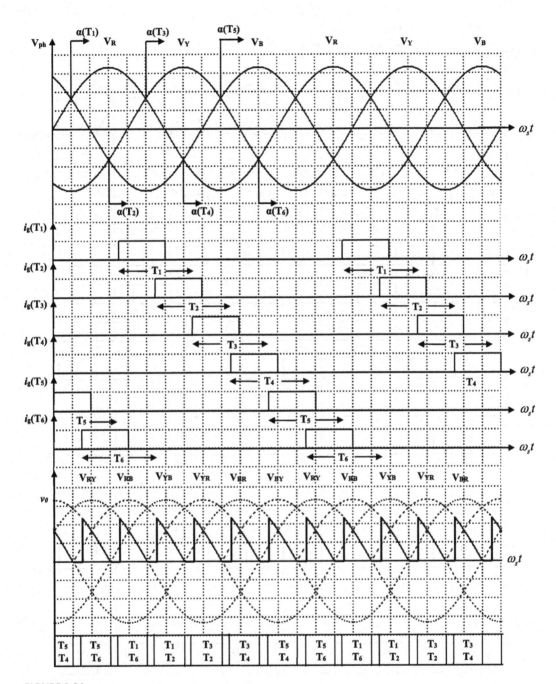

FIGURE 3.36
Three-phase full-converter waveforms for R-load with $\alpha = 75°$.

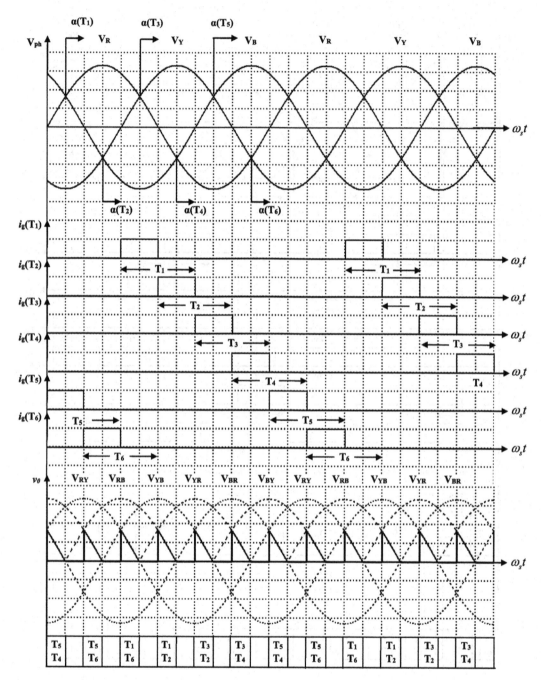

FIGURE 3.37
Three-phase full-converter waveforms for R-load with $\alpha = 90°$.

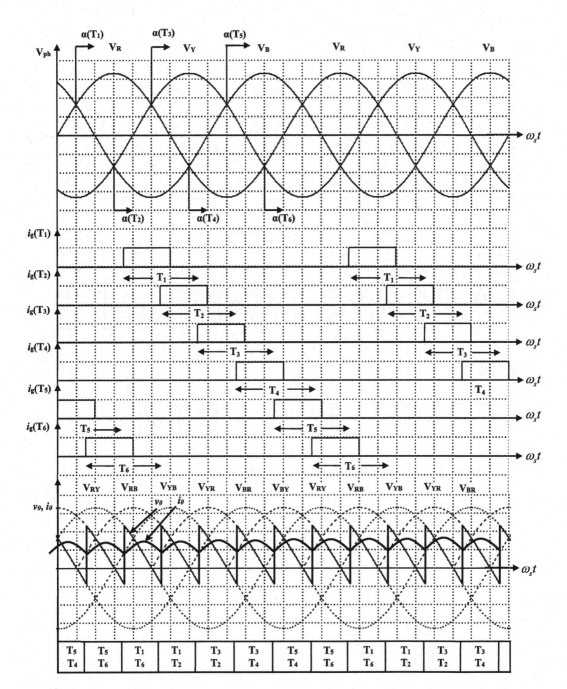

FIGURE 3.38

Three-phase full-converter waveforms for R-L load with $\alpha = 75°$ and current continuity.

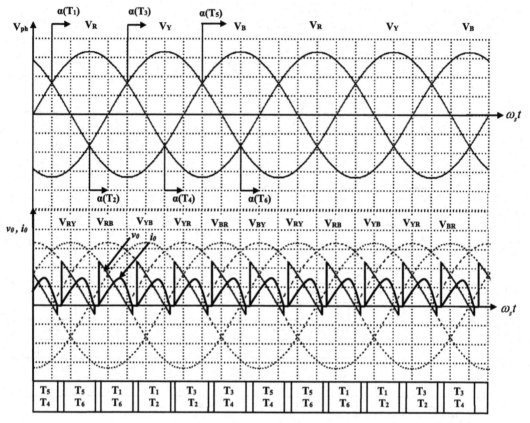

FIGURE 3.39
Three-phase full-converter waveforms for $\alpha = 75°$ with R-L load with current discontinuity.

An expression for the average value of the output voltage is derived below:

$$V_{o(av)} = \left(\frac{1}{2\pi/3}\right)\left[\int_{\alpha}^{\pi/3} v_{RY} d\omega_s t + \int_{\pi/3}^{2\pi/3+\alpha} v_{RB} d\omega_s t\right]$$

Taking $\alpha(T_1)$ as the reference point,

$v_{RY} = V_{Lm} \sin\left(\omega_s t + \pi/3\right)$ and

$v_{RB} = V_{Lm} \sin\omega_s t$ where, V_{Lm} is the maximum value of line voltage.

$$V_{o(av)} = \frac{1}{2\pi/3}\left[\int_{\alpha}^{\pi/3} V_{Lm} \sin\left(\omega_s t + \frac{\pi}{3}\right) d\omega_s t + \int_{\pi/3}^{2\pi/3+\alpha} V_{Lm} \sin(\omega_s t) d\omega_s t\right]$$

$$V_{o(av)} = \frac{3V_{Lm}}{2\pi}\left\{\left[-\cos\left(\omega_s t + \frac{\pi}{3}\right)\right]_{\alpha}^{\pi/3} + \left[-\cos(\omega_s t)\right]_{\pi/3}^{2\pi/3+\alpha}\right\}$$

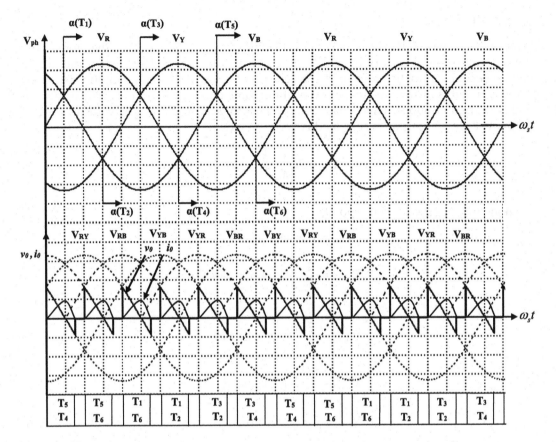

FIGURE 3.40
Three-phase full-rectifier waveforms for R-L load with $\alpha = 90°$ with current discontinuity.

$$V_{o(av)} = \frac{3V_{Lm}}{2\pi}\left[1 + \cos\left(\alpha + \frac{\pi}{3}\right) - \cos\left(\frac{2\pi}{3} + \alpha\right)\right]$$

$$V_{o(av)} = \frac{3V_{Lm}}{2\pi}\left[1 + \cos\alpha \cdot \cos\left(\frac{\pi}{3}\right) - \sin\alpha \cdot \sin\left(\frac{\pi}{3}\right) - \cos\left(\frac{2\pi}{3}\right)\cos\alpha + \sin\alpha \cdot \sin\left(\frac{2\pi}{3}\right)\right]$$

$$V_{o(av)} = \frac{3V_{Lm}}{2\pi}[1 + 0.5\cos\alpha - (-0.5\cos\alpha)]$$

$$V_{o(av)} = \frac{3V_{Lm}}{2\pi}[1 + \cos\alpha] \tag{3.24}$$

3.6.2 For $\alpha \geq 60°$

For SCR firing angles larger than or equal to 60°, the output voltage contains only one line voltage, which repeats in a sequence. For $\alpha = 60°$, the output voltage is only continuous. For $\alpha > 60°$, the nature of the output voltage waveform depends largely on the nature of the load, such as R, R-L, or R-L-E_b. Each case is examined below.

3.6.2.1 R Load

When a three-phase full bridge converter is feeding a purely resistive load, then for $\alpha > 60°$, the output voltage and current will have identical waveforms. Thus, when the output voltage becomes zero, the current also reaches zero value. The conducting thyristors are turned off at this instant due to natural commutation, making the output voltage and current both discontinuous. This is depicted in Fig. 3.35, 3.36, and 3.37. The value of average output voltage for resistance type load is derived below:

$$V_{o(av)} = \frac{1}{2\pi/3} \int_{\alpha}^{\pi} v_{RY} d\omega_s t$$

Taking $\alpha(T_1)$ as the reference point,

$$v_{RY} = V_{Lm} \sin(\omega_s t)$$

$$V_{o(av)} = \frac{1}{\frac{2\pi}{3}} \int_{\alpha}^{\pi} V_{Lm} \sin(\omega_s t) d\omega_s t$$

$$V_{o(av)} = \frac{3}{2\pi} \int_{\alpha}^{\pi} V_{Lm} \sin(\omega_s t) d\omega_s t$$

$$V_{o(av)} = \frac{3V_{Lm}}{2\pi} \left[-\cos(\omega_s t) \right]_{\alpha}^{\pi}$$

$$V_{o(av)} = \frac{3V_{Lm}}{2\pi} (1 + \cos\alpha) \qquad (3.25)$$

3.6.2.2 R-L Load

With this type of load, the output current can be either continuous or discontinuous. Here, load current continues to exist even after output voltage goes to zero value, and the input voltage waveform therefore contains negative portions also. Three such cases for $\alpha = 75°$ and 90° are shown in Fig. 3.38, Fig. 3.39, and Fig. 3.40.

For $\alpha > \pi/3$ and with a continuous load current, the average output voltage of the converter is derived below:

$$V_{o(av)} = \frac{1}{\pi/3} \int_{\alpha}^{\pi/3 + \alpha} v_{RY} d\omega_s t$$

Taking $\alpha(T_1)$ as the reference point,

$$v_{RY} = V_{Lm} \sin\left(\omega_s t + \frac{\pi}{3} \right)$$

$$V_{o(av)} = \frac{1}{\frac{\pi}{3}} \int_{\alpha}^{\pi/3 + \alpha} V_{Lm} \sin\left(\omega_s t + \frac{\pi}{3} \right) d\omega_s t$$

$$V_{o(av)} = \frac{3}{\pi} \int_{\alpha}^{\pi/3+\alpha} V_{Lm} \sin\left(\omega_s t + \frac{\pi}{3}\right) d\omega_s t$$

$$V_{o(av)} = \frac{3V_{Lm}}{\pi} \left[-\cos\left(\omega_s t + \frac{\pi}{3}\right)\right]_{\alpha}^{\pi/3+\alpha}$$

$$= \frac{3V_{Lm}}{\pi} \left[\cos\left(\alpha + \frac{\pi}{3}\right) - \cos\left(2\frac{\pi}{3} + \alpha\right)\right]$$

$$= \frac{3V_{Lm}}{\pi} \left[\cos\alpha\cos\left(\frac{\pi}{3}\right) - \sin\alpha\sin\left(\frac{\pi}{3}\right) - \cos\left(2\frac{\pi}{3}\right)\cos(\alpha) + \sin\left(2\frac{\pi}{3}\right)\sin\alpha\right]$$

$$V_{o(av)} = \frac{3V_{Lm}}{\pi}\cos\alpha \qquad (3.26)$$

Questions

1. A single-phase semi-converter is feeding a resistive load. Sketch the output voltage and current. Also derive an expression for the average value of output voltage.

2. A single-phase bridge-type semi-converter is triggered at an angle of α and is feeding a separately excited dc motor drive. If the armature current is constant, sketch the input voltage and current waveforms. Also compute the rms value of input current.

3. A single-phase semi-converter is feeding a purely resistive load. If the SCR firing angle is 45°, name the commutation with which the SCRs are turned off.

4. With a neat circuit diagram and relevant waveforms, explain the operation of a single-phase full converter with continuous load current.

5. Derive an expression for the average value of output voltage of a single-phase full converter if output current is continuous.

6. A single-phase full converter is feeding an R-L load with the SCR fired at α. If the load current is continuous, sketch the supply voltage, load voltage, and load current. Also derive an expression for the average output voltage.

Unsolved Problems

1. A single-phase semi-converter is triggered at 65°. The converter is supplied from a 220-V, 50-Hz, single-phase supply. The load consists of a pure resistance of 50 Ω. Compute the following: (i) average output load voltage, (ii) average load current, (iii) duration of current conduction, and (iv) maximum positive and negative voltage across the SCR.

2. A 250-V, 60-Hz source supplies a single-phase semi-converter feeding a load with a resistance of 3 Ω, inductance of 10 mH, and back emf of 50 V. The firing angle of the SCR is 30°. Find the minimum steady-state load current.

3. A 220-V, 50-Hz source supplies a single-phase full converter feeding a load with a resistance of 1 Ω, inductance of 0.02 H, and emf of 68 V. Find the minimum steady-state load current.

4. An RL load is fed from a three-phase full converter with an input line voltage of 230 V. Calculate the average output voltages for firing angles of 30°.

5. A single-phase fully controlled thyristor bridge converter supplies an RL load. Assuming that the output current is virtually constant at 10 A, determine the following if the supply source is 230 V and 50 Hz with a firing angle of 60°.

   i. Average output voltage of the converter

   ii. Supply rms current, total

   iii. Supply rms current, fundamental

   iv. Supply power factor

6. A single-phase, fully controlled thyristor bridge converter is supplied from a 230-V, 50-Hz source. Find the output power and power factor for the following cases of loading. In each case, the load current can be assumed to be constant by the addition large inductance in series with the load.

   i. A dc motor armature running at a speed of 1,000 rpm developing an induced emf of 120 V. The armature resistance is 1 Ω. The firing angle is 45°.

   ii. A battery with an emf of 130 V. The internal resistance is 1.5 Ω, and the firing angle is set at 30°.

7. A single-phase fully controlled thyristor bridge convertor is used as a line-commutated inverter (LCI) for feeding power to the ac grid (one-phase, 230 V, 50 Hz) from solar photovoltaic panels. At certain operating conditions, the solar panels produced 1,200 W with a generated voltage of 120 V. Determine the following: (i) the delay angle α such that this power (1,200 W) is supplied to the ac grid from the solar panels, (ii) the power transferred to the ac grid, and (iii) the losses in the inductor resistance. Assume that the dc current is maintained virtually constant by connecting a large inductance between the bridge terminals and solar panels with a resistance of 0.25 Ω.

8. A single-phase fully controlled thyristor bridge converter is supplied from a 230-V, 50-Hz source. Find the output power and power factor for the following cases of loading:

   i. A dc motor armature running at a speed of 1,000 rpm developing an induced emf of 120 V. The armature resistance is 1 Ω. The firing angle is 45°.

   ii. A battery with a emf of 130 V. The internal resistance is 1.5 Ω, and the firing angle is set at 30°.

   In each case, the load current can be assumed to be constant by the addition large inductance in series with the load.

9. A three-phase full-wave bridge circuit with a controlled rectifier is to be used for rectification and inversion. The leakage inductance of each phase of the input transformer windings is 2.0 mH. The input supply is 230 rms per phase of frequency of 50 Hz. The load current on the dc signal is 15 A. Calculate the firing angle required for the SCR to get a dc output voltage of 200 V.

Answers

1. $V_{o(av)} = 140.88$ V, $I_{o(av)} = 2.81$ A, duration of conduction = 115°, $V_{max} = +281.97$ V, −311 V
2. $I_\pi = 68.9$ A, $I_{min} = 21.83$ A
3. 0 A
4. 268.99 V
5. 103.5 V, 9 A, 10 A, 0.45
6. 0.64, 0.78
7. 124.57°, 1,175 W
8. 0.64, 0.78
9. $\alpha = 49.91°$

4

Direct Current Motor Modeling and Control Aspects

4.1 Introduction

Direct current (dc) motors have been used in the industry for the past several years. There are several classifications among dc motors, and separately excited dc motors and series motors present excellent speed-torque characteristics, suitable for many industrial utilizations. As such, dc motors are suitable for a wide range of variable-speed operation, braking, and speed reversal. For successful implementation of closed-loop speed control, a dc motor needs to be modeled either in state-space or in the transfer-function form. This chapter introduces of power conversion, state-space, and transfer-function models. Measurement of various motor parameters is also included.

4.2 Voltage Equation

A simple representation of a dc motor is shown in Fig. 4.1. The field system consists of a pair of electromagnets excited from the field voltage V_f. The field current is indicated as I_f. The armature conductors are assumed to carry the current as given in Fig. 4.1. The field flux is constant and stationary in space; furthermore, this flux is perpendicular to the armature current at any instant. This is one of the most interesting features of dc motors because such a position produces the maximum torque. Applying Fleming's left-hand rule, the armature flux can be obtained, and it can be seen that the introduction of field flux with armature flux gives clockwise rotational torque to the motor.

The induced voltage in the armature of a dc motor is equivalent to the generated voltage in a dc generator. The induced emf, generally termed back-emf, is labeled as E_b and is given by

$$E_b = \frac{\Phi_m Z_a N_r}{60} \cdot P_a \Big/ A \tag{4.1}$$

where,

Φ_m = Field flux
Z_a = Total number of armature conductors
N_r = Rotor speed in rpm
P_a = Number of poles
A = 2 for wave winding
  = P_a for lap winding

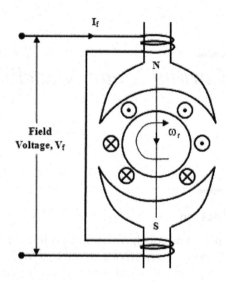

FIGURE 4.1
Sketch of a dc motor structure.

Putting

$$\omega_r = \frac{2\pi N_r}{60},$$

$$E_b = \frac{\Phi_m Z_a}{60} \cdot \frac{60\,\omega_r}{2\pi} \cdot P_a\!\!\Big/\!A$$

$$= \frac{\Phi_m Z_a \omega_r}{2\pi} \cdot P_a\!\!\Big/\!A$$

Making

$$K = \frac{Z_a}{2\pi} \cdot P_a\!\!\Big/\!A, \text{ then}$$

$$E_b = K\Phi_m\omega_r$$

In a separately excited dc motor, field flux Φ_m is kept constant, and hence

$$E_b = K_b\omega_r \tag{4.2}$$

where $K_b = K\Phi_m$ and is called the back-emf constant. The unit of K_b is V/(rad/s).

4.3 Torque Equation

A separately excited dc motor can be represented in R-L-back emf form and is shown in Fig. 4.2. Here, r_a and L_a represent armature resistance and inductance, respectively. Armature voltage and current are shown as v_a and i_a in Fig. 4.2. Writing Kirchhoff's voltage law (KVL),

$$v_a = e_b + r_a i_a + L_a\frac{di_a}{dt} \tag{4.3}$$

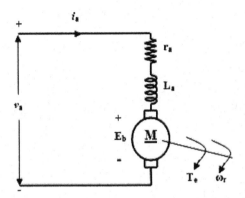

FIGURE 4.2
Sketch of a dc motor equivalent circuit.

Multiplying both sides using i_a, the power equation can be written as

$$v_a i_a = r_a i_a^2 + e_b i_a + L_a i_a \frac{di_a}{dt}$$

Under steady-state conditions, $\frac{di_a}{dt} \approx 0$ and hence

$$V_a I_a = r_a I_a^2 + E_b I_a$$

where $I_a^2 r_a$ represents armature copper loss. Hence, the air gap power is $E_b I_a$. If T_e represents the electromagnetic torque, then

$$\text{Output power} = T_e \omega_r$$

Neglecting friction and windage losses, this should be equal to air gap power.

$$\text{i.e., } E_b I_a = T_e \omega_r$$

$$\text{i.e., } T_e = \frac{E_b I_a}{\omega_r} \tag{4.4}$$

From Equation (4.2),

$$K_b = \frac{E_b}{\omega_r}$$

and substituting in Equation (4.4), we get

$$T_e = K_b I_a \tag{4.5}$$

It is evident that the torque constant is equivalent to the back-emf constant for a fixed field excited dc motor.

EXAMPLE 4.1

A permanent magnet dc commutator motor has no load speed of 5,000 rpm when connected to a 115-V dc supply. The armature resistance is 2.8 Ω, and other losses may be neglected. Find the speed of the motor at supply voltage of 80 V and developing a torque of 0.7 N-m.

SOLUTION:

As an assumption, under no-load, armature current $I_a = 0$.

$$\therefore E_b = V_a = 115 \text{ V}$$

$$K = \frac{115}{\left(\dfrac{2\pi \times 5,000}{60}\right)} \Rightarrow 0.22 \text{ V/rad/s}$$

Electromagnetic torque, $T_e = K_b I_a$

$$I_a = \frac{0.7}{0.22} \Rightarrow 3.18 \text{ A}$$

$$E_{b2} = V - I_a R_a \Rightarrow 80 - (3.18 \times 2.8)$$
$$= 71.09 \text{ V}$$

Now, $\quad \dfrac{N_2}{N_1} = \dfrac{E_{b2}}{E_{b1}}$

$$N_2 = 5000 \times \left(\frac{71.09}{115}\right) = 3090 \text{ rpm}$$

EXAMPLE 4.2

A separately excited dc motor runs at 1,500 rpm at no load with 220-V supply at the armature. The voltage is maintained at its rated value. The speed of the motor when it delivers a torque of 5 N-m is 1,400 rpm. The rotational losses and armature reaction losses are neglected. Find (a) armature resistance of the motor and (b) voltage applied to the armature for the motor to deliver a torque of 2.5 N-m at 1,350 rpm.

SOLUTION:

(a)

$$I_a = 0 \text{ (at no load)}$$

$$E_{b1} = V_a = 220 \text{ V}$$

$$E_{b1} = K\omega \Rightarrow 220 = K\left(\frac{2\pi \times 1500}{60}\right) \Rightarrow K = 1.401$$

Torque, $T_e = K_b I_a$

$$5 = 1.401 \times I_a \Rightarrow I_a = 3.568 \text{ A}$$

$$E_{b2} = V_a - I_a r_a$$

$$\frac{N_2}{N_1} = \frac{E_{b2}}{E_{b1}} \Rightarrow E_{b2} = \frac{1400 \times 220}{1500} = 205.33 \text{ V}$$

$$r_a = \frac{V - E_{b2}}{I_a} = \frac{220 - 205.33}{3.568}$$

$$\therefore r_a = 4.1 \, \Omega$$

(b) $T = K_bI_a \Rightarrow 2.5 = 1.401 \times I_a$

     $\Rightarrow I_a = 1.784$ A

$E_b = K_b\omega_r \Rightarrow 1.401 \times \left(\dfrac{2\pi \times 1350}{60}\right) \Rightarrow 198.06$ V

$V_a = E_b + I_a r_a$

        $= 198.03 + (1.784 \times 4.1) = 205.34$ V

EXAMPLE 4.3

A separately excited dc motor has the parameters 220 V, 25 A, 1,500 rpm, J = 0.6 kg-m^2, $K_b = 0.567$ V/rad/s, and friction is negligible. If the motor starts from rest, find the time taken by the motor to reach a speed of 1,000 rpm with no load. The armature current is maintained constant at its rated value during starting.

SOLUTION:

$$T_e = K_bI_a$$
$$= 0.567 \times 25$$
$$= 14.175 \text{ N-m}$$

During starting, armature current is constant, and hence torque remains constant at rated value.

$$J\frac{d\omega}{dt} = 14.175$$

Integrating on either side,

$$\int_{\omega_1}^{\omega_2} d\omega = \frac{14.175}{J}\int_{t_1}^{t_2} dt = 23.625\int_{t_1}^{t_2} dt$$

$$\omega_2 - \omega_1 = 23.625(t_2 - t_1)$$

$$\omega_1 = 0, \ \omega_2 = 104.71 \text{ rad/s}$$

And hence

$$t_1 - t_2 = 4.432 \text{ s}$$

The time taken by the motor to reach a speed of 1,000 rpm = 4.432 s.

EXAMPLE 4.4

A variable speed drive rated for 1,500 rpm, 60N-m is reversing to 1,000 rpm under no load. The motor torque is 20 N-m and reversing time is 0.5 s. Find the moment of inertia of the drive.

SOLUTION:

$$T_e - T_L = J\frac{d\omega}{dt}$$

$$T_L = 0 \text{ (Given)}$$

$$T_e = 20 \text{ N-m(Given)}$$

$$\Delta t = 0.5 \text{ sec}$$

$$\Delta\omega = (1500 - (-1000)) \times \frac{2\pi}{60}$$

$$= 261.799 \text{ rad/s}$$

$$J = \frac{20 \times 0.5}{261.799}$$

$$J = 0.03819 \text{ kg-m}^2$$

EXAMPLE 4.5

In a speed-controlled dc motor drive, the load torque is 30 N-m. At time $t = 0$, the motor is running at 500 rpm and the generated torque is 90 N-m. The inertia of the drive is 0.01 N-ms^2/rad. The friction is negligible. Evaluate the time taken for the speed to reach 1,000 rpm.

SOLUTION:

$$J\frac{d\omega}{dt} = T_e - T_L$$

$$\Delta\omega = 1,000 - 500 = 500 \text{ rpm} = 52.35 \text{ rad/s}$$

$$T_e - T_L = 90 - 30 = 60 \text{ N-m}$$

$$J = 0.01$$

$$\Delta t = \frac{J\Delta\omega}{T_e - T_L}$$

$$= \frac{0.01 \times 52.35}{60}$$

$$\Delta t = 8.726 \text{ ms}$$

EXAMPLE 4.6

An electric motor is developing a starting torque of 20 N-m, and starts with a load torque of 8 N-m on its shaft. If the acceleration at start is 100 rad/sec^2, what is the value of moment of inertia?

SOLUTION:

$$T_e - T_L = J\frac{d\omega}{dt}$$

$$20 - 8 = J\frac{d\omega}{dt}$$

$$\because \frac{\Delta\omega}{\Delta t} = 100\ \text{rad}/\sec^2\,(\text{Given})$$

$$12 = 100\,J$$

$$J = 0.12\ \text{kg-m}^2$$

4.4 State-Space Model

Referring to Fig. 4.2, the voltage equation of a separately excited dc motor is rewritten as

$$v_a = L_a\frac{di_a}{dt} + r_a i_a + e_b$$

Re-arranging the terms and setting $e_b = K_b\omega_r$,

$$\frac{di_a}{dt} = -\frac{r_a i_a}{L_a} - \frac{K_b\omega_r}{L_a} + \frac{v_a}{L_a} \tag{4.6}$$

Let J represent the moment of inertia in kg-m^2 and B the friction coefficient in N-m/(rad/s). Now

$$T_e = B\omega_r + J\frac{d\omega_r}{dt} + T_L$$

$$\text{i.e., } K_b i_a = B\omega_r + J\frac{d\omega_r}{dt} + T_L \tag{4.7}$$

$$\text{i.e., } \frac{d\omega_r}{dt} = -\frac{B\omega_r}{J} - \frac{T_L}{J} + \frac{K_b i_a}{J}$$

Equations (4.6) and (4.7) completely describe the dynamics of a dc motor and can be put in matrix form as

$$\begin{bmatrix} \dfrac{di_a}{dt} \\ \dfrac{d\omega_r}{dt} \end{bmatrix} = \begin{bmatrix} \dfrac{-r_a}{L_a} & \dfrac{-K_b}{L_a} \\ \dfrac{K_b}{J} & \dfrac{-B}{J} \end{bmatrix} \begin{bmatrix} i_a \\ \omega_r \end{bmatrix} + \begin{bmatrix} \dfrac{+1}{L_a} & 0 \\ 0 & \dfrac{-1}{J} \end{bmatrix} \begin{bmatrix} v_a \\ T_L \end{bmatrix} \tag{4.8}$$

This is equivalent to state-space form

$$\frac{dx}{dt} = Ax + Bu$$

where

$$x = \begin{bmatrix} i_a \\ \omega_r \end{bmatrix}$$

$$A = \begin{bmatrix} \dfrac{-r_a}{L_a} & \dfrac{-K_b}{L_a} \\ \dfrac{K_b}{J} & \dfrac{-B}{J} \end{bmatrix}$$

$$B = \begin{bmatrix} \dfrac{1}{L_a} & 0 \\ 0 & \dfrac{-1}{J} \end{bmatrix}$$

where u is the input vector and is $\begin{bmatrix} v_a \\ T_L \end{bmatrix}$.

The Eigenvalues of the drive system can be found from A matrix by writing:

$$|sI - A| = 0 \qquad (4.9)$$

i.e., $$\left| s \begin{bmatrix} 1 & 0 \\ 0 & 1 \end{bmatrix} - \begin{bmatrix} \dfrac{-r_a}{L_a} & \dfrac{-K_b}{L_a} \\ \dfrac{K_b}{J} & \dfrac{-B}{J} \end{bmatrix} \right| = 0$$

$$\left| \begin{bmatrix} s + \dfrac{r_a}{L_a} & \dfrac{K_b}{L_a} \\ \dfrac{-K_b}{J} & s + \dfrac{B}{J} \end{bmatrix} \right| = 0$$

i.e., $\left(s + \dfrac{r_a}{L_a} \right)\left(s + \dfrac{B}{J} \right) + \dfrac{K_b^2}{JL_a} = 0$

$$s^2 + \dfrac{sB}{J} + \dfrac{sr_a}{L_a} + \dfrac{r_a B}{L_a J} + \dfrac{K_b^2}{JL_a} = 0$$

$$s^2 + s\left(\dfrac{B}{J} + \dfrac{r_a}{L_a} \right) + \dfrac{K_b^2}{JL_a} + \dfrac{r_a B}{L_a J} = 0 \qquad (4.10)$$

$$s = \dfrac{-\left(\dfrac{r_a}{L_a} + \dfrac{B}{J} \right) \pm \sqrt{\left(\dfrac{r_a}{L_a} + \dfrac{B}{J} \right)^2 - 4\left(\dfrac{K_b^2 + r_a B}{JL_a} \right)}}{2}$$

EXAMPLE 4.7

A separately excited dc motor has the following parameters:

$$r_a = 0.5\,\Omega, \quad L_a = 0.003\,H, \quad k_b = 0.8\,v/rad/sec$$

$$J = 0.0167\,kg\text{-}m^2, \quad B = 0.01\,N\text{-}m/rad/sec$$

Find the Eigen value and asses the stability of the system.

SOLUTION:

$$s = \frac{-\left(\dfrac{r_a}{L_a} + \dfrac{B}{J}\right) \pm \sqrt{\left(\dfrac{r_a}{L_a} + \dfrac{B}{J}\right)^2 - 4 \times \left(\dfrac{Br_a}{JL_a} + \dfrac{k_b^2}{JL_a}\right)}}{2}$$

$$s = \frac{-167.27 \pm \sqrt{-23519.28}}{2}$$

$$= -83.635 \pm i76.68$$

The system is stable because the roots are placed in the left side of the s-plane.

EXAMPLE 4.8

A separately excited dc motor has the following parameters:

$$r_a = 0.34\,\Omega, \quad L_a = 1.13\,mH, \quad k_b = 1.061\,v/rad/sec$$

$$J = 0.035\,kg\text{-}m^2, \quad B = 0$$

Find the Eigen value and assess the stability of the system.

SOLUTION:

$$s = \frac{-\left(\dfrac{r_a}{L_a} + \dfrac{B}{J}\right) \pm \sqrt{\left(\dfrac{r_a}{L_a} + \dfrac{B}{J}\right)^2 - 4 \times \left(\dfrac{Br_a}{JL_a} + \dfrac{k_b^2}{JL_a}\right)}}{2}$$

$$s = \frac{-300.89 \pm \sqrt{-23317.29}}{2}$$

$$s = -150.45 \pm i76.35$$

The system is stable because the roots are placed in the left side of the s-plane.

EXAMPLE 4.9

A separately excited dc motor has the following parameters:

$$r_a = 1.39\,\Omega, \quad L_a = 0.00182\,H, \quad k_b = 0.331\,v/rad/sec$$

$$J = 0.002\,kg\text{-}m^2, \quad B = 0.005\,N\text{-}m/rad/sec$$

Find the Eigen value and assess the stability of the systems.

SOLUTION:

$$s = \frac{-\left(\dfrac{r_a}{L_a} + \dfrac{B}{J}\right) \pm \sqrt{\left(\dfrac{r_a}{L_a} + \dfrac{B}{J}\right)^2 - 4\times\left(\dfrac{Br_a}{JL_a} + \dfrac{k_b^2}{JL_a}\right)}}{2}$$

$$s^2 + 766.24s + 32008.52 = 0$$

$$s = \frac{-766.24 \pm \sqrt{766.24^2 - 4\times1\times32008.52}}{2}$$

$$s = -44.4, \ -721.9$$

The system is stable because the roots are placed in the left side of the s-plane.

4.5 Transfer Function Model

The voltage equation of a dc motor is

$$v_a = \frac{L_a di_a}{dt} + r_a i_a + K_b \omega_r$$

Taking Laplace transform,

$$\begin{aligned}
V_a(s) &= L_a s I_a(s) + r_a I_a(s) + K_b \omega_r(s) \\
&= I_a(s)\{L_a s + r_a\} + K_b \omega_r(s) \\
\therefore I_a(s) &= \frac{V_a(s) - K_b \omega_r(s)}{r_a + L_a s}
\end{aligned} \tag{4.11}$$

The speed-torque equation is rewritten from Equation (4.7) as

$$\frac{d\omega_r}{dt} + \frac{B}{J}\omega_r = \frac{-T_L}{J} + \frac{K_b i_a}{J}$$

Taking Laplace transform,

$$\begin{aligned}
s\omega_r(s) + \omega_r(s)\cdot\frac{B}{J} &= -\frac{T_L(s)}{J} + \frac{K_b I_a(s)}{J} \\
\omega_r(s) &= \frac{K_b I_a(s) - T_L(s)}{Js + B}
\end{aligned} \tag{4.12}$$

Equations (4.11) and (4.12) can be rearranged to obtain the block diagram in Fig. 4.3. Neglecting the load torque $T_L(s)$, the no-load transfer function is obtained as

$$\frac{\omega_r(s)}{V_a(s)} = \frac{K_b}{JL_a s^2 + (BL_a + r_a J)s + Br_a + K_b^2} \tag{4.13}$$

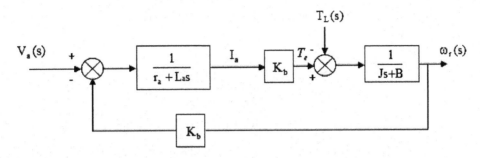

FIGURE 4.3
Transfer function of a dc motor.

4.6 Closed-Loop Control Design

To explain the closed-loop control of a dc motor, consider a proportional-integral (PI) controller be incorporated as given in Fig. 4.4. There are several methods to find the values of PI constants, and here the Routh-Hurwitz method is employed.

The characteristic equation for the above closed loop control system is

$$1 + G_c(s) \cdot G(s) = 0 \tag{4.14}$$

where $G_c(s)$ is the transfer function of PI controller and hence

$$G_c(s) = K_P + \frac{K_I}{S} \tag{4.15}$$

The dc motor is represented by $G(s)$ and is

$$G(s) = \frac{K_b}{JL_a s^2 + (BL_a + r_a J)s + Br_a + K_b^2}$$

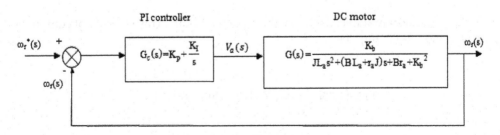

FIGURE 4.4
Closed-loop speed control using a proportional-integral (PI) controller.

Substituting $G_c(s)$ and $G(s)$ in the characteristic equation, we get

$$1+\left[K_P+\frac{K_I}{S}\right]\times\left[\frac{K_b}{JL_aS^2+(BL_a+r_aJ)S+Br_a+K_b^2}\right]=0$$

$$1+\left[\frac{(K_ps+K_I)K_b}{JL_as^3+(BL_a+r_aJ)s^2+(Br_a+K_b^2)s}\right]=0 \qquad (4.16)$$

$$JL_as^3+(BL_a+r_aJ)s^2+(Br_a+K_b^2)s+(K_ps+K_I)K_b=0$$

$$JL_as^3+(BL_a+r_aJ)s^2+(Br_a+K_b^2+K_PK_b)s+K_IK_b=0$$

This is the final characteristic equation for the above control system.

EXAMPLE 4.10

Consider a dc motor with the following parameters: L_a = 1.13 H, J = 0.035 kg-m^2, B = 0.1 N-m/rad/s, r_a = 0.3 Ω, K_b = 1.061 V/rad/s. Calculate the range of K_p and K_I values for the closed-loop operation of the motor.

SOLUTION:

The characteristic equation for the above case is obtained using Equation (4.16):

$$0.039\,s^3+0.123\,s^2+(1.156+1.061\,K_P)s+1.061\,K_I=0.$$

To find the range of controller parameters (proportional gain K_P and integral gain K_I), apply the Routh-Hurwitz criteria.

s^3	0.039	$1.156+1.061\,K_P$
s^2	0.123	$1.061\,K_I$
s^1	$\dfrac{0.123(1.156+1.061\,K_P)-0.0395*1.061\,K_I}{0.123}$	
s^0	$1.061\,K_I$	

To make the system stable, the first column of the Routh array should not contain any sign changes, which implies

$$1.061\,K_I>0$$

$$\frac{0.123(1.156+1.061\,K_p)-0.0395\times1.061\,K_I}{0.123}>0$$

$$0.1421+0.130\,K_P-0.041\,K_I>0$$

$$1.1+K_P-0.315\,K_I>0$$

As an illustration, let $K_I = 5$ (which is greater than zero). Substituting this value in the above equation,

$$1.1 + K_P - 0.315 \times 5 > 0$$
$$\therefore K_P > 0.4769$$

Here for $K_I = 5$, K_p should be greater than 0.4769.

It is important to mention that the above values of controller constants always guarantee stability, but they need not provide optimal dynamic stability. Improved dynamic response can be obtained by fine-tuning the values of K_p and K_I. In several cases, an inner current loop is also added, which helps limit the armature current to permissible values.

4.7 The dc Series Motor

In the dc series motor, the field circuit is connected in series with the armature coil as shown in Fig. 4.5. The back emf, E_b, is

$$E_b = K_b \Phi_m \omega_r$$

Armature current produces the flux, Φ_m and hence,

$$\Phi_m \alpha i_a$$
$$\Phi_m = k_f i_a$$

Hence,

$$E_b = K_b \cdot k_f i_a \omega_r \tag{4.17}$$

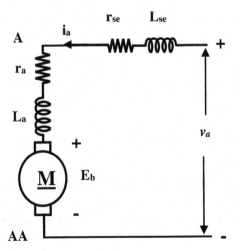

FIGURE 4.5
Sketch of a dc series motor.

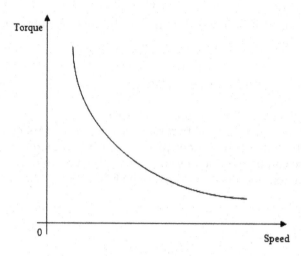

FIGURE 4.6
Torque-speed characteristics of dc series motor.

In the above equation, k_f is the series field constant. Reproducing Equation (4.4) yields

$$T_e = \frac{E_b i_a}{\omega_r} = \frac{K_b \Phi_m \omega_r i_a}{\omega_r}$$
$$= K_b \Phi_m i_a$$
$$= K_b \cdot k_f i_a^2$$

(4.18)

Let $K_b k_f = k_T$, be as the torque constant, and its unit is N-m/A^2. Thus,

$$T_e = k_T i_a^2$$

Furthermore, Equation (4.17) becomes

$$E_b = k_T i_a \omega_r$$

(4.19)

$$v_a = i_a r_a + k_T i_a \omega_r$$

(4.20)

In the above equation, r_a includes armature resistance together with series field resistance. The torque speed characteristic curve is shown in Fig. 4.6.

EXAMPLE 4.11

A dc series motor has the following parameters: 220 V, 2.5 hp, 5,000 rpm, armature resistance = 1.26 Ω (includes armature resistance and series field resistance). Torque constant is $k_T = 0.035$ N-m/A^2. Under rated condition, find (a) motor current and (b) torque.

SOLUTION:

(a)

$$\omega_r = \frac{2\pi N_r}{60} = \frac{2 \times 3.14 \times 5,000}{60} = 523.3 \text{ rad/s}$$

$$v_a = i_a r_a + k_T i_a \omega_r$$

$$220 = i_a \times 1.26 + 0.035 \times i_a \times 523.3$$

$$\therefore i_a = 11.24 \text{ A}$$

(b)

$$T_e = k_T i_a^2$$

$$= 0.035 \times 11.24^2$$

$$= 4.4213 \text{ N-m}$$

4.8 Determination of r_a and L_a

The armature resistance r_a of the motor is measured by applying a low dc voltage to armature terminals. The value of r_a can be taken as the ratio of applied armature voltage to the armature current. In case, the exact value is required, the brush voltage drop is to be subtracted from applied voltage.

To measure L_a, a low ac voltage (probably through a variac) is applied to the armature terminals, and the ratio of voltage to current is taken as Z_a. Then the armature inductance is computed as

$$L_a = \frac{\sqrt{Z_a^2 - r_a^2}}{2\pi f_1}$$

where f_1 is the frequency of ac supply in Hz.

4.9 Determination of K_b

The field current is adjusted to the rated value and the motor is rotated by a prime mover (another motor usually) at its rated speed, ω_r. The armature terminals are open circuited and the induced voltage across the armature is measured as E_b. The back-emf constant is then determined as

$$K_b = \frac{E_b}{\omega_r} \tag{4.21}$$

4.10 Determination of the Moment of Inertia of a Drive System

The moment of inertia of the dc drive system can be determined by retardation or a running-down test. The dc motor is started under no load and, once the no-load speed is reached, the input power is noted. Then the motor is switched OFF; as the armature

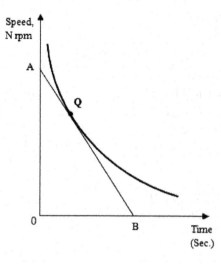

FIGURE 4.7
Speed versus time curve of dc motor.

slows down, its kinetic energy is drawn up to supply the various losses produced by rotation. Now, the variation of motor speed against time t is plotted as shown in Fig. 4.7.

If J is the moment of inertia of the armature and ω_r its angular velocity at any instant, then the kinetic energy of the armature is

$$K \cdot E = \frac{1}{2} J \omega_r^2$$

Rotational losses P_r = rate of loss of kinetic energy:

$$P_r = \frac{d}{dt}\left[\frac{1}{2} J \omega_r^2\right]$$

(4.22)

$$P_r = J \omega_r \frac{d\omega_r}{dt}$$

To calculate P_r it is therefore necessary to determine the curve of ω_r against time t.

Referring to Fig. 4.7, to find the gradient $\dfrac{dN}{dt}$ at any point Q, it is usual to draw the tangent to the curve and to measure the intercepts OA and OB.

$$\frac{dN}{dt} = \frac{OA}{OB}$$

Now, if the moment of inertia, J, is expressed in kg-m^2 and ω_r in rad/s, then the losses will be given by Equation (4.22). However,

$$\omega_r = \frac{2\pi N_r}{60}$$

$$\frac{d\omega_r}{dt} = \frac{2\pi}{60}\frac{dN_r}{dt}$$

Hence,

$$P_r = \left(\frac{2\pi}{60}\right)^2 JN_r \frac{dN_r}{dt} = 0.0109\, JN_r \frac{dN_r}{dt}$$

Thus,

$$J = \frac{P_r}{0.0109 \times N_r \times \left(\dfrac{dN_r}{dt}\right)}$$

$$J = \frac{91.74 \times P_r}{N_r \times \left(\dfrac{dN_r}{dt}\right)} \tag{4.23}$$

Questions

1. Starting from the fundamental equations, derive a state space model for a separately excited dc motor drive.
2. Explain a practical method for determining the back-emf constant of a separately excited dc motor drive.
3. Give a brief introduction about the construction of dc motors. Derive the necessary equations and develop a state-space model of the drive.
4. Explain how to account for field circuit saturation of a dc motor drive in its state-space modeling.
5. Deduce a state-space model for an armature voltage–controlled dc motor drive system.

Unsolved Problems

1. A separately excited dc motor has the following parameters: 220 V, 10 A, 1,400 rpm, armature resistance = 2.3 Ω, field excited at rated current. When the armature is open-circuited and rotated at 1,500 rpm by a prime mover, the armature voltage was 250 V. If the motor is now supplied at rated voltage and if it carries 5.6 A armature current, what will be the approximate load torque on the shaft?
2. A separately excited dc motor has the following specifications: 1,500 kW, 2,650 A, 600 rpm, brush drop of 2.0 V, armature resistance of 0.003645 Ω, armature inductance of 0.1 mH, and a machine frictional torque coefficient of 15 N-m/rad/s. If field current is maintained constant at its normal value, compute the steady-state input voltage for rated armature current and rated speed.
3. A separately excited dc motor has the following parameters: armature resistance = 0.5 Ω, armature inductance = 0.003 H, K_b = 0.8 V/rad/s, J = 0.0167 kg-m^2, and B = 0.01 N-m/rad/s. If the motor is supplied at rated field current and armature is supplied at 220 V, find the steady-state motor speed at 100N-m load.

4. A separately excited dc motor has the following parameters: 220 V, 10 A, 1,400 rpm, armature resistance = 2.3 Ω and field excited at rated current. When the armature terminals are open-circuited and externally driven at 1,500 rpm, the armature voltage is 250 V. If the motor is now supplied at rated voltage and runs at 1,000 rpm, what electromagnetic torque will be developed?

5. A separately excited dc motor has the following parameters: 220 V, 10 A, 1,400 rpm, armature resistance = 2.3 Ω, and field excited at rated current. When the armature terminals are open-circuited and externally driven at 1,500 rpm, the armature voltage is 250 V. If the motor is now supplied at rated voltage and runs at 1,000 rpm, what electromagnetic torque will be developed?

6. A separately excited dc motor driving a fan type has the following constants: armature resistance = 0.2 Ω, back-emfconstant = 0.8 V/(rad/s), load constant = 75×10^{-6} N-m/rad^2. If the armature current required to deliver a load torque of 0.75 N-m is 10 A, find the armature voltage. Neglect friction and windage losses.

7. An electric motor takes 2.33 s to start from a standstill to no-load speed of 1,490 rpm. If losses are neglected and electromagnetic torque required for starting is 23.77 N-m, compute moment of inertia of the drive.

Answers

1. K_b = 1.5915 V/rad/s, T_L = 8.9124 N-m
2. E_b = 586.22 V, V_a = 597.88 V
3. ω_r = 198.42 rad/s, N = 1,894.77 rpm
4. K_b = 1.5915 V/rad/s, E_b = 166.67 V, I_a = 23.19 A, T_e = 36.9 N-m
5. K_b = 1.5915 V/rad/s, T_e = 36.9 N-m, I_a = 23.19 A, E_b = 166.67 V
6. ω = 100 rad/s, E_b = 80 V, Va = 82 V
7. 0.354 kg-m^2

5

ac/dc Converter-Fed dc Motor Drives

5.1 Introduction

A wide range of speed control of dc motors is traditionally achieved using armature voltage control and field current control. Line commutated silicon-controlled rectifier (SCR) converters are simple, economical, and reliable, and they are extensively employed for single-quadrant, two-quadrant, or four-quadrant operation of variable-speed dc motors. However, line-current pollution is the major challenge associated with SCR-fed dc motor systems. This chapter introduces circuit operation of converter fed dc motors, and then it investigates line-current harmonics in semi-converters and full converters. Improved switching schemes for performance enhancement are discussed. The chapter ends with a vivid explanation of the four-quadrant operation of dc motors.

5.2 Single-Phase Converter Circuits

Let v_a and i_a indicate instantaneous values, and let V_a and I_a denote average values. The operation of converter-fed dc motor can be explained with V_a and I_a as independent variables plotted along vertical and horizontal axes and motor operation can be viewed in first, second, third, or fourth quadrants as the case may be. Different configuration of ac/dc converters employing SCRs are available and are summarized in Table 5.1. The motor performance with each circuit is explained in the following sections.

5.2.1 Operation and Analysis of Half-Wave Controlled Converter-Fed dc Motor Drive

The schematic of this drive is shown in Fig. 5.1(a). The half-wave controlled converter is capable of rectification of power in the positive half-cycle alone, and hence this circuit is generally employed for motors of rating below half horsepower.

During the positive half-cycle, SCR is triggered at an angle α and current builds up in motor. At $\omega_s t = \beta$, the current falls to zero. The variations of v_s, v_a and i_a are shown in Fig. 5.1(b).

Applying Kirchhoff's voltage law,

$$\frac{L_a di_a}{dt} + i_a r_a + e_b = V_m \sin \omega_s t, \ \alpha \leq \omega_s t \leq \beta \tag{5.1}$$

TABLE 5.1

Configurations of ac/dc converters employing SCRs.

Circuit	Name	Operating Quadrant
	Half-wave controlled converter	
	Semi converter	
	Full converter	
	Dual converter	

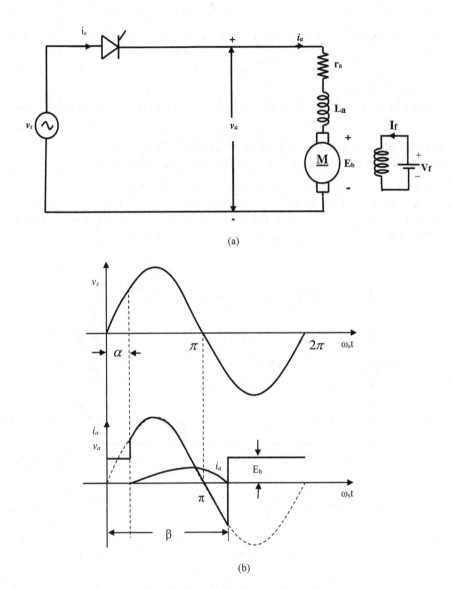

(a)

(b)

FIGURE 5.1

(a) Half-wave controller-fed dc motor drive. (b) Waveforms of v_s, v_a and i_a.

Because $e_b = K_b \omega_r$, we get

$$\frac{di_a}{dt} + \frac{i_a r_a}{L_a} + \frac{K_b \omega_r}{L_a} = \frac{V_m}{L_a} \sin \omega_s t \quad \alpha \leq \omega_s t \leq \beta \tag{5.2}$$

During starting or speed change, ω_r changes and hence dynamics of speed-torque characteristic should be taken into account. Then

$$T_e = \frac{J d\omega_r}{dt} + B\omega_r + T_L$$

Assume that the motor is started on no-load and further by putting $T_e = K_b i_a$, we get

$$\frac{J d\omega_r}{dt} + B\omega_r = K_b i_a \qquad (5.3)$$

Solution of Equations (5.2) and (5.3) gives the dynamic behavior of converter-fed motor.

5.2.2 Analysis of a Separately Excited dc Motor Supplied from a Half-Wave Converter with a Freewheeling Diode

With the introduction of the freewheeling diode (FD), the half-wave controlled rectifier-fed drive system is shown in Fig. 5.2(a). Here, at $\omega_s t = \alpha$, SCR is triggered, making $v_a = V_m \sin \omega_s t$. At $\omega_s t = \pi$, FD is forward-biased and carries the armature current. Hence SCR is turned OFF at $\omega_s t = \pi$. Typical waveforms are shown in Fig. 5.2(b).

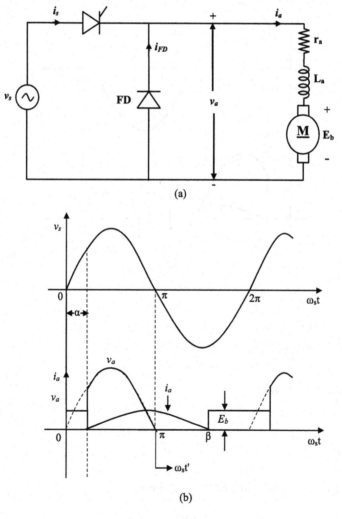

FIGURE 5.2
(a) Half-wave converter-fed dc motor. (b) Waveforms of a half-wave converter-fed dc motor.

At $\omega_s t = \pi$, FD starts carrying the armature current and decays to zero at $\omega_s t = \beta$. It is rare that i_a becomes continuous. The analysis is carried out below:

$$\frac{L_a di_a}{dt'} + r_a i_a + K_b \omega_r = 0, \ 0 \le \omega_s t' \le \beta - \pi \tag{5.4}$$

where $\omega_s t' = \omega_s t - \pi$.

Equation (5.4) along with Equation (5.3) completely define the drive dynamics.

5.3 Analysis and Operation of a Single-Phase Bridge-Type Semi-Converter-Fed dc Motor Drive

The circuit of a single-phase bridge type semi-converter-fed separately excited dc motor drive system is shown in Fig. 5.3(a). Here, there are two SCRs and two diodes in the bridge configuration; the freewheeling diode (FD) is connected across the motor terminals. The typical motor terminal voltage and current are shown in Fig. 5.3(b). During the

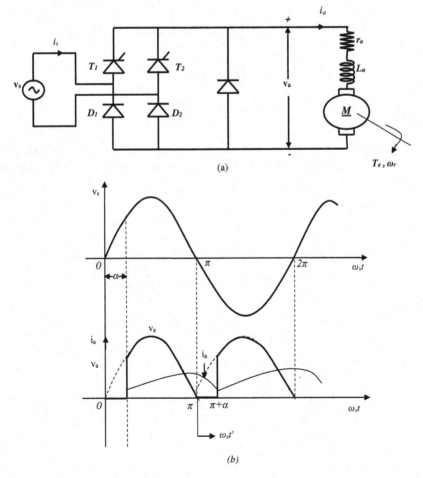

FIGURE 5.3
(a) Semi-converter-fed dc motor. (b) Typical motor terminal voltage and current waveforms for continuous conduction.

positive half-cycle, T_1 is triggered at α such that T_1 and D_2 conduct from α to π, making $v_a = V_m \sin \omega_s t$. At $\omega_s t = \pi$, FD is forward-biased and carries i_a. Thus T_1 and D_2 are switched OFF. Assume voltage and current waveforms are continuous in nature.

For the motor current to be continuous, FD carries load current from π to $\pi + \alpha$. At $\pi + \alpha$, when T_2 is triggered, FD is reverse-biased and the motor current builds up through T_2 and D_1. The analytical equation can be written as

$$\frac{L_a di_a}{dt'} + i_a r_a + K_b \omega_r = V_m \sin \omega_s t, \ \alpha \le \omega_s t \le \pi \tag{5.5}$$

$$\frac{L_a di_a}{dt'} + i_a r_a + K_b \omega_r = 0, \ 0 \le \omega_s t' \le \alpha \tag{5.6}$$

where $\omega_s t' = \omega_s t - \pi$.

The drive speed can be obtained by combining Equation (5.3) with Equations (5.5) and (5.6). The average value of the motor terminal voltage with continuous armature current is derived below.

$$\begin{aligned} V_{a(av)} &= \frac{1}{\pi} \int_{\alpha}^{\pi} V_m \sin \omega_s t \ d\omega_s t \\ &= \frac{V_m}{\pi} (-\cos \omega_s t)_{\alpha}^{\pi} \\ &= \frac{V_m}{\pi} (1 + \cos \alpha) \end{aligned} \tag{5.7}$$

It may be noted that the current need not be continuous, in which case the waveforms look like the one shown in Fig. 5.4. Here, during freewheeling, the current drops to zero at

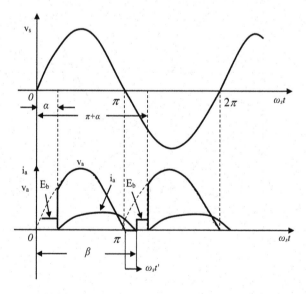

FIGURE 5.4
Waveforms with discontinuous conduction.

$\omega_s t = \beta$. Hence the voltage in Equation (5.5) remains the same, whereas Equation (5.6) is modified as Equation (5.8).

$$\frac{L_a di_a}{dt'} + i_a r_a + K_b i_a = 0, \quad 0 \le \omega_s t' \le \beta - \pi \tag{5.8}$$

The average value of the motor terminal voltage with a discontinuous armature current was derived in section 3.2.3 and is reproduced in Equation (5.9).

$$V_{a(av)} = \frac{V_m}{\pi}(1 + \cos\alpha) + E_b(\pi + \alpha - \beta) \tag{5.9}$$

EXAMPLE 5.1

A 220-V, 10-A, 1,440-rpm separately excited dc motor is operating at half load, and the armature is supplied from a single-phase semi-converter fired at 60°. The input to the converter is $230\sqrt{2}\sin(314t)$. The armature resistance is 1.82 Ω. The back-emf constant is 1.39 V/rad/s. Compute the motor speed in rpm.

SOLUTION:

$$V_{o(avg)} = \frac{V_m}{\pi}(1 + \cos\alpha)$$

$$= \frac{230\sqrt{2}}{\pi}(1 + 1/2)$$

$$= 155.35 \text{ V}$$

$$I_a = 5 \text{ A}(\because \text{ half load})$$

$$I_a r_a = 5 \times 1.8 = 9 \text{ V}$$

$$E_b = V_{o(av)} - I_a R_a = 155.3 - 9$$

$$= 146.35 \text{ V}$$

$$\omega_r = \frac{E_b}{K_b} = \frac{146.35}{1.39} = 105.28 \text{ rad/s}$$

EXAMPLE 5.2

A single-phase half-wave controlled rectifier is driving a separately excited dc motor. The dc motor has a back-emf constant of 0.5 V/rpm. The armature current is 5 A without any ripple. The armature resistance is 2 Ω. The converter is supplied from a 240-V, single-phase ac source with a firing angle of 38°. Find the speed of the motor.

SOLUTION:

$$I_a = 5 \text{ A (Without ripple)}$$

$$K_b = 0.5 \text{ V/rpm}$$

$$r_a = 2 \Omega$$

$$V_i = 240 \text{ V}_{rms} \text{(Single phase AC)}$$

$$\alpha = 38°$$

$$V_0 = \frac{V_m}{\pi}(1 + \cos\alpha)$$

$$= \frac{240\sqrt{2}}{\pi}(1 + \cos 38°)$$

$$= 193.2 \text{ V}$$

$$E_b = V_0 - I_a r_a$$

$$= 193.2 - (5 \times 2)$$

$$= 183.20 \text{ V}$$

$$\text{Speed } N = \frac{E_b}{K_b}$$

$$= \frac{183.2}{0.5}$$

$$N = 366.4 \text{ rpm}$$

EXAMPLE 5.3

A single-phase, 230-V, 50-Hz supply feeds a separately excited dc motor through two single-phase semi-converters, one for the field and the other for the armature. The firing angle for the semi-converter for the field is 30°. The field resistance is 150 Ω, and the armature resistance is 0.2 Ω. The load torque is 40 N-m at 800 rpm. The back-emf constant is 0.7 V/rad/s and is equal to the torque constant. Assume that armature and field current are continuous and constant, and neglect losses. Find the firing angle of the converter in the armature circuit.

SOLUTION:

The output voltage of a single-phase semi-converter connected to field is

$$V_f = \frac{V_m}{\pi}(1 + \cos\alpha_f)$$

$$V_f = \frac{230\sqrt{2}}{\pi}(1 + \cos 30) = 193.20 \text{ V}$$

$$\text{Field current} = \frac{193.20}{150} = 1.288 \text{ A}$$

$$\text{Torque} = K_t I_f I_a$$

$$I_a = \frac{40}{0.7 \times 1.288} = 44.365 \text{ A}$$

$$\omega_r = \frac{800 \times 2\pi}{60} = 83.775 \text{ rad/s}$$

$$\text{Back-emf, } E_b = K_b I_f \omega_r$$

$$= 0.7 \times 83.775 \times 1.288 = 75.531 \text{ V}$$

$$V_a = 75.531 + 44.365 \times 0.2 = 84.404 \text{ V}$$

$$84.404 = \frac{230\sqrt{2}}{\pi}(1 + \cos\alpha_a)$$

$$\alpha_a = 100.648°$$

5.4 Analysis of a Single-Phase, Bridge-Type Full Converter-Fed dc Motor Drive

The dc motor drive supplied from a single-phase bridge-type full converter is shown in Fig. 5.5(a). Here, there are four SCRs, labeled as T_1, T_2, T_3, and T_4. Thyristors T_1 and T_3 are turned ON at $\omega_s t = \alpha$ during the positive half-cycle, whereas T_2 and T_4 are gated at $\omega_s t = \pi + \alpha$ during the negative half-cycle of the supply voltage. Typical motor voltage and current are shown in Fig. 5.5(b), where the motor current is assumed to be continuous.

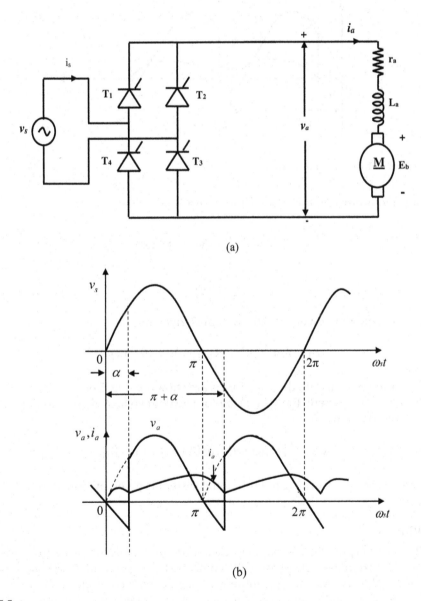

(a)

(b)

FIGURE 5.5
(a) Full converter-fed dc drive. (b) Waveforms.

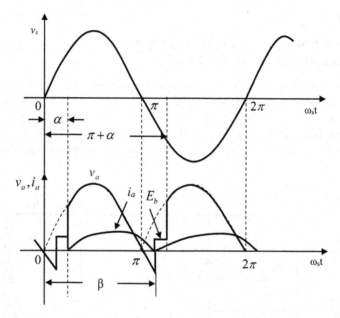

FIGURE 5.6
Motor waveforms with discontinuous current.

With continuous current, the average motor terminal voltage is

$$V_{o(av)} = \frac{1}{\pi} \int_{\alpha}^{\pi+\alpha} V_m \sin \omega_s t \ d\omega t$$

$$= \frac{2V_m}{\pi} \cos \alpha$$

(5.10)

The transient equation governing the system is given by

$$r_a i_a(t) + L \frac{di_a(t)}{dt} + K_b \omega_r = V_m \sin (\omega_s t)$$

(5.11)

With discontinuous current, the waveforms are shown in Fig. 5.6.

With discontinuous current, the average motor terminal voltage is given in Equation (3.15) and is reproduced below.

$$V_{av} = \frac{1}{\pi} \left[V_m (\cos \alpha - \cos \beta) + E_b (\pi + \alpha - \beta) \right]$$

The torque equation governing the rotational system is given by Equation (5.3).

EXAMPLE 5.4

(a) A 210-V, 1,500-rpm, 20-A, separately excited dc motor has an armature resistance of 1 Ω. It is fed from a-single phase fully controlled bridge rectifier with an ac source voltage of 230 V (rms). Assuming continuous load current, compute (a) motor speed at a firing angle of 30° and torque of 5 N-m, and (b) developed torque at a firing angle of 51° and speed of 900 rpm.

SOLUTION:

(a) Under rated conditions,

$$N = 1{,}500 \text{ rpm}$$

$$I_a = 20 \text{ A}$$

$$E_b = V_a - I_a R_a = 210 - (20 \times 1)$$
$$= 190 \text{ V}$$

$$E_b \propto \phi_m N$$

$$E_b = K_b \phi_m N$$

$$\omega_r = \frac{2\pi N}{60}$$
$$= 157.07 \text{ rad/s}$$

$$K_b \phi_m = \frac{190}{157.07}$$
$$= 1.21 \text{ rad/s}$$

$$T\omega_r = E_b I_a$$

$$T = \frac{E_b}{\omega_r} I_a$$

$$T = 1.21 \, I_a$$

$$I_a = \frac{5}{1.21}$$
$$= 4.133 \text{ A}$$

$$V_{o(av)} = \frac{2V_m}{\pi} \cos\alpha$$

$$= \frac{2 \times 230\sqrt{2}}{\pi} \cos 30$$

$$= 179.33 \text{ V}$$

$$E_b = V_{o(av)} - I_a r_a$$
$$= 179.33 - (4.133 \times 1)$$
$$= 175.196 \text{ V}$$

$$\frac{E_{b2}}{E_{b1}} = \frac{N_2}{N_1}$$

$$N_2 = \frac{175.196}{190} \times 1500$$

$$N = 1{,}383.12 \text{ rpm}$$

(b) $\alpha = 51°$, $N = 900$ rpm

$$V_{o(av)} = \frac{2 \times 230\sqrt{2}}{\pi} \cos 51$$

$$= 130.31 \text{ V}$$

$$\frac{E_{b2}}{E_{b1}} = \frac{N_2}{N_1}$$

$$E_b = \frac{900}{1500} \times 190$$

$$= 114 \text{ V}$$

$$I_a = \frac{130.31 - 114}{1}$$

$$= 16.31 \text{ A}$$

$$T = \frac{E_b}{\omega_r} I_a$$

$$T = 19.728 \text{ N-m}$$

EXAMPLE 5.5

A separately excited dc motor is controlled by varying its armature voltage using a single-phase full converter bridge. The field current is kept constant at the rated value. The motor has an armature resistance of 0.2 Ω, and the motor voltage constant is 2.33 V/rad/s. The motor is driving a mechanical load having a constant torque of 150 N-m. The triggering angle of the converter is 60°. The armature current can be assumed to be continuous and ripple-free. (a) Calculate the motor armature current. (b) Evaluate the motor speed in rad/s. (c) Calculate the rms value of the fundamental component of the input current to the bridge.

SOLUTION:

(a) Armature current:

$$E_b I_a = T\omega_r$$

$$K_b \cdot \omega_r I_a = T\omega_r$$

$$I_a = \frac{T}{K_b}$$

$$= \frac{150}{2.33}$$

$$I_a = 64.37 \text{ A}$$

(b) Motor speed:

$$V_{o(av)} = \frac{2V_m}{\pi}\cos\alpha$$

$$= \frac{2 \times 250\sqrt{2}}{\pi}\cos 60$$

$$= 112.54 \text{ V}$$

$$E_b = V_{o(av)} - I_a r_a$$

$$= 112.54 - (56 \times 0.2)$$

$$= 101.34 \text{ V}$$

$$\omega_r = \frac{101.34}{2.33}$$

$$\omega_r = 43.43 \text{ rad/s}$$

(c) rms value of I_1

$$I_{1rms} = \frac{2\sqrt{2}I_{dc}}{\pi}$$

$$I_{1rms} = 50.417 \text{ A}$$

EXAMPLE 5.6

A single-phase fully controlled thyristor bridge converter is supplied by a 230-V, 50-Hz source. Find the input power and power factor for a dc motor armature running at 1,000 rpm developing an induced emf of 120 V. The armature resistance is 1 Ω. The firing angle is 45°.

SOLUTION:

$$V_{o(avg)} = \frac{2V_m}{\pi}\cos\alpha$$

$$= \frac{2 \times 230\sqrt{2}}{\pi}\cos 45$$

$$= 146.4 \text{ V}$$

$$I_o = I_a = \frac{V_{dc} - E_b}{r_a}$$

$$= 26.4 \text{ A}$$

output power $P_o = V_o I_o = 146.4 \times 26.4$

$$= 3.865 \text{ kW}$$

Assuming $P_{in} = P_o = 3.865$ kW

$$\text{Input VA} = V_{rms} \times I_{rms}$$

$$I_{rms} = 24.6 \text{ A (refer to section 5.6)}$$

$$\therefore \text{Input VA} = 230 \times 24.6 = 5,658 \text{ VA}$$

$$\text{power factor} = \frac{3.865 \text{ kW}}{5658 \text{ VA}}$$

$$= 0.683$$

EXAMPLE 5.7

A 230-V, single-phase, 50-Hz supply feeds the armature and field circuits of a separately excited dc motor through two full converters. The firing angle of the converter in the field circuit is 30°. The armature and field resistance are 0.38 Ω and 175 Ω, respectively. The torque and voltage constants are 1.0 each. The firing angle of the converter in the armature circuit is 45°, and the armature current is 40 A. Find the torque developed and the motor speed.

SOLUTION:

Output of full converter in field circuit

$$\text{Average field voltage } V_f = \frac{2V_m}{\pi}(\cos\alpha_f)$$

$$V_f = \frac{2 \times 230\sqrt{2}}{\pi}(\cos 30) = 179.33 \text{ V}$$

$$\text{Field current} = \frac{179.33}{175} = 1.025 \text{ A}$$

Given torque constant = 1, and armature current = 40 A:

$$\text{Torque} = K_t I_f I_a$$

$$= 1 \times 40 \times 1.025 = 41 \text{ N-m}$$

$$V_{o(av)} = \frac{2v_m}{\pi}(\cos\alpha_a)$$

$$= \frac{2 \times 230\sqrt{2}}{\pi}(\cos 45) = 146.40 \text{ V}$$

$$\text{Back-emf} = V_a - I_a r_a$$

$$= 146.4 - 40 \times 0.38 = 131.2 \text{ V}$$

$$\omega_r = \frac{131.2}{1 \times 1.025} = 128 \text{ rad/s}$$

$$= 1,222.31 \text{ rpm}$$

EXAMPLE 5.8

A dc shunt motor operating from a single-phase semi-controlled bridge operating at 1,500 rpm has an input voltage (v_s) = 330 sin 314t and back-emf of 75 V. The SCRs are fired symmetrically at $\alpha = 45°$ in every half-cycle, and $r_a = 5\ \Omega$. Neglecting armature inductance, calculate the average armature current and the torque.

SOLUTION:

Given $v_s = 330 \sin 314t$, $E_b = 75$ V, $\alpha = 45°$, and $r_a = 5\ \Omega$:

$$V_{o(av)} = \frac{V_m}{\pi}[\cos\alpha + 1]$$

$$= \frac{330}{\pi}[\cos 45 + 1]$$

$$= 179.318\ V$$

$$I_a = \frac{V_{o(av)} - E_b}{r_a}$$

$$= \frac{179.318 - 75}{5}$$

$$= 20.8636\ A$$

Mechanical power output:

$$P_{mech} = E_b I_a$$

i.e. $$\frac{2\pi N_r T}{60} = 75 \times 20.8636$$

$$\therefore T = 10.305\ \text{N-m}$$

5.5 Performance of a Converter-Fed dc Motor Drive System

When dc motors are powered from ac/dc converters, the motor terminal voltage is not steady dc, nor is the supply current sinusoidal. This brings noticeable performance difference on supply lines as well as on a dc motor drive system. Various factors that require important attention on the supply performance are listed below.

5.5.1 Input Power Factor

In the case of an ac system with purely sinusoidal voltage and current waveforms, the power factor is traditionally defined as the cosine of the angle between voltage and current. However, if the voltage or current or both are non-sinusoidal, this definition becomes invalid. Therefore, the power factor as seen from the ac source side is derived from the fundamental definition as

$$\text{Input power factor} = \frac{\text{Average value of input power}}{\text{Input volt amps}}$$

In the case of an ac/dc converter-fed dc motor, the input voltage is sinusoidal, while the input current is non-sinusoidal. In other words, the source current contains a component having a sinusoidal waveform with the same frequency as that of the source voltage, which is called a fundamental component, and the sinusoidal waves of multiple frequencies, which are called harmonics. Thus source current can be split into a fundamental component and harmonics. The useful power comes from the fundamental, while harmonics produce a heating effect. Let V, I_1, and ϕ_1 be the rms value of supply voltage, the rms value of the fundamental component of source current, and the phase angle between supply voltage and fundamental component of source current, respectively. Then

$$\text{Input power factor} = \frac{V_{rms}I_1 \cos\phi_1}{V_{rms}I_{srms}}$$

$$= \frac{I_1 \cos\phi_1}{I_{srms}} \tag{5.12}$$

where I_{srms} is the rms value of source current, which includes fundamental and harmonic currents.

5.5.2 Displacement Factor

This is defined as

$$DF = \cos\phi_1 \tag{5.13}$$

5.5.3 Total Harmonic Distortion

Total harmonic distortion (THD) is sometimes referred as the harmonic factor and is defined as

$$THD = \frac{\text{rms value of harmonic content}}{\text{rms value of fundamental current}}$$

Let I_h represent rms value of harmonic current. Then

$$THD = \frac{I_h}{I_1} = \frac{\sqrt{I_{srms}^2 - I_1^2}}{I_1} \tag{5.14}$$

5.6 Investigations on Line Current Harmonics in Single-Phase Full Converter-Fed Drives

One important parameter with power converter-fed drives is the magnitude of line current harmonics. This aspect with full and semi-converters is discussed below.

5.6.1 Single-Phase Full Converter-Fed Drive

Assuming that the motor current is ripple-free and its amplitude is I_a, the supply voltage, v_s, and source current, i_s, are plotted in Fig. 5.7(a). Referring to the power circuit in Fig. 5.5(a),

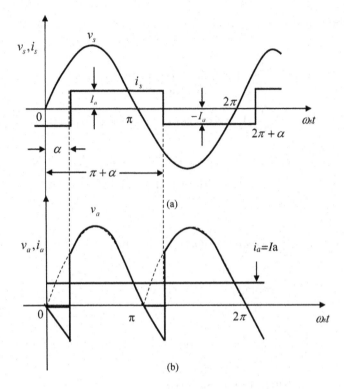

FIGURE 5.7
(a) Supply voltage and current with full converter-fed dc motor. (b) Waveforms of v_a and i_a.

the source current is I_a from α to $\pi+\alpha$, whereas it is $-I_a$ from $\pi+\alpha$ to $2\pi+\alpha$. The supply current can be split up using Fourier series as shown below:

$$i_s = I_{dc} + \sum_{n=1}^{\infty}(a_n \cos(n\omega_s t) + b_n \sin(n\omega_s t)) \tag{5.15}$$

$$= I_{dc} + \sum_{n=1}^{\infty}\sqrt{2}I_n \sin(n\omega_s t + \phi_n) \tag{5.16}$$

where $I_{dc} = \dfrac{1}{T}\displaystyle\int_0^T i_s dt$

$$a_n = \frac{2}{T}\int_0^T i_s \cos(n\omega_s t)dt \tag{5.17}$$

$$b_n = \frac{2}{T}\int_0^T i_s \sin(n\omega_s t)dt \tag{5.18}$$

$$I_n = \left[\left[\frac{a_n}{\sqrt{2}}\right]^2 + \left[\frac{b_n}{\sqrt{2}}\right]^2\right]^{1/2} \tag{5.19}$$

$$\phi_n = \tan^{-1}\left(\frac{a_n}{b_n}\right) \tag{5.20}$$

Referring to the waveform of i_s in Fig. 5.7, the Fourier coefficient can be computed as

$$I_{dc} = \frac{1}{2\pi}\left\{\int_{\alpha}^{\pi+\alpha} I_a d\omega_s t + \int_{\pi+\alpha}^{2\pi+\alpha}(-I_a)d\omega_s t\right\}$$

$$= \frac{1}{2\pi}\{I_a\{\pi+\alpha-\alpha+(-2\pi-\alpha)+\pi+\alpha\}\} = 0$$

(5.21)

$$a_n = \frac{2}{2\pi}\left\{\int_{\alpha}^{\pi+\alpha} I_a \cos(n\omega_s t)d\omega_s t + \int_{\pi+\alpha}^{2\pi+\alpha}(-I_a)\cos(n\omega_s t)d\omega_s t\right\}$$

$$= \frac{I_a}{\pi}\left\{\int_{\alpha}^{\pi+\pi} I_a \cos(n\omega_s t)d\omega_s t + \int_{\pi+\alpha}^{2\pi+\pi}(-I_a)\cos(n\omega_s t)d\omega_s t\right\}$$

$$= \frac{I_a}{\pi}\left\{\left(\frac{\sin(n\omega_s t)}{n}\right)_{\alpha}^{\pi+\alpha} + \left(\frac{-\sin(n\omega_s t)}{n}\right)_{\pi+\alpha}^{2\pi+\alpha}\right\}$$

(5.22)

$$= \frac{I_a}{n\pi}\{\sin n(\pi+\alpha) - \sin(n\alpha) - \sin n(2\pi+\alpha) + \sin n(\pi+\alpha)\}$$

$$= \frac{2I_a}{n\pi}\{\sin n(\pi+\alpha) - \sin(n\alpha)\}$$

In Equation (5.21), $a_n = 0$ for $n = 2, 4, 6\ldots$

$$a_n = \frac{-4I_a}{n\pi}\sin n\alpha, \; n = 1,3,5$$

(5.23)

Further,

$$b_n = \frac{2}{2\pi}\left\{\int_{\alpha}^{\pi+\alpha} I_a \sin(n\omega_s t)d\omega_s t + \int_{\pi+\alpha}^{2\pi+\alpha}(-I_a)\sin(n\omega_s t)d\omega_s t\right\}$$

$$= \frac{I_a}{\pi}\left\{\left(\frac{-\cos(n\omega_s t)}{n}\right)_{\alpha}^{\pi+\alpha} + \left(\frac{\cos(n\omega_s t)}{n}\right)_{\pi+\alpha}^{2\pi+\alpha}\right\}$$

$$= \frac{I_a}{n\pi}\{-\cos n(\pi+\alpha) + \cos(n\alpha) + \cos n(2\pi+\alpha) - \cos n(\pi+\alpha)\}$$

(5.24)

$$= \frac{2I_a}{n\pi}\{\cos(n\alpha) - \cos n(\pi+\alpha)\}$$

$$= 0 \text{ for } n = 2, 4, 6\ldots\ldots$$

$$= \frac{4I_a}{n\pi}\cos(n\alpha) \text{ for } n = 1, 3, 5\ldots..$$

From Equations (5.23) and (5.24), n^{th} harmonic current is calculated as

$$I_n = \left[\left[\frac{a_n}{\sqrt{2}}\right]^2 + \left[\frac{b_n}{\sqrt{2}}\right]^2\right]^{1/2}$$

$$I_n = \left(\frac{a_n^2 + b_n^2}{2}\right)^{1/2}$$

$$= \left(\frac{16I_a^2 \sin^2(n\alpha)}{2n^2\pi^2} + \frac{16I_a^2 \cos^2(n\alpha)}{2n^2\pi^2}\right)^{1/2}$$

$$= \frac{2\sqrt{2}I_a}{n\pi}$$

(5.24a)

And

$$\phi_n = \tan^{-1}\left(\frac{a_n}{b_n}\right)$$

$$= \tan^{-1}\left(\frac{\dfrac{-4I_a}{n\pi}\sin(n\alpha)}{\dfrac{4I_a}{n\pi}\cos(n\alpha)}\right)$$

$$= \tan^{-1}[\tan(n\alpha)]$$

$$= -n\alpha$$

$$\text{Input power factor} = \frac{I_1}{I_{srms}}\cos\phi_1$$

(5.25)

From the wave shape of i_s in Fig. 5.7, the rms value of i_s is I_a. Furthermore, putting $n = 1$ in Equations (5.24a) and (5.25) yields

$$I_1 = \frac{2\sqrt{2}I_a}{\pi} \quad \text{and} \quad \phi_n = -\alpha$$

Thus,

$$\text{input power factor} = \frac{2\sqrt{2}I_a \cos\alpha}{\pi \cdot I_a}$$

$$= \frac{2\sqrt{2}}{\pi}\cos\alpha$$

(5.26)

and the

$$\text{displacement factor} = \cos\phi_1 = \cos\alpha$$

(5.27)

The total harmonic distortion is now computed as

$$\text{THD} = \frac{I_h}{I_1} = \frac{\left(I_{srms}^2 - I_1^2\right)^{1\backslash 2}}{I_1}$$

$$= \frac{\left(I_a^2 - \frac{8I_a^2}{\pi^2}\right)^{1/2}}{\frac{2\sqrt{2}I_a}{\pi}} = \frac{I_a\left[1 - \frac{8}{\pi^2}\right]^{1/2}}{\frac{2\sqrt{2}I_a}{\pi}} \tag{5.28}$$

$$= \frac{\left[\pi^2 - 8\right]^{1/2}}{2\sqrt{2}}$$

$$= 0.4834 \ (48.34\%)$$

5.6.2 Single-Phase Semi-Converter-Fed Drive

In a single-phase semi-converter, freewheeling is inherent, and during this interval, the source current is zero. With a constant armature current, I_a, the supply voltage and current of a single-phase semi-converter is sketched in Fig. 5.8.

The Fourier coefficients can be obtained as given below:

$$I_{dc} = \frac{1}{2\pi}\left\{\int_\alpha^\pi I_a d\omega_s t + \int_{\pi+\alpha}^{2\pi} (-I_a) d\omega_s t\right\} \tag{5.29}$$

$$I_{dc} = \frac{I_a}{2\pi}\{(\pi - \alpha) - (2\pi - \pi - \alpha)\} = 0$$

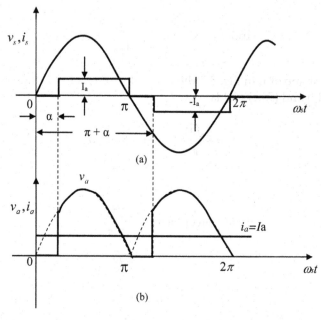

FIGURE 5.8
(a) Supply voltage and current waveforms with a semi-converter. (b) Waveforms of v_a and i_a.

$$a_n = \frac{2}{2\pi} \left\{ \int_{\alpha}^{\pi} I_a \cos(n\omega_s t) d\omega_s t + \int_{\pi+\alpha}^{2\pi} (-I_a) \cos(n\omega_s t) d\omega_s t \right\}$$

$$= \frac{I_a}{\pi} \left\{ \left(\frac{\sin(n\omega_s t)}{n} \right)_{\alpha}^{\pi} - \left(\frac{\sin(n\omega_s t)}{n} \right)_{\pi+\alpha}^{2\pi} \right\}$$

$$= \frac{I_a}{n\pi} \{ \sin(n\pi) - \sin(n\alpha) - \sin(n2\pi) + \sin n(\pi+\alpha) \}$$

$$= \frac{-2I_a}{n\pi} \sin(n\alpha) \quad n = 1, 3, 5\ldots\ldots$$

$$= 0 \text{ for } n = 2, 4,\ldots\ldots\ldots$$

(5.30)

$$b_n = \frac{2}{2\pi} \left\{ \int_{\alpha}^{\pi} I_a \sin(n\omega_s t) d\omega_s t + \int_{\pi+\alpha}^{2\pi} (-I_a) \sin(n\omega_s t) d\omega_s t \right\}$$

$$= \frac{I_a}{\pi} \left\{ \left(\frac{-\cos(n\omega_s t)}{n} \right)_{\alpha}^{\pi} + \left(\frac{\cos(n\omega_s t)}{n} \right)_{\pi+\alpha}^{2\pi} \right\}$$

$$= \frac{I_a}{n\pi} \{ -\cos(n\pi) + \cos(n\alpha) + \cos(n2\pi) - \cos n(\pi+\alpha) \}$$

$$= \frac{2I_a}{n\pi} (1 + \cos(n\alpha)), \quad n = 1, 3, 5\ldots$$

(5.31)

Now, the nth harmonic current is given by

$$I_n = \left(\frac{a_n^2 + b_n^2}{2} \right)^{1/2}$$

$$= \left(\frac{4I_a^2}{2n^2\pi^2} \sin^2(n\alpha) + \frac{4I_a^2}{2n^2\pi^2} (1 + \cos(n\alpha))^2 \right)^{1/2}$$

$$= \frac{\sqrt{2}I_a}{n\pi} \left\{ \sin^2(n\alpha) + \left(1 + \cos^2(n\alpha) + 2\cos(n\alpha) \right) \right\}^{1/2}$$

$$= \frac{\sqrt{2}I_a}{n\pi} \{ 1 + 1 + 2\cos(n\alpha) \}^{1/2}$$

(5.32)

$$= \frac{2I_a}{n\pi} \{ 1 + \cos(n\alpha) \}^{1/2}$$

$$= \frac{2I_a}{n\pi} \left\{ 2\cos^2\left(\frac{n\alpha}{2} \right) \right\}^{1/2}$$

$$= \frac{2\sqrt{2}I_a}{n\pi} \left(\cos\left(\frac{n\alpha}{2} \right) \right)$$

$$\phi_n = \tan^{-1}\left(\frac{a_n}{b_n}\right)$$

$$= \tan^{-1}\left(\frac{\dfrac{-2I_a}{n\pi}\sin(n\alpha)}{\dfrac{2I_a}{n\pi}(1+\cos(n\alpha))}\right)$$

$$= \tan^{-1}\left(\frac{-\sin(n\alpha)}{1+\cos(n\alpha)}\right)$$

$$= \tan^{-1}\left(\frac{-2\sin\left(\dfrac{n\alpha}{2}\right)\cos\left(\dfrac{n\alpha}{2}\right)}{2\cos^2\left(\dfrac{n\alpha}{2}\right)}\right) \qquad (5.33)$$

$$= \tan^{-1}\left(\frac{-\sin\left(\dfrac{n\alpha}{2}\right)}{\cos\left(\dfrac{n\alpha}{2}\right)}\right)$$

$$= -\frac{n\alpha}{2}$$

Referring to i_s in Fig. 5.8(a), the rms value of i_s is computed below, where the rms value of i_s is denoted.

$$I_{srms}^2 = \frac{1}{2\pi}\left\{\int_\alpha^\pi I_a^2 d\omega_s t + \int_{\pi+\alpha}^{2\pi}(-I_a)^2 d\omega_s t\right\}$$

$$= \frac{I_a^2}{2\pi}\left\{(\omega_s t)_\alpha^\pi + (\omega_s t)_{\pi+\alpha}^{2\pi}\right\}$$

$$= \frac{I_a^2}{2\pi}\{\pi - \alpha + 2\pi - (\pi + \alpha)\} \qquad (5.34)$$

$$= \frac{I_a^2}{2\pi}\{2\pi - 2\alpha\}$$

$$= I_a^2\left(\frac{\pi - \alpha}{\pi}\right)$$

$$I_{srms} = I_a\left(\frac{\pi - \alpha}{\pi}\right)^{1/2}$$

$$\text{Input power factor} = \frac{V_{rms}I_1\cos\phi_1}{V_{rms}I_{srms}}$$

$$= \frac{I_1\cos\phi_1}{I_{srms}}$$

Putting n = 1 in Equations (5.32) and (5.33), $I_1 = \dfrac{2\sqrt{2}I_a}{\pi}\left(\cos\frac{\alpha}{2}\right)$ and $\cos\phi_1 = \cos\left(\frac{-\alpha}{2}\right) = \cos\frac{\alpha}{2}$. Substituting these values in Equation (5.34) yields

$$\text{Input power factor} = \frac{2\sqrt{2}I_a \cos\frac{\alpha}{2}\cdot\cos\frac{\alpha}{2}}{\pi\cdot I_a\left(\dfrac{\pi-\alpha}{\pi}\right)^{1/2}}$$

$$= \frac{2\sqrt{2}\cos^2\frac{\alpha}{2}}{\pi\left(\dfrac{\pi-\alpha}{\pi}\right)^{1/2}} \tag{5.35}$$

$$\text{Displacement factor, } DF = \cos\frac{\alpha}{2} \tag{5.36}$$

$$\text{THD} = \frac{I_h}{I_1}$$

$$= \frac{\left(I_{srms}^2 - I_1^2\right)^{1/2}}{I_1}$$

$$= \frac{\left(\dfrac{I_a^2(\pi-\alpha)}{\pi} - \dfrac{8I_a^2}{\pi^2}\cos^2\frac{\alpha}{2}\right)^{1/2}}{2\sqrt{2}\dfrac{I_a}{\pi}\cos\frac{\alpha}{2}} \tag{5.37}$$

$$= \frac{\left[\left(\dfrac{\pi-\alpha}{\pi}\right) - \left(\dfrac{8}{\pi^2}\cos^2\frac{\alpha}{2}\right)\right]^{1/2}}{\dfrac{2\sqrt{2}}{\pi}\cos\frac{\alpha}{2}}$$

5.6.3 Comparison of Input Performance Parameters of Single-Phase Full Converter and Semi-Converter-Fed dc Motor Drives

From the expressions derived in sections 5.6.1 and 5.6.2, now it is possible to compare the input performance of a dc motor drive with full and semi-converters. The comparison is performed by taking the output voltage of the converter as an independent variable because this value is different for the two converters for the same α.

The average value of output voltage of a full converter with continuous motor current is taken from Equation (5.10):

$$V_{o(av)} = \frac{2V_m}{\pi}\cos\alpha$$

Taking $\dfrac{2V_m}{\pi}$ as one per unit (p.u.) of voltage,

$$V_{o(av)} = \cos\alpha \text{ p.u.} \tag{5.38}$$

From Equation (5.7), the average value of the semi-converter in p.u. is

$$V_{o(av)} = \frac{\dfrac{V_m}{\pi}(1 + \cos\alpha)}{\dfrac{2V_m}{\pi}} \qquad (5.39)$$

$$= \frac{1}{2}(1 + \cos\alpha) \text{ p.u.}$$

For different values of α, p.u. output voltages of full and semi-converters are evaluated using Equations (5.38) and (5.39) and marked along X-axis. For the same α, input performance parameters of the two converters are then evaluated using Equations (5.26), (5.27), (5.28), (5.35), (5.36), and (5.37). The variation of performance parameters of the two converters is shown in Fig. 5.9. It is evident that the semi-converter exhibits improved power factor and displacement factors. It possesses higher THD at reduced voltage (up to 0.5 p.u.), but THD falls thereafter.

EXAMPLE 5.9

A single-phase full converter is fired at 45° and is supplying power to a separately excited dc motor. If the armature current is held constant at 10 A, compute the input power factor.

SOLUTION:

$$\text{input p.f} = \frac{2\sqrt{2}}{\pi}\cos\alpha$$

$$= \frac{2\sqrt{2}}{\pi}\cos 45°$$

$$= 0.63 \text{ (lagging)}$$

5.7 Power Factor Improvement of Bridge-Type, Single-Phase Converters

As seen in Fig. 5.9, the performance of a full converter is poor in comparison with that of a semi-converter. Two schemes to improve the power factor of bridge-type converters are explained below.

5.7.1 Adopting Freewheeling in a Full Converter

We have seen that a semi-converter exhibits good performance compared to a full converter, but a semi-converter is not capable of regenerative braking. On the other hand, a full converter has poor performance but is capable of two-quadrant operation, i.e., motoring and regenerative braking modes. Hence, to integrate advantages of both converters, a full converter with a modified firing scheme is developed and explained in this section. This is shown in Fig. 5.10. In the modified gating signals, SCRs T_3 and T_4 are triggered to act as diodes in the respective half-cycles, and gating pulses $i_g(T_1)$ and $i_g(T_2)$ are extended for one half-cycle period. This leads to freewheeling of load current, making the operation same as that with a semi-converter. This improves the input power factor and the displacement factor of full converters.

A little change in firing pattern is now incorporated into the full converter circuit so that it can retain its regenerative capability with an improved power factor. This is shown in Fig. 5.11.

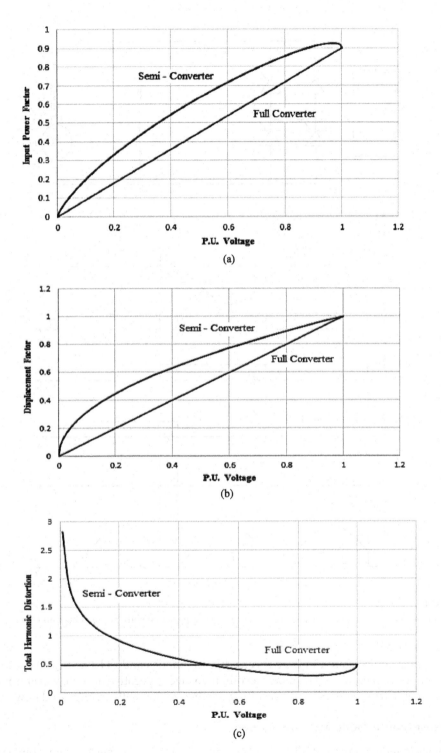

FIGURE 5.9
Plot between performance parameters and per-unit voltage for semi-converter and full converter. (a) Input power factor versus P.U. Voltage. (b) Displacement factor versus P.U. Voltage (c) Total Harmonic Distortion versus P.U. Voltage.

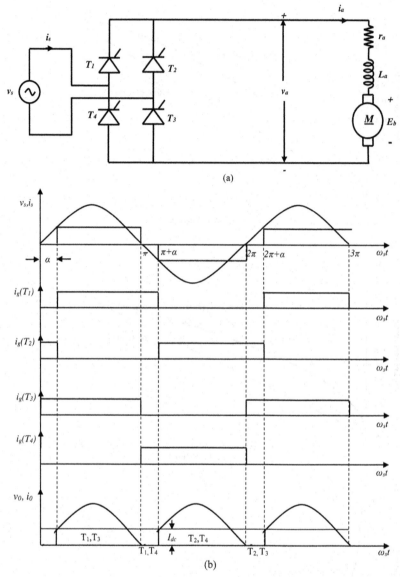

FIGURE 5.10
Modified operation of a full converter. (a) Supply voltage. (b) Gating signal with output voltage and source current.

Here, T_1 and T_2 are triggered in their reverse-biasing modes and $i_g(T_3)$ and $i_g(T_4)$ controls the average output voltage, $V_{o(av)}$. The motor terminal voltage is negative and the waveform of i_s is similar to that with a semi-converter, resulting in regeneration with an improved power factor.

5.7.2 Pulse-Width Modulation Technique

By employing self-commutating devices such as power MOSFETs, insulated-gate bipolar transistors (IGBTs), etc., pulse-width modulation strategies can be incorporated with bridge-type converters employing a single IGBT as shown in Fig. 5.12. The IGBT is switched ON and OFF many times in a half-cycle, such that effective voltage control and harmonic elimination can

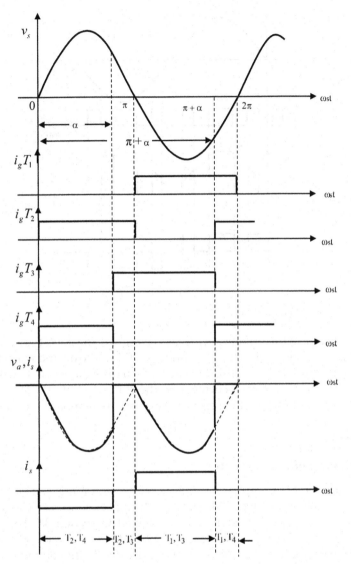

FIGURE 5.11
Regenerative braking in the semi-converter mode.

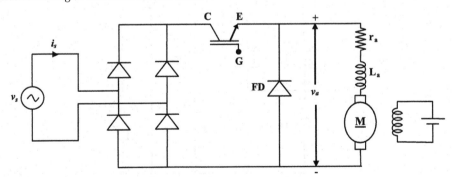

FIGURE 5.12
Pulse-width modulation technique in an ac/dc converter feeding a dc motor drive.

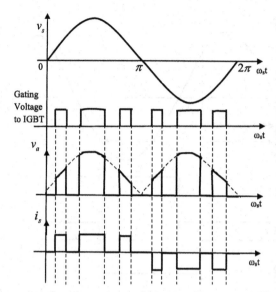

FIGURE 5.13
Waveforms with pulse-width modulation.

be met simultaneously. If the armature current is assumed constant, typical waveforms are shown in Fig. 5.13 simultaneously. The diode bridge rectifies ac to dc, and when the IGBT is triggered, the motor is connected to rectified voltage. When the gating signal to the IGBT is withdrawn, the IGBT turns OFF and load current freewheels through the FD. The switching signal to the IGBT can be varied such that lower-order line current harmonics can be eliminated.

5.8 Three-Phase Semi-Converter

The power circuit of a three-phase semi-converter supplying power to a separately excited dc motor is shown in Fig. 5.14. Here the firing sequence of SCRs as well as conduction modes of all devices are similar, as explained in section 3.5. For any SCR firing angle $\alpha \le 60°$, the output voltage waveform is identical to the one with R load. However, for $\alpha > 60°$, the nature of the output voltage waveform is decided by the continuity of the armature current. If the armature current is continuous, then the freewheeling mode exists, the output

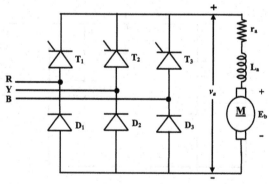

FIGURE 5.14
Three-phase semi-converter.

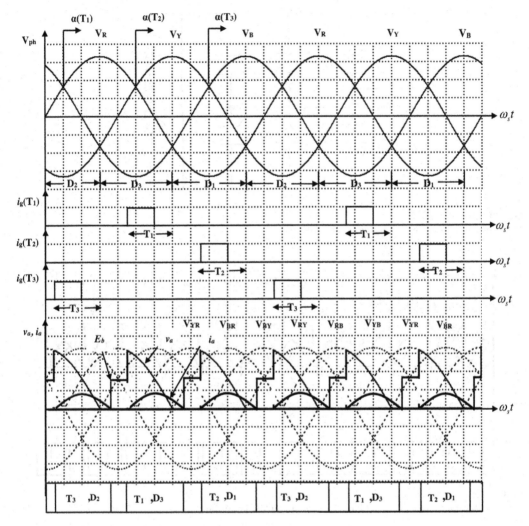

FIGURE 5.15
Waveforms for $\alpha = 105°$ with a dc motor load.

voltage waveforms resemble that of R load, and the output voltage expression remains same as given in Equation (3.22). Once the armature current becomes discontinuous, the back-emf E_b appears across the motor terminals. A typical case for $\alpha = 105°$ is given in Fig. 5.15.

5.9 Three-Phase Full-Converter

The circuit diagram of a three-phase full converter feeding a dc motor load is depicted in Fig. 5.16. The firing sequence of SCRs are the same as with R load in this case. Up to $\alpha = 60°$, the output voltage waveform is same as with R load, but for $\alpha > 60°$ the wave shape of the output voltage is decided by the continuity of the armature current. Two typical cases (i.e., $\alpha = 75°$ and $\alpha = 90°$) for discontinuous armature current are plotted in Fig. 5.17 and Fig. 5.18, respectively. As a special case, the motor terminal voltage and current waveforms during the regenerative braking mode are given in Fig. 5.19, where $\alpha = 120°$.

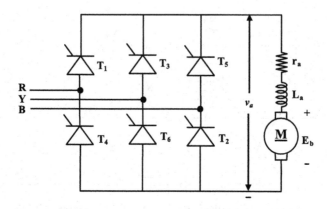

FIGURE 5.16
Three-phase full-converter supplying a dc motor load.

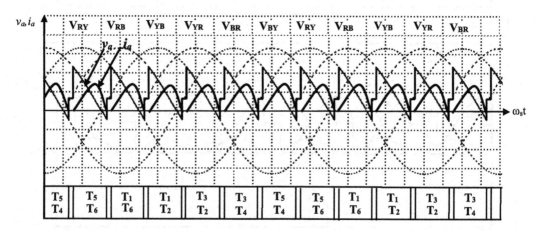

FIGURE 5.17
Waveforms for a dc motor load with $\alpha = 75°$ and current discontinuity.

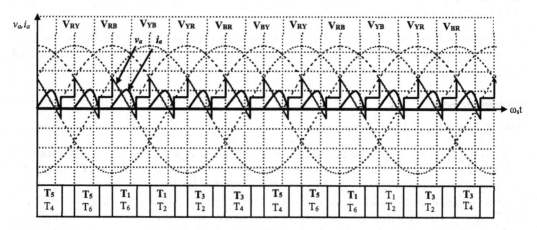

FIGURE 5.18
Waveforms for a dc motor load with $\alpha = 90°$ and current discontinuity.

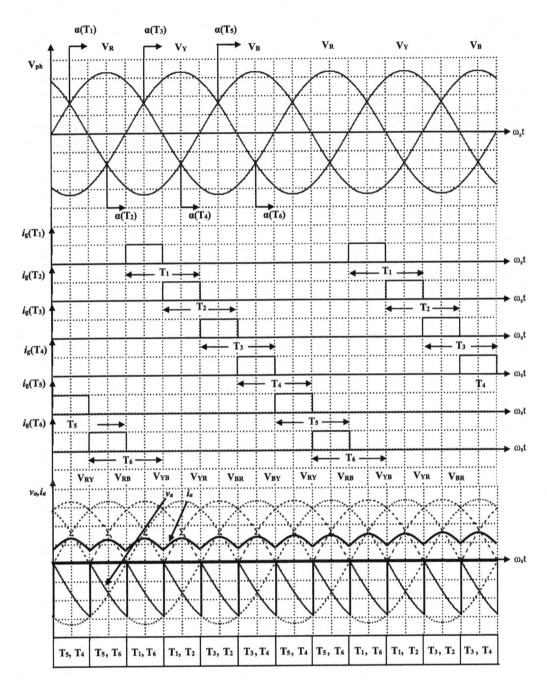

FIGURE 5.19
Waveforms during regenerative braking mode with $\alpha = 120°$ and current continuity.

EXAMPLE 5.10

The speed of a 125-horsepower, 600-V, 1,800-rpm, separately excited dc motor is controlled by a three-phase full converter. The converter is operated from a three-phase, 480-V, 50-Hz supply. The rated armature current at the motor is 165 A. The motor parameters are $r_a = 0.0874\ \Omega$, $L_a = 6.5$ mH, and $K_a\Phi_m = 0.33$ V/rpm. (a) Find the no-load speeds at firing angles $\alpha = 0°$ and $\alpha = 30°$, assuming that, at no load, the armature current is 10% at the rated current and is continuous. (b) Find the firing angle to obtain the rated speed of 1,800 rpm at the rated motor current.

SOLUTION:

(a) No-load condition:

The supply phase voltage is $V_{ph} = \dfrac{480}{\sqrt{3}}$

The motor terminal voltage is

$$V_a = \frac{3V_{LM}}{\pi}\cos\alpha = \frac{3\sqrt{3}\ V_m}{\pi}\cos\alpha$$

$$= \frac{3\sqrt{3}\times\sqrt{2}\times V_{ph}}{\pi}\cos\alpha$$

$$V_a = \frac{3\sqrt{3}\times\sqrt{2}\times 277}{\pi}\cos\alpha$$

For $\alpha = 0°$:

$$V_a = 648\ \text{V}$$
$$E_b = V_a - I_a r_a$$
$$= 648 - 16.5\times 0.0874$$
$$E_b = 646.6\ \text{V}$$

No-load speed is

$$N_0 = \frac{E_b}{K_a\phi_m} = \frac{646.6}{0.33} = 1,959\ \text{rpm}$$

For $\alpha = 30°$:

$$V_a = 648\ \cos 30°$$
$$E_b = 561.2 - (16.5\times 0.0874) = 555.8\ \text{V}$$

(b) Full-load condition:

$$N_{fL} = \frac{E_b}{K_a\phi_m}$$

To find E_b at full load, motor back-emf at 1,800 rpm is

$$E_b = 0.33\times 1800 = 594\ \text{V}$$

∴ Motor terminal voltage at rated current of 165 A is

$$V_a = E_b + I_a r_a$$
$$= 594 + 165 \times 0.0874$$
$$V_a = 608.4 \text{ V}$$

$$\therefore \frac{3\sqrt{6}}{\pi} V_{ph} \cos\alpha = 608.4$$

$$\alpha = 20.1342°$$

EXAMPLE 5.11

A 220-V, 1,500-rpm, 50-A, separately excited dc motor with $r_a = 0.5\ \Omega$, is fed from a three-phase fully controlled rectifier. An available ac voltage of 440 V, 50 Hz is used. Determine the value of α (a) when the motor is running at 1,200 rpm and at the rated torque, and (b) when the motor is running at −800 and at twice the rated torque. Assume continuous conduction.

SOLUTION:

Given the motor rating at 220 V, 1,500 rpm, 50 A, and $R_a = 0.5\ \Omega$:

$$220 = \frac{3\sqrt{6}V_{ph}}{\pi}$$
$$V_{ph} = 94$$

$$\text{Supply line voltage} = 440 \text{ V}$$

$$\text{Supply phase voltage} = \frac{440}{\sqrt{3}} = 254.034$$

$$V = 220 \text{ V}$$
$$I_a = 50 \text{ A}$$

$$\text{Back emf } E_b = V - I_a r_a = 220 - (50 \times 0.5) = 195 \text{ V}$$

(a) At 1,200 rpm

$$\frac{E_{b1}}{E_{b2}} = \frac{N_1}{N_2}$$

$$E_{b2} = \frac{N_2}{N_1} \times E_{b1} = \frac{1200}{1500} \times 195 = 156 \text{ V}$$

$$V_a = 156 + (50 \times 0.5) = 181 \text{ V}$$

$$181 = \frac{3\sqrt{6}V_{ph}}{\pi}\cos\alpha = \frac{3\sqrt{6} \times 94}{\pi}\cos\alpha$$

$$\cos\alpha = 0.8232$$
$$\alpha = 34.6°$$

(b) At –800 rpm

$$\frac{E_{b1}}{E_{b2}} = \frac{N_1}{N_2}$$

$$\frac{195}{E_{b2}} = \frac{1500}{-800}$$

$$E_{b2} = \frac{-800}{1500} \times 195 = -104 \text{ V}$$

If torque is twice the rating, then I_a doubles as well, so $I_a = 100$ A.

$$V_a = -104 + (50 \times 0.5) = -54 \text{ V}$$

$$-54 = \frac{3\sqrt{6}V_{ph}}{\pi}\cos\alpha = \frac{3\sqrt{6} \times 94}{\pi}\cos\alpha$$

$$\cos\alpha = -0.2456$$

$$\alpha = 104.21°$$

EXAMPLE 5.12

A 440-V, three-phase ac supply feeds a separately excited dc motor through two three-phase full converters, one for armature and the other for field. The firing angle of the field converter is 30°; $r_a = 0.2$ Ω and $r_f = 200$ Ω. The armature current is equal to the rated value of 40 A. The motor voltage constant is 1.2 V/rad/s. If the motor runs at 1,250 rpm, find the firing angle of the converter in the armature circuit.

SOLUTION:

Output average voltage of the three-phase full converter in the field circuit is

$$V_f = \frac{3V_{LM}}{\pi}\cos\alpha = \frac{3\sqrt{3}\ V_m}{\pi}\cos\alpha$$

$$= \frac{3\sqrt{3} \times 440\sqrt{2}}{\sqrt{3} \times \pi}(\cos 30) = 514.60 \text{ V}$$

$$\text{Field current} = \frac{514.60}{200} = 2.573 \text{ A}$$

$$\omega_r = \frac{1,250 \times 2\pi}{60} = 130.90 \text{ rad/s}$$

Back-emf $E_b = K_b I_f \omega_r$

$$= 1.2 \times 2.573 \times 130.90 = 404.166 \text{ V}$$

$$V_a = E_b + I_a r_a$$

$$V_a = 404.165 + (40 \times 0.2) = 412.165 \text{ V}$$

$$V_a = \frac{3\sqrt{3}\ V_m}{\pi}\cos\alpha$$

$$412.165 = \frac{3\sqrt{3} \times 440\sqrt{2}}{\sqrt{3} \times \pi}(\cos\alpha_a)$$

$$\alpha_a = 40.27°$$

5.10 Four-Quadrant Operation of Single-Phase Full Converter-Fed dc Motor Drive

The circuit schematic for speed reversal or two-quadrant operation of a dc motor drive system is shown in Fig. 5.20. The circuit contains a full bridge converter and four mechanical contactors suitably connected for regenerative braking and then speed reversal.

The drive rotates in the forward direction with $\alpha < \pi/2$, and F_1 and F_2 are closed. With the use of R_1 and R_2, regenerative braking together with speed reversal is achieved. The complete scheme of operation is detailed below.

The time variations of various motor variables are plotted in Fig. 5.21. At $t = 0$, α is less than $\pi/2$ and F_1 and F_2 are closed, while R_1 and R_2 remain opened. It should be noted that simultaneous closing of F_1-R_1 or F_2-R_2 causes a short circuit of dc voltage and should be avoided. The motor drives a finite amount of load in the forward direction. At $t = t_1$, the speed reversal command reaches the control unit and, as a first step, the motor is unloaded. This increases ω_r to its no-load value, and i_1 and i_a fall to corresponding no-load amplitudes. The firing pulses to the converter are now withdrawn, and i_1 and i_a soon fall to zero. Once the current goes to zero, F_1 and F_2 are opened. This makes $V_a = E_b$, and ω_r continues at its no-load value. From t_1 onward, a dead time of the order of a few tens of milliseconds is allowed, during which all contactors are kept opened; this prevents simultaneous closing of F_1-R_1 and F_2-R_2. During this period, the motor is disconnected from the power supply and is coasting.

Once the allowable delay is over at $t = t_2$, the contactors R_1 and R_2 are closed and the firing angle α is set above $\pi/2$. This makes the polarities of v_0 and v_a compatible with each other so that regeneration can be performed. At this interval, the firing angle is adjusted to make $V_0 = V_a$, and the motor current i_a is given by

$$i_a = \frac{V_a - E_b}{r_a} \tag{5.40}$$

where

$$V_a = \frac{2v_m}{\pi}\cos\alpha$$

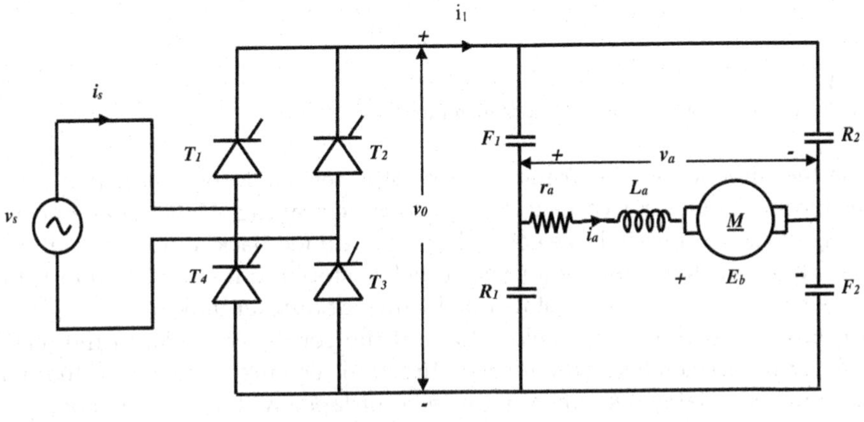

FIGURE 5.20
Two-quadrant dc motor drive for speed reversal.

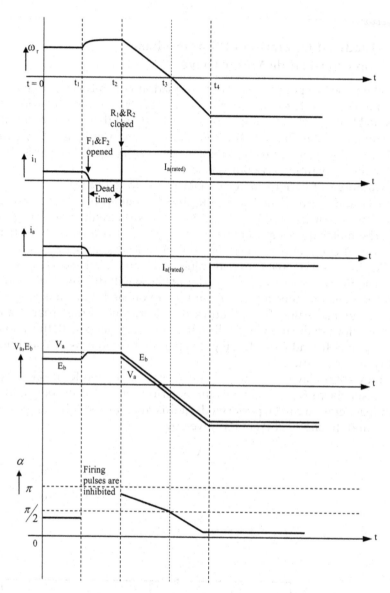

FIGURE 5.21
Regenerative braking and speed reversal of a dc motor drive.

Because the direction of the armature current is reversed, the polarity of the electro-magnetic torque, T_e, is negative, causing regenerative braking. Thus the kinetic energy of the rotating parts is converted to electrical energy and fed back to the mains through the converter. This decelerates the motor very quickly, stopping it at $t = t_3$. From t_2 to t_3, α is continuously decreased to keep i_a at its rated value for quicker braking.

From t_3 onward, α is reduced below $\pi/2$ and the converter works in the rectification mode. The motor accelerates in the reverse direction. The motor current is maintained at its rated value by suitably decrementing α in accordance with the rise in speed. At $t = t_4$, the motor speed reaches its no-load value in the opposite direction, and speed control for different loading conditions can be achieved by changing α.

5.11 Dual Converter

Before discussing the theory of dual converters, the four-quadrant operation of a separately excited dc motor driving a lift is explained with the help of the $V_a - I_a$ diagram shown in Fig. 5.22.

Consider that a separately excited dc motor driving a lift. The lift cabin is connected to motor shaft through steel ropes. When the lift has to move up from ground floor, say to second floor, the motor drives the lift upwards against the action of gravity. Here, both V_a and I_a are positive and the motor operates in the first quadrant. The motor is rotating in clockwise direction and electromagnetic torque is positive. As the cabin reaches second floor, motor current is reversed. Electromagnetic torque, T_e of the motor is given by

$$T_e \propto \phi_m \, I_a$$

For a separately excited motor ϕ_m is constant and hence

$$T_e \propto I_a$$

With the reversal of I_a, the electromagnetic torque is reversed, applying a braking action to the cabin. During this interval, V_a is positive but I_a is negative, indicating that the motor converts the kinetic energy of the load to electrical energy that feeds power back to the supply lines. Thus second-quadrant operation is termed regenerative braking. When the cabin stops, the regenerative braking ends. When the lift has to go down, the motor has to rotate in the counter-clockwise direction. This requires both V_a and I_a to be negative, leading to third-quadrant operation. In this quadrant, power is positive, but both electromagnetic torque and speed are negative. To stop the cabin at the desired floor, regenerative braking is applied by reversing I_a alone. Thus fourth-quadrant operation results in deceleration of drive system. It should be mentioned that in quadrants 1 and 3, electrical energy is supplied to the drive system, while energy is fed back from the mechanical system to the mains in quadrants 2 and 4.

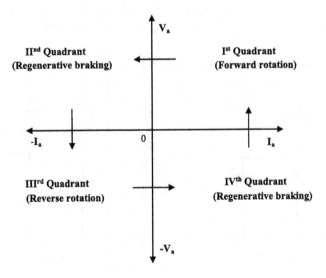

FIGURE 5.22
A $V_a - I_a$ diagram of a reversible-speed dc motor drive.

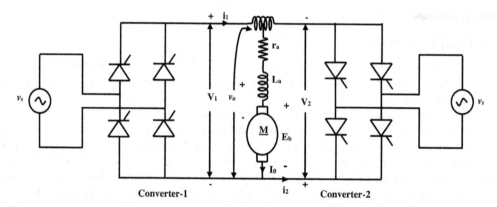

FIGURE 5.23
Power circuit of a dual converter with single-phase converters.

From the theory of semi-converters, it is evident that its output current and voltage are always positive. Thus semi-converters are capable of operating in the first quadrant of the $V_a - I_a$ diagram as shown in Fig. 5.22. In other words, regenerative braking is not possible with a semi-converter-fed dc motor drive. In the case of a full converter, the output current is always positive, but the output voltage can be either positive or negative depending on the SCR firing angle, α. Thus, in principle, a full converter-fed dc motor can operate either in first or fourth quadrant. Hence, the drive system can rotate in one direction, and it can be stopped with regenerative braking; however, speed reversal is not possible with a full converter.

If a power converter is capable of operating in all four quadrants, it is called a four-quadrant converter. A dual converter falls to this category. The circuit of a dual converter is shown in Fig. 5.23. It consists of two full converters connected across the load in an anti-parallel manner. The converters can be either single-phase or three-phase as shown in Fig. 5.23 and Fig. 5.24, respectively.

A reactor coil is inserted between the converters to avoid short circuit of converter voltages. Converter 1 provides first- and fourth-quadrant operation, while converter 2 is used to operate the motor in the second and third quadrants. The output voltage and current

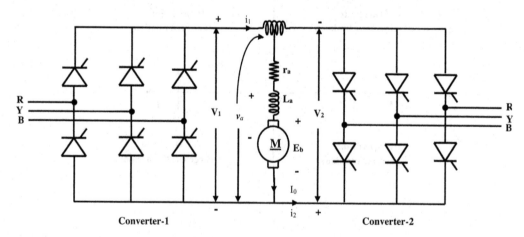

FIGURE 5.24
Power circuit of a dual converter for three-phase operation.

of converter 1 are indicated as V_1 and I_1, respectively; notations V_2 and I_2 are used for converter 2. For the circuit topology, the average value and polarity of V_1 and V_2 should be the same. For the indicated polarity of V_1 and V_2, we can write,

$$V_1 = -V_2$$
$$V_1 + V_2 = 0$$

If α_1 and α_2 represent firing angles of converter 1 and converter 2, respectively, then

$$V_1 = K_1 \cos\alpha_1 \text{ and } V_2 = K_1 \cos\alpha_2$$

In the above, $K_1 = \dfrac{2V_m}{\pi}$ for single-phase converters, and $K_1 = \dfrac{3V_{Lm}}{\pi}$ for three-phase converters.

$$\therefore K_1 \cos\alpha_1 + K_1 \cos\alpha_2 = 0$$
$$\cos\alpha_1 + \cos\alpha_2 = 0$$
$$2\cos\left(\frac{\alpha_1 + \alpha_2}{2}\right)\cos\left(\frac{\alpha_1 - \alpha_2}{2}\right) = 0$$

The solution of the above equation yields

$$\frac{\alpha_1 + \alpha_2}{2} = \frac{\pi}{2} \quad \text{or} \quad \frac{\alpha_1 - \alpha_2}{2} = \frac{\pi}{2}$$
$$\alpha_1 + \alpha_2 = \pi \quad \text{or} \quad \alpha_1 - \alpha_2 = \pi$$
$$\alpha_1 = \pi - \alpha_2 \quad \text{or} \quad \alpha_1 = \pi + \alpha_2$$

Because the maximum value of α_1 or α_2 is π, the feasible solution is

$$\alpha_1 = \pi - \alpha_2 \tag{5.41}$$

This implies that a dual converter-supplied dc motor drive system can be operated only in the closed-loop mode so that the control law given by Equation (5.41) is satisfied. For further investigation, consider the plots of V_1 and V_2 of a dual converter with a single-phase supply as shown in Fig. 5.25. It is clear that the average value of V_1 is positive whereas that of V_2 is negative. The instantaneous variation of V_1 and V_2 with a single-phase supply is shown in Fig. 5.25. Hence, converter 1 is supplying power to the motor as well as to converter 2. Converter 2 is operating in the inverting mode and the power supplied to it is fed back to the mains. Furthermore, Fig. 5.25 shows that instantaneous values of V_1 and V_2 differ; hence, to avoid a short circuit between the two converters, a reactor is inserted between them.

The four-quadrant operation of a dual converter can be carried out using two schemes, namely a non-circulating current scheme and a circulating current scheme. A brief description of the two schemes is given below.

5.11.1 Non-Circulating Current Scheme

In this scheme, only one converter operates at a time. Therefore, there is no cause for a short circuit between the converters, and the reactor can be dispensed with. However, the control law given by Equation (5.41) is strictly followed. The sequential stages of non-circulating current can be described with the help of Fig. 5.26.

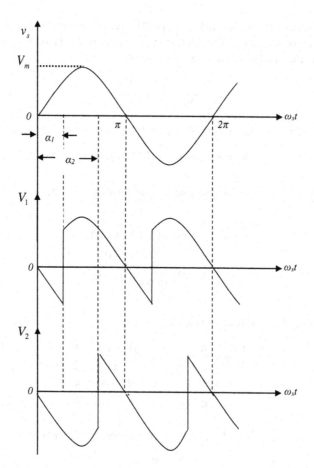

FIGURE 5.25
Instantaneous voltage waveforms of a dual converter.

Initially, at $t = 0$, converter 1 is operating at a finite firing angle of $\alpha_1 < \frac{\pi}{2}$ and drives the motor. The motor drives a constant load at a fixed speed. At $t = t_1$, the speed reversal command reaches the controller. The load is removed and motor speed increases to no-load value; in addition, i_a falls to the no-load current and E_b increases. This situation prevails until $t = t_2$. At t_2, firing pulses to converter 1 are withdrawn. This makes i_a and i_1 fall to zero, and V_a becomes E_b. The motor coasts, but due to its inertia, speed and E_b remain almost the same as the last values. The duration $t_3 - t_2$ is a few milliseconds to ensure that converter 2 is switched ON only after converter 1 is completely turned OFF.

At $t = t_3$, converter 2 is triggered at slightly less than $\pi - \alpha_1$ so that both E_b and V_2 are of the same polarity but E_b is little higher than V_2. The current is given by

$$(E_b - V_2) / r_a = i_a = I_{a\ rated} \tag{5.42}$$

where $I_{a\ rated}$ indicates rated armature current. Thus, E_b acts as a generated voltage due to the kinetic energy of the running parts of the motor and energy flows from the motor to supply mains resulting in regenerative braking of the motor. As the energy transfer takes place, speed and E_b decrease. Hence, to maintain the constant armature current at rated value, α_2 is successively reduced. At $t = t_4$, motor speed becomes zero and regenerative braking ends at this instant. From t_4 onward, α_2 is less than $\frac{\pi}{2}$ and hence V_2 is negative,

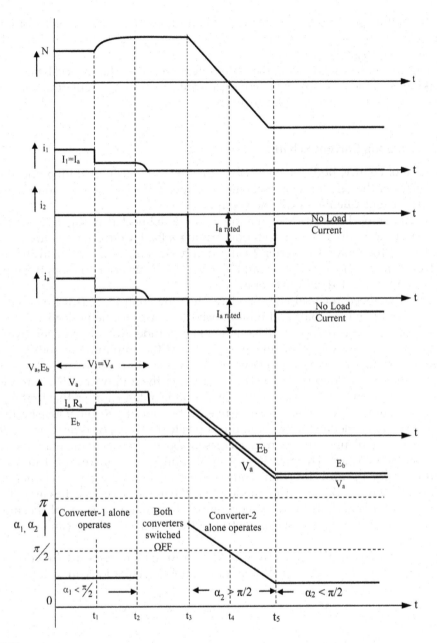

FIGURE 5.26
Dynamic response of the drive system during speed reversal in the non-circulating current mode.

accelerating the motor in the reverse direction. Motor start-up in the opposite direction is performed at a constant armature current by controlling α_2 to satisfy the following equation:

$$(V_2 - E_b) / r_a = i_a = I_{a\,rated} \qquad (5.43)$$

At t = t_5, motor speed reaches no-load speed and I_a and I_2 fall to no-load motor current. The firing angle of converter 2 is maintained at $\alpha_2 = \alpha_m$. After t_4, load can be applied and α_2 can be varied to get desired speed.

Referring to Fig. 5.26, it is evident that a definite time of $t_3 - t_2$ has lapsed before regenerative braking is applied. Hence, the major disadvantage of a non-circulating scheme is that increased time is required for speed reversal.

5.11.2 Circulating Current Scheme

In this type of control, both the converters are simultaneously operating and the speed reversal is faster than the previous scheme. The speed reversal of a dc motor drive using circulating current scheme is explained below.

The variations of different motor parameters are delineated in Fig. 5.27. At t = 0, the motor is running in the forward direction driving a load. Converter 1 is fired at $\alpha_1 < \pi/2$ and is driving the motor. Converter 2 is fired at $\pi - \alpha_1$ and is operating in the inverting mode. Hence, $I_1 = I_a + I_2$, $V_1 = V_2 = V_a$, and $E_b = V_a - I_a r_a$. Converter 1 is supplying power to the motor, and $V_2 I_2$ is fed back to supply lines.

At t = t_1, the speed reversal command reaches the controller; as a first step, load is removed from the motor shaft and motor speed increases to the no-load value together with back-emf, E_b. This reduces I_a to its no-load amplitude. At t = t_2, regenerative braking is initiated by increasing α_1. This makes α_2 fall, and the amplitude of both V_1 and V_2 is now lower than E_b. This forces I_a to flow in the reverse direction. Because I_a cannot pass through converter 1, it flows to converter 2 and reaches the supply lines. The variation of α_1 and α_2 is continued in such a way that I_a remains at its rated value so that braking can be faster. At t = t_3, motor speed reaches zero value, and so does E_b. Regenerative braking ends at this instant. To accelerate the motor in the opposite direction, α_2 is decreased below $\pi/2$, and to provide circulating current, α_1 is increased above $\pi/2$ so as to maintain the equation $\alpha_1 + \alpha_2 = \pi$ at any instant. The polarities of V_1, V_2, and V_a are now reversed, and I_a continues to flow in the upward direction. Thus converter 2 is working as a rectifier and converter 1 is operating in the inverting mode. Once the motor speed reaches its full value in the reverse direction, I_a falls to no-load current and motor speed reversal is completed at t = t_4. After this instant, motor speed control in the reverse direction can be achieved by changing α_2.

It is worth mentioning that when the motor is running in the forward direction (positive value), converter 1 is driving the motor and the following equations are valid.

$$V_a = E_b + I_a r_a, \quad \text{where} \quad V_a = \frac{2V_m}{\pi} \cos\alpha_1 \tag{5.44}$$

$$\alpha_1 < \frac{\pi}{2} \quad \text{and} \quad \alpha_2 > \frac{\pi}{2} \tag{5.45}$$

During regenerative braking, motor speed is positive and

$$V_a = E_b - I_a r_a, \quad \text{where} \quad V_a = \frac{2V_m}{\pi} \cos\alpha_1 \tag{5.46}$$

$$\alpha_1 < \frac{\pi}{2} \quad \text{and} \quad \alpha_2 > \frac{\pi}{2} \tag{5.47}$$

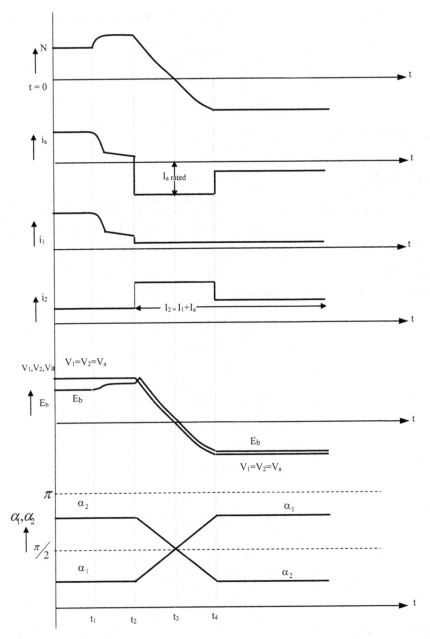

FIGURE 5.27
Dynamic response of the drive system during speed reversal with circulating current mode.

When the motor runs in the opposite direction, speed will be taken as negative; converter 2 is driving the motor, and hence the following equations are valid.

$$V_a = E_b + I_a r_a, \quad \text{where} \quad V_a = \frac{2V_m}{\pi} \cos \alpha_2 \tag{5.48}$$

$$\alpha_1 > \frac{\pi}{2} \quad \text{and} \quad \alpha_2 < \frac{\pi}{2} \tag{5.49}$$

During regenerative braking with negative speed, the following equations are employed.

$$V_a = E_b - I_a r_a, \quad \text{where} \quad V_a = \frac{2V_m}{\pi} \cos \alpha_2 \tag{5.50}$$

$$\alpha_1 > \frac{\pi}{2} \quad \text{and} \quad \alpha_2 < \frac{\pi}{2} \tag{5.51}$$

These concepts are illustrated below:

EXAMPLE 5.13

A 220-V, 22.7-A, 1,500-rpm separately excited dc motor has an armature resistance of 6.6 Ω. The motor is running in the reverse direction at 600 rpm at rated torque. The motor is driven by a single-phase dual converter powered from $230\sqrt{2} \sin 314t$ source. Compute the converter firing angles.

SOLUTION:

Back-emf at rated conditions is given by

$$E_{b(rated)} = V_a - I_a r_a$$
$$= 220 - 22.7 \times 6.6$$
$$= 136.2 \text{ V}$$

Back-emf at 600 rpm at rated torque is given by

$$\frac{E_{b(600)}}{E_{b(rated)}} = \frac{600}{1500}$$

$$\frac{E_{b(600)}}{136.2} = \frac{600}{1500}$$

$$E_{b(600)} = \frac{600}{1500} \times 136.2 = 54.48 \text{ V}$$

$$V_{a(at\ 600\ rpm)} = E_b + I_a r_a$$
$$= 54.48 + 22.7 \times 6.6$$
$$= 204.3 \text{ V}$$

The motor is running in the reverse direction. Converter 2 is supplying power.

$$V_a = \frac{2V_m}{\pi} \cos \alpha_2 = 204.3 \text{ V}$$

$$\cos(\alpha_2) = \frac{204.3 \times \pi}{2 \times 230\sqrt{2}} = 0.98$$

$$\alpha_2 = 11.47°$$

$$\therefore \alpha_1 = 180 - \alpha_2 = 168.52°$$

EXAMPLE 5.14

A 220-V, 22.7-A, 1,500 rpm separately excited dc motor has an armature resistance of 6.6 Ω. The motor is driving a load in the reverse direction at 500 rpm and 60% of rated torque. The motor is driven by a single-phase dual converter powered from a $230\sqrt{2}\sin 314t$ source. Compute the converter firing angles.

SOLUTION:

Back-emf at rated conditions is given by

$$E_{b(rated)} = V_a - I_a r_a$$
$$= 220 - 22.7 \times 6.6$$
$$= 136.2 \text{ V}$$

Back-emf at 500 rpm at 60% torque is given by

$$\frac{E_{b(500)}}{E_{b(rated)}} = \frac{500}{1500}$$

$$\frac{E_{b(500)}}{136.2} = \frac{500}{1500}$$

$$E_{b(500)} = \frac{500}{1500} \times 136.2 = 45.39 \text{ V}$$

$$V_{a(at\ 500\ rpm)} = E_b + I_a r_a$$
$$= 45.39 + (0.6 \times 22.7) \times 6.6$$
$$= 135.282 \text{ V}$$

$$V_a = \frac{2V_m}{\pi}\cos\alpha_2 = 135.3 \text{ V}$$

$$\cos(\alpha_2) = \frac{135.3 \times \pi}{2 \times 230\sqrt{2}} = 0.65$$

$$\alpha_2 = 49.20°$$

$$\therefore \alpha_1 = 180 - \alpha_2 = 130.79°$$

EXAMPLE 5.15

A 220-V, 50-A, 1,500-rpm, separately excited dc motor has an armature resistance of 0.5 Ω and is supplied from a dual converter powered from $220\sqrt{2}\sin(314t)$. If the motor runs clockwise during regenerative braking, speed is 800 rpm at rated torque. Compute the converter firing angles.

SOLUTION:

Under rated conditions,

$$E_{b(rated)} = V_{a(rated)} - I_{a(rated)} \times r_a$$
$$= 220 - (50 \times 0.5) = 195 \text{ V}$$

Back-emf at 800 rpm is given by

$$\frac{E_{b(800)}}{E_{b(rated)}} = \frac{800}{1500}$$

$$E_{b(800)} = \frac{800}{1500} \times 195 = 104 \text{ V}$$

During regenerative braking

$$V_a = E_b - I_a \times r_a$$
$$= 104 - (50 \times 0.5) = 79 \text{ V}$$

Because converter 1 is feeding power,

$$V_a = \frac{2V_m}{\pi} \cos\alpha_1$$

$$\text{i.e., } 79 = \frac{2 \times 220\sqrt{2}}{\pi} \cos\alpha_1$$

$$\alpha_1 = 66.49°$$

$$\alpha_2 = 180 - \alpha_1 = 180 - 66.49 = 113.51°$$

Questions

1. Sketch the circuit of a single-phase half-wave controlled rectifier supplying a separately excited dc motor. Show the motor terminal voltage and current wave forms.

2. Draw the circuit and relevant wave forms of a separately excited dc motor supplied from a single-phase semi-converter.

3. Repeat the above question with full converter.

4. Sketch four prominent ac/dc converters used for speed control of a separately excited dc motor with V-I diagrams.

5. Define input performance of an ac/dc converter-fed dc motor drive system.

6. Using Fourier series analysis, derive expressions for input pf, displacement factor, and total harmonic distortion (THD) of a single-phase semi-converter-fed dc motor drive.

7. Prepare a semi-converter comparison table of performance between a separately excited dc motor drive driven by a semi-converter and by a full converter.

8. With the necessary waveforms and circuits, explain how you would operate a single-phase full converter as a single-phase semi-converter.

9. With a neat circuit diagram and relevant waveform, explain the operation of a pulse-width modulation ac/dc converter.

10. With a neat circuit diagram and the necessary waveforms, explain the process of regenerative braking and speed reversal of a separately excited dc motor drive.

11. Sketch the circuit of a dual converter-fed dc motor drive, and derive the necessary conditions for the firing angles for the two converters.

12. A reactor is generally connected between the two converters in a dual converter-fed dc motor drive. Explain the reason.

13. Compare circulating and non-circulating current modes of operation of a dual converter.

14. Using Fourier series, derive an expression for the input power factor of a single-phase full converter-fed armature voltage-controlled dc motor drive. Assume that the motor armature current is constant at I_{dc}.

15. A dual converter in the circulating current mode uses an inductance, L, between the two converters. Derive an expression for the average inductor current.

16. A single-pulse width-modulated ac/dc converter, employing symmetrical switching, drives a dc motor at constant current. Find the third harmonic component of the line current when the motor input voltage is 0.5 p.u.

Unsolved Problems

1. A dc motor is supplied from a single-phase semi-converter. The motor draws a constant current of 10 A when the converter is fired at 35°. Calculate (a) rms input current, (b) input power factor, (c) displacement factor, and (d) THD.

2. A single-phase full converter is used to drive a dc motor at a constant armature current of 20 A. If the firing angle of the converter is 50°, compute the performance parameters (i.e., input pf, displacement factor, THD).

3. A dc shunt motor operating from a single-phase half-wave controlled bridge at 1,450 rpm has an input $V_s = 330\sin 314t$ and a back-emf of 75 V. The SCRs are fired symmetrically at $\alpha = \Pi/4$ every half-cycle, and the armature has a resistance of 5 Ω. Neglecting armature inductance, calculate the average armature current and the torque.

4. A single-phase half-wave controlled rectifier is supplied from a 220-V, 50-Hz source. What is the value of the dc voltage for a firing angle of 60°? The dc current is constant at 20 A. Calculate the rms value of the fundamental component of input current and the input power factor.

5. A 220-V, 20-A, 1,500-rpm, separately excited dc motor has an armature resistance of 4.5 Ω. The motor is running in the reverse direction at 900 rpm at half the rated torque. The motor is driven by single-phase dual converter powered from a $230\sqrt{2}\sin 314t$ source. Compute the converter firing angles.

6. A 220-V, 1,500-rpm, 50-A, separately excited dc motor with $r_a = 0.5$ Ω is fed from a three-phase fully controlled rectifier. An available ac voltage of 440 V, 50 Hz is used. Interpose a suitable transformer and draw the circuit diagram of the given system. Determine the value of α (a) when the motor is running at 1,200 rpm and at rated torque; and (b) when the motor is running at −800 and at twice the rated torque. Assume continuous conduction.

7. A 200-V, 25-A, 1,500-rpm, separately excited dc motor has an armature resistance of 1 Ω and is supplied from a dual converter powered from a 230-V, 50-Hz supply. If the motor runs clockwise during regenerative braking, the speed is 500 rpm at rated torque. Compute the firing angles.

Answers

1. $I_{rms} = 8.975$ A, pf $= 0.9124$, DF $= 0.9537$, THD $= 0.3043$
2. pf $= 0.5787$, DF $= 0.6427$, THD $= 0.4834$
3. $I_{a(avg.)} = 20.86$ A, torque $= 10.27$ N-m
4. $V_{dc} = 148.62$ V, $I_1 = 15.6$ A, pf $= 0.827$
5. $\alpha_1 = 164.87°$, $\alpha_2 = 15.13°$
6. $\alpha(1{,}200 \text{ rpm}) = 34.6°$, $\alpha(-800 \text{ rpm}) = 104.21°$
7. $\alpha_1 = 80.73°$, $\alpha_2 = 99.27°$

6

dc/dc Converters

6.1 Introduction

If the existing input is dc voltage, then a chopper circuit is employed to get a variable dc voltage at the output. A chopper is a power electronic circuit connected between a constant dc voltage source and load. Chopper circuits are also called dc/dc converters. In principle, a chopper is equivalent to a mechanical switch connecting and disconnecting the load from the source at desired intervals. Chopper circuits, usually employing silicon-controlled rectifiers (SCRs), are widely used for speed control of dc motors in industries, electric traction, and battery-powered vehicles. Chopping at high frequency is used in dc power supplies such as buck, boost, and Ćuk converters.

6.2 Principle of Operation

Figure 6.1 illustrates the principle of a chopper. The chopper is represented by a switch that is shown inside the dotted square as given in Fig. 6.1(a). The switch may be a gate turn-off thyristor, a SCR, a metal-oxide semiconductor field-effect transistor (MOSFET), or an insulated-gate bipolar transistor. It is triggered periodically and is kept conducting for a period of T_{on} and is blocked for a period $T - T_{on}$. The chopped load voltage waveform is shown in Fig. 6.1(b). During the period T_{on}, when the chopper is ON, the supply terminals are connected to the load terminals. During the OFF period of $T - T_{on}$, when the chopper is OFF, load is disconnected from the supply. With inductive-type loads, generally a freewheeling diode (FD) is connected across the load, which allows the load current to freewheel through it and makes the output voltage zero during $T - T_{on}$.

The average value of load voltage can be derived now.

$$V_{o(av)} = \frac{1}{T} \int_0^T V_o \, dt = \frac{1}{T} \int_0^{T_{on}} V_{dc} \, dt$$

$$= \frac{V_{dc}}{T} [t]_0^{T_{on}} \tag{6.1}$$

$$V_{o(av)} = V_{dc} \frac{T_{on}}{T}$$

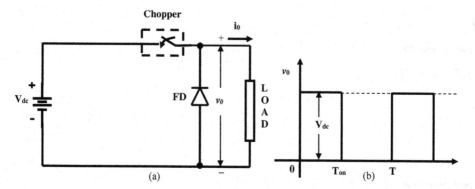

FIGURE 6.1
(a) Schematic of a chopper circuit. (b) Output voltage.

The ratio $\dfrac{T_{on}}{T}$ is known as the duty ratio and denoted as D. Further, it is evident that $V_{o(av)} < V_{dc}$, and hence this configuration is called as a step-down chopper.

The output voltage of a chopper can be controlled in two ways.

a. Pulse-Width Modulation Control: In this control strategy, the ON time, T_{on}, is varied, but the chopping frequency, f $\left(f = \dfrac{1}{T} \right)$, and hence the chopping period(T) is kept constant.

b. Frequency Modulation Control: In this type, the chopping frequency, f, is varied and either ON time, T_{on}, or OFF time $(T - T_{on})$ is kept constant.

Pulse width modulation is commonly employed for dc motor speed control whereas frequency modulation is suitable for switching power supplies.

EXAMPLE 6.1

A dc motor with a rated terminal voltage of 110 V is to be fed from a 220 V dc mains with the help of a dc/dc converter. Find the duty ratio required for the converter.

SOLUTION:

$$V_{in} = 220 \text{ V}, \ V_o = 110 \text{ V}$$
$$V_{o(av)} = D \cdot V_{in}$$
$$D = \frac{110}{220} \times 100 = 50\%$$

EXAMPLE 6.2

A separately excited dc motor fed from a 250-V mains through a dc/dc converter is running at 1,500 rpm and produces a back-emf of 195 V. Given an armature resistance 1 Ω, and the duty ratio of the chopper is 80%. Calculate the armature current.

SOLUTION:

$$V_{in} = 250 \text{ V} \quad E_b = 195 \text{ V} \quad D = 80\%$$

$$V_{o(av)} = D \cdot V_{in} = 0.8 \times 250 = 200 \text{ V}$$

$$I_a = \frac{V_0 - E_b}{R_a} = \frac{200 - 195}{1} = 5 \text{ A}$$

6.3 Steady-State Analysis of First-Quadrant Chopper-Fed dc-Motor Drive

The power circuit of a first-quadrant chopper feeding a dc motor load is shown in Fig. 6.2(a). The output voltage of the chopper along with the motor current is shown in Fig. 6.2(b). The magnitude and continuity of load current depends on the duty cycle, $\frac{T_{on}}{T}$, and load torque, T_L. In general, it can be said that, for $\frac{T_{on}}{T}$ and T_L becoming larger, motor current will be continuous.

The operation of chopper is analyzed with continuous and discontinuous modes of load current separately.

6.3.1 Continuous Current Operation

The variation of motor current is shown in Fig. 6.2(b). There are two modes of operation for the circuit and is given below.

Mode 1: $(0 < t < T_{on})$

During this mode, S_1, is closed, and Kirchhoff's voltage law can be written as

$$L_a \frac{di_a}{dt} + r_a i_a + E_b = V_{dc}$$

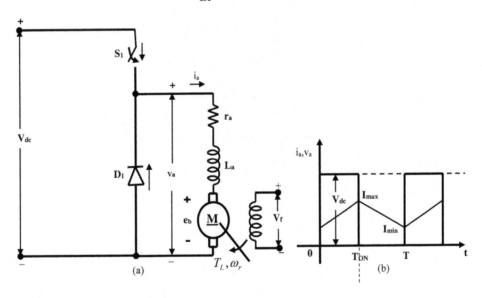

FIGURE 6.2
(a) Type-A chopper feeding the motor load. (b) Motor terminal voltage and current (continuous mode).

The resulting current will have steadystate as well as transient components and can be written as

$$i_a = K_1 e^{\frac{-r_a}{L_a}t} + \frac{V_{dc} - E_b}{r_a}$$

At $t = 0$, $i_a = I_{min}$ and we get

$$I_{min} = K_1 + \frac{V_{dc} - E_b}{r_a}$$

$$K_1 = I_{min} - \left[\frac{V_{dc} - E_b}{r_a} \right]$$

Hence,

$$i_a = \left[I_{min} - \left(\frac{V_{dc} - E_b}{r_a} \right) \right] e^{\frac{-r_a}{L}t} + \frac{V_{dc} - E_b}{r_a}$$

$$i_a = I_{min} e^{\frac{-r_a}{L}t} + \left[\frac{V_{dc} - E_b}{r_a} \right] \left(1 - e^{\frac{-r_a}{L_a}t} \right)$$

At $t = T_{on}$, $i_a = I_{max}$ and substituting,

$$I_{max} = I_{min} e^{\frac{-r_a}{L}T_{on}} + \left[\frac{V_{dc} - E_b}{r_a} \right] \left(1 - e^{\frac{-r_a}{L_a}T_{on}} \right) \tag{6.2}$$

Mode 2: Freewheeling Mode $(T_{on} < t < T - T_{on})$

During this mode, S_1 is opened and D_1 short circuits the load. Load current freewheels through D_1.

Consider a new time, $t' = t - T_{on}$; the relevant equation is

$$L_a \frac{di_a}{dt'} + i_a r_a + E_b = 0$$

The solution is

$$i_a = K_2 e^{\frac{-r_a}{L_a}t'} - \frac{E_b}{r_a}$$

At $t' = 0$, $i_a = I_{max}$ and the above equation becomes

$$I_{max} = K_2 - \frac{E_b}{r_a}$$

$$K_2 = I_{max} + \frac{E_b}{r_a}$$

Hence, $i_a = \left(I_{max} + \frac{E_b}{r_a} \right) e^{\frac{-r_a}{L_a}t'} - \frac{E_b}{r_a}$

$$i_a = I_{max} e^{\frac{-r_a}{L_a}t'} + \frac{E_b}{r_a} \left(e^{\frac{-r_a}{L_a}t'} - 1 \right)$$

At $t' = T - T_{on}$, $i_a = I_{min}$ and putting these values in the above equation yields

$$I_{min} = I_{max}e^{-\frac{r_a}{L_a}(T-T_{on})} + \frac{E_b}{r_a}\left[e^{-\frac{r_a}{L_a}(T-T_{on})} - 1\right] \qquad (6.3)$$

Now substituting I_{max} from Equation (6.2) into Equation (6.3),

$$I_{min} = \left[I_{min}e^{\frac{-r_a}{L_a}T_{on}} + \left(\frac{V_{dc} - E_b}{r_a}\right)\left(1 - e^{\frac{-r_a}{L_a}T_{on}}\right)\right]e^{\frac{-r_a}{L_a}(T-T_{on})} + \frac{E_b}{r_a}\left(e^{\frac{-r_a}{L_a}(T-T_{on})} - 1\right)$$

$$= I_{min}e^{\frac{-r_a}{L_a}T} + \left(\frac{V_{dc} - E_b}{r_a}\right)\left(e^{\frac{-r_a}{L_a}(T-T_{on})} - e^{\frac{-r_a}{L_a}T}\right) + \frac{E_b}{r_a}\left(e^{\frac{-r_a}{L_a}(T-T_{on})} - 1\right)$$

$$I_{min} - I_{min}e^{\frac{-r_a}{L_a}T} = \frac{V_{dc}}{r_a}\left[e^{\frac{-r_a}{L_a}(T-T_{on})} - e^{\frac{-r_a}{L_a}T}\right] - \frac{E_b}{r_a}\left(e^{\frac{-r_a}{L_a}(T-T_{on})} - e^{\frac{-r_a}{L_a}T} - e^{\frac{-r_a}{L_a}(T-T_{on})} + 1\right)$$

$$I_{min}\left(1 - e^{\frac{-r_a}{L_a}T}\right) = \frac{V_{dc}}{r_a}\left[e^{\frac{-r_a}{L_a}(T-T_{on})} - e^{\frac{-r_a}{L_a}T}\right] + \frac{E_b}{r_a}\left(1 - e^{\frac{-r_a}{L_a}T}\right)$$

$$I_{min} = \frac{V_{dc}}{r_a}\left[\frac{e^{\frac{-r_a}{L_a}(T-T_{on})} - e^{\frac{-r_a}{L_a}T}}{\left(1 - e^{\frac{-r_a}{L_a}T}\right)}\right] + \frac{E_b}{r_a}$$

Eliminating $e^{\frac{-r_a}{L_a}T}$ from the numerator and the denominator of the first term gives,

$$I_{min} = \frac{V_{dc}}{r_a}\left[\frac{e^{\frac{r_a}{L_a}T_{on}} - 1}{e^{\frac{r_a}{L_a}T} - 1}\right] - \frac{E_b}{r_a} \qquad (6.4)$$

Substituting Equation (6.3) in Equation (6.2) results in

$$I_{max} = \left[I_{max}e^{-\frac{r_a}{L_a}(T-T_{on})} + \frac{E_b}{r_a}\left(e^{\frac{-r_a}{L_a}(T-T_{on})} - 1\right)\right]e^{-\frac{r_a}{L_a}T_{on}} + \frac{V_{dc} - E_b}{r_a}\left(1 - e^{-\frac{r_a}{L_a}T_{on}}\right)$$

$$I_{max} = \left[I_{max}e^{-\frac{r_a}{L_a}T} + \frac{E_b}{r_a}\left(e^{-\frac{r_a}{L_a}T} - e^{-\frac{r_a}{L_a}T_{on}}\right)\right] + \frac{V_{dc}}{r_a}\left(1 - e^{-\frac{r_a}{L_a}T_{on}}\right) - \frac{E_b}{r_a}\left(1 - e^{-\frac{r_a}{L_a}T_{on}}\right)$$

$$I_{max}\left(1 - e^{-\frac{r_a}{L_a}T}\right) = \frac{V_{dc}}{r_a}\left(1 - e^{-\frac{r_a}{L_a}T_{on}}\right) - \frac{E_b}{r_a}\left(1 - e^{-\frac{r_a}{L_a}T}\right)$$

Hence

$$I_{max} = \frac{V_{dc}}{r_a}\left[\frac{1 - e^{-\frac{r_a}{L_a}T_{on}}}{1 - e^{-\frac{r_a}{L_a}T}}\right] - \frac{E_b}{r_a} \qquad (6.5)$$

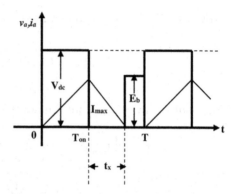

FIGURE 6.3
Motor terminal voltage and current (discontinuous mode).

6.3.2 Discontinuous Current Operation

The motor current need not be continuous as discussed in the previous section. For increased OFF time or reduced load torque, motor current becomes discontinuous. Typical motor terminal voltage and current are plotted in Fig. 6.3. During the freewheeling period, motor current becomes zero at $t = T_{on} + t_x$. When current falls to zero, D_1 stops conducting and v_a becomes E_b.

Then, $I_{min} = 0$ and from Equation (6.2),

$$I_{max} = \frac{V_{dc} - E_b}{r_a}\left(1 - e^{-\frac{r_a}{L_a}T_{on}}\right) \tag{6.6}$$

During the freewheeling period, $0 < t' \le t_x$,

$$L\frac{di_a}{dt'} + i_a r_a + E_b$$

where t_x is indicated in Fig. 6.3.

The equation for motor current is

$$i_a = Ke^{-\frac{r_a}{L_a}t'} - \frac{E_b}{r_a}$$

$$i_a = I_{max} \text{ at } t' = 0$$

$$I_{max} = K - \frac{E_b}{r_a}$$

$$K = I_{max} + \frac{E_b}{r_a}$$

$$i_a = \left(I_{max} + \frac{E_b}{r_a}\right)e^{-\frac{r_a}{L_a}t'} - \frac{E_b}{r_a}$$

$$i_a = 0, \ t' = t_x$$

$$0 = \left(I_{max} + \frac{E_b}{r_a}\right)e^{-\frac{r_a}{L_a}t^x} - \frac{E_b}{r_a}$$

$$\left(I_{max} + \frac{E_b}{r_a}\right)e^{-\frac{r_a}{L_a}t^x} = \frac{E_b}{r_a}$$

$$e^{-\frac{r_a}{L_a}t^x} = \frac{\dfrac{E_b}{r_a}}{\left(I_{max} + \dfrac{E_b}{r_a}\right)}$$

$$\frac{-r_a}{L_a}t_x = \ln\left[\frac{\dfrac{E_b}{r_a}}{\left(I_{max} + \dfrac{E_b}{r_a}\right)}\right]$$

$$= \ln\left[\frac{\dfrac{E_b}{r_a}}{\left(\dfrac{V_{dc} - E_b}{r_a}\right)\left(1 - e^{-\frac{r_a}{L_a}T_{on}}\right) + \dfrac{E_b}{r_a}}\right]$$

$$= \ln\left[\frac{\dfrac{E_b}{r_a}}{\dfrac{V_{dc}}{r_a}\left(1 - e^{-\frac{r_a}{L_a}T_{on}}\right) - \dfrac{E_b}{r_a}\left(1 - e^{-\frac{r_a}{L_a}T_{on}}\right) + \dfrac{E_b}{r_a}}\right]$$

$$t^x = -\frac{L_a}{r_a}\ln\left[\frac{\dfrac{E_b}{r_a}}{\dfrac{V_{dc}}{r_a}\left(1 - e^{-\frac{r_a}{L_a}T_{on}}\right) + \dfrac{E_b}{r_a}\left(1 - e^{-\frac{r_a}{L_a}T_{on}}\right)}\right]$$

Hence,

$$t^x = -\frac{L_a}{r_a}\ln\left[\frac{E_b}{V_{dc}\left(1 - e^{-\frac{r_a}{L_a}T_{on}}\right) + E_b\left(e^{-\frac{r_a}{L_a}T_{on}}\right)}\right] \tag{6.7}$$

For a given set of values of V_{dc}, E_b, T_{on}, T, R, and L, initially it is not known whether the current is continuous or discontinuous. One way to identify this is to compute t_x using Equation (6.7); if it is more than $T - T_{on}$, then the current is continuous. Then Equations (6.4) and (6.5) can be used to find I_{min} and I_{max}. If t_x is lower than $T - T_{on}$, then the current is discontinuous and $I_{min} = 0$ and I_{max} is obtained from Equation (6.6).

An alternative method to check whether the current is continuous or otherwise is to find I_{min} using Equation (6.4). If it is a positive value, then the current is continuous. If it is negative, then the current is discontinuous.

If the current is discontinuous, the chopper output voltage contains E_b during the current's zero interval. The average motor terminal voltage with discontinuous current is derived below:

$$V_{o(av)} = \frac{1}{T}\left[\int_0^{T_{on}} V_{dc}\,dt + \int_{T_{on}}^{T_{on}+t^x} 0.dt + \int_{T_{on}+t^x}^{T} E_b\,dt\right]$$

$$V_{o(av)} = \frac{1}{T}\left[V_{dc}T_{on} + E_b\left(T-\left(T_{on}+t^x\right)\right)\right] \tag{6.8}$$

$$V_{o(av)} = V_{dc}\frac{T_{on}}{T} + E_b\left(1-\frac{T_{on}}{T}-\frac{t^x}{T}\right)$$

EXAMPLE 6.3

A first-quadrant chopper drives a separately excited dc motor with the following details: $V_{dc} = 110$ V, $L_a = 1$ mH, $r_a = 0.25\ \Omega$, $E_b = 11$ V, $T = 4{,}500\ \mu s$, $T_{on} = 1{,}000\ \mu s$. Find I_{min}, I_{max}, and $V_{o(av)}$.

SOLUTION:

$$I_{min} = \frac{V_{dc}}{r_a}\left[\frac{e^{\frac{r_a}{L_a}T_{on}}-1}{e^{\frac{r_a}{L_a}T}-1}\right] - \frac{E_b}{r_a}$$

$$I_{min} = \frac{110}{0.25}\left[\frac{e^{\frac{0.25}{1\times10^{-3}}1000\times10^{-6}}-1}{e^{\frac{0.25}{1\times10^{-3}}4500\times10^{-6}}-1}\right] - \frac{11}{0.25}$$

$$I_{min} = 16.0759\,\text{A}$$

If I_{min} is positive, then the motor current is continuous.

$$I_{max} = \frac{V_{dc}}{r_a}\left[\frac{1-e^{\frac{-r_a}{L_a}T_{on}}}{1-e^{\frac{-r_a}{L_a}T}}\right] - \frac{E_b}{r_a}$$

$$I_{min} = \frac{110}{0.25}\left[\frac{e^{-\frac{0.25}{1\times10^{-3}}1000\times10^{-6}}-1}{e^{-\frac{0.25}{1\times10^{-3}}4500\times10^{-6}}-1}\right] - \frac{11}{0.25}$$

$$I_{max} = 100.11\,\text{A}$$

$$V_{o(av)} = V_{dc}\frac{T_{on}}{T}$$

$$= 110\times\frac{1000\times10^{-6}}{4500\times10^{-6}}$$

$$V_{o(av)} = 24.44\,\text{V}$$

EXAMPLE 6.4

A first-quadrant chopper drives a separately excited dc motor with the following details: $V_{dc} = 200$ V, $L_a = 2$ mH, $r_a = 1.0\ \Omega$, $E_b = 40$ V, $T = 10$ ms, $T_{on} = 7$ ms. Calculate I_{min}, I_{max}, and $V_{o(av)}$.

SOLUTION:

$$I_{min} = \frac{V_{dc}}{r_a} \left[\frac{e^{\frac{r_a}{L_a}T_{on}} - 1}{e^{\frac{r_a}{L_a}T} - 1} \right] - \frac{E_b}{r_a}$$

$$= \frac{200}{1} \left[\frac{e^{\frac{1}{2\times10^{-3}}\times7\times10^{-3}} - 1}{e^{\frac{1}{2\times10^{-3}}\times10\times10^{-3}} - 1} \right] - \frac{40}{1}$$

$$I_{min} = 3.572 \text{ A}$$

Positive sign of I_{min} indicates the motor current is continuous.

$$I_{max} = \frac{V_{dc}}{r_a} \left[\frac{1 - e^{\frac{-r_a}{L_a}T_{on}}}{1 - e^{\frac{-r_a}{L_a}T}} \right] - \frac{E_b}{r_a}$$

$$= \frac{200}{1} \left[\frac{1 - e^{-\frac{1}{2\times10^{-3}}\times7\times10^{-3}}}{1 - e^{-\frac{1}{2\times10^{-3}}\times10\times10^{-3}}} \right] - \frac{40}{1}$$

$$I_{max} = 155.276 \text{ A}$$

$$V_{o(av)} = V_{dc} \frac{T_{on}}{T}$$

$$= 200 \times \frac{7 \times 10^{-3}}{10 \times 10^{-3}}$$

$$V_{o(av)} = 140 \text{ V}$$

EXAMPLE 6.5

A first-quadrant chopper drives a separately excited dc motor with the following details: $V_{dc} = 200$ V, $L_a = 0.8$ mH, $r_a = 0.65$ Ω, $E_b = 120$ V, $T = 3$ ms, $T_{on} = 0.25$ ms. Find I_{min}, I_{max}, and $V_{o(av)}$.

SOLUTION:

$$I_{min} = \frac{V_{dc}}{r_a} \left[\frac{e^{\frac{r_a}{L_a}T_{on}} - 1}{e^{\frac{r_a}{L_a}T} - 1} \right] - \frac{E_b}{r_a}$$

$$= \frac{200}{0.65} \left[\frac{e^{\frac{0.65}{0.8\times10^{-3}}0.25\times10^{-3}} - 1}{e^{\frac{0.65}{0.8\times10^{-3}}3\times10^{-3}} - 1} \right] - \frac{120}{0.65}$$

$$I_{min} = -177.9857 \text{ A}$$

A negative sign indicates that the motor current is discontinuous.

Hence, $I_{min} = 0$. From Equation (6.6),

$$I_{max} = \frac{V_{dc} - E_b}{r_a}\left(1 - e^{-\frac{r_a}{L_a}T_{on}}\right)$$

$$= \frac{200 - 120}{0.65}\left(1 - e^{-\frac{0.65}{0.8\times10^{-3}}0.25\times10^{-3}}\right)$$

$$I_{max} = 22.6245 \text{ A}$$

$$t^x = -\frac{L_a}{r_a}\ln\left[\frac{E_b}{V_{dc}\left(1 - e^{-\frac{r_a}{L_a}T_{on}}\right) + E_b\left(e^{-\frac{r_a}{L_a}T_{on}}\right)}\right]$$

$$= -\frac{0.8\times10^{-3}}{0.65}\ln\left[\frac{120}{200\left(1 - e^{\frac{-0.65}{0.8\times10^{-3}}0.25\times10^{-3}}\right) + 120\left(e^{\frac{-0.65}{0.8\times10^{-3}}0.25\times10^{-3}}\right)}\right]$$

$$t^x = 0.1423 \text{ ms}$$

$$V_{o(av)} = V_{dc}\frac{T_{on}}{T} + E_b\left(1 - \frac{T_{on}}{T} - \frac{t^x}{T}\right)$$

$$= 200\frac{0.25\times10^{-3}}{3\times10^{-3}} + 120\left(1 - \frac{0.25\times10^{-3}}{3\times10^{-3}} - \frac{0.1423\times10^{-3}}{3\times10^{-3}}\right)$$

$$V_{o(av)} = 120.9812 \text{ V}$$

EXAMPLE 6.6

A first-quadrant chopper drives a separately excited dc motor with the following details: $V_{dc} = 110$ V, $L_a = 0.2$ mH, $r_a = 0.25$ Ω, $E_b = 40$ V, $T = 2,500$ μs, $T_{on} = 1,250$ μs. Find I_{min}, I_{max}, and $V_{o(av)}$.

SOLUTION:

$$I_{min} = \frac{V_{dc}}{r_a}\left[\frac{e^{\frac{r_a}{L_a}T_{on}} - 1}{e^{\frac{r_a}{L_a}T} - 1}\right] - \frac{E_b}{r_a}$$

$$I_{min} = \frac{110}{0.25}\left[\frac{e^{\frac{0.25}{0.2\times10^{-3}}1250\times10^{-6}} - 1}{e^{\frac{0.25}{0.2\times10^{-3}}2500\times10^{-6}} - 1}\right] - \frac{40}{0.25}$$

$$I_{min} = -83.754 \text{ A}$$

A negative sign indicates that the motor current is discontinuous.

Hence $I_{min} = 0$. Now, using Equation (6.6),

$$I_{max} = \left(\frac{V_{dc} - E_b}{R}\right)\left(1 - e^{-\frac{R}{L}T_{on}}\right)$$

$$= \left(\frac{110 - 40}{025}\right)\left(1 - e^{-\frac{0.25}{0.2 \times 10^{-3}}1250 \times 10^{-6}}\right)$$

$$I_{max} = 221.312 \text{ A}$$

$$t^x = -\frac{L_a}{r_a}\ln\left[\frac{E_b}{V_{dc}\left(1 - e^{-\frac{r_a}{L_a}T_{on}}\right) + E_b\left(e^{-\frac{r_a}{L_a}T_{on}}\right)}\right]$$

$$t^x = -\frac{0.2 \times 10^{-3}}{0.25}\ln\left[\frac{40}{110\left(1 - e^{\frac{-0.25}{0.2 \times 10^{-3}}1250 \times 10^{-6}}\right) + 40\left(e^{\frac{-0.25}{0.2 \times 10^{-3}}1250 \times 10^{-6}}\right)}\right]$$

$$t^x = 0.6948 \text{ ms}$$

$$V_{o(av)} = V_{dc}\frac{T_{on}}{T} + E_b\left(1 - \frac{T_{on}}{T} - \frac{t_x}{T}\right)$$

$$= 110\frac{1250 \times 10^{-6}}{2500 \times 10^{-6}} + 40\left(1 - \frac{1250 \times 10^{-6}}{2500 \times 10^{-6}} - \frac{0.6948 \times 10^{-3}}{2500 \times 10^{-6}}\right)$$

$$V_{o(av)} = 63.882 \text{ V}$$

6.4 Commutation Process in SCR-Based dc/dc Converters

The switching device employed in the dc/dc converter can be either SCR or self commutating devices such as insulated-gate bipolar transistors or power MOSFETs. When a SCR is used, an additional circuit that mainly consists of inductor L_c and capacitor C_c is employed. Furthermore, the commutation process can be either voltage commutation or current commutation. These two commutation processesas they occur in a dc/dc converter are explained in this section.

6.4.1 Voltage-Commutated First-Quadrant Chopper

In this section, the commutation of a first-quadrant chopper employing SCRs using voltage commutation is explained. The power circuit of a first-quadrant chopper together with

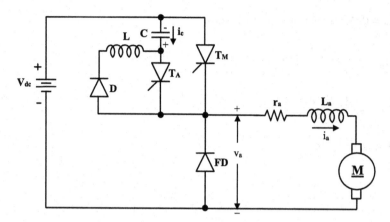

FIGURE 6.4
Circuit of SCR-based type-A chopper employing voltage commutation.

commutating elements is shown in Fig. 6.4. For clarity, the following notations are used in the circuit:

T_M: The main SCR carrying the load current

FD: Freewheeling diode

r_a, L_a: Resistance and self-inductance of dc motor

T_A: The auxiliary thyristor; commutation of T_M is initiated by triggering this SCR

D: Diode employed in the commutation network

C,L: Capacitance and inductance of commutation circuit; generally termed as commutating components.

The explanation of turning OFF of the main thyristor is carried out in a sequential way. It is important to mention that the duration of the whole commutation process is of the order of a few microseconds, which is very small in comparison to the electrical time constant of motor. Therefore, during the course of commutation, the load current is assumed to be constant. For better understanding, the whole process is split up into various modes, and variations of pertinent variables are plotted in Fig. 6.5. Initially the capacitor is charged with a voltage of V_{dc} with bottom plate positive and top plate negative as indicated in Fig. 6.4. This polarity is treated as negative, and the direction of capacitor current, i_c, as given in this figure is taken as positive. Various modes are described below.

Mode I: This mode refers to the instant just prior to the commutation of T_M and is shown in Fig. 6.6. This refers to the ON period of the chopper, and during this interval T_M is conducting and $v_a = V_{dc}$ as shown in Fig. 6.5.

Mode II: The commutation starts at the end of T_{on}. This is considered to be the reference time for the starting of commutation, such that $t = 0$ starts at T_{on}. At $t = 0$, the auxiliary thyristor T_A is triggered; this results in the application of capacitance voltage across T_M, reverse-biasing it and instantaneously turning it OFF. Once T_M is turned OFF by voltage commutation, the circuit then appears as given in the Fig. 6.7. The FD is reverse-biased due to the presence of V_{dc} together with the polarity of v_c and hence will not conduct. Thus the load current, which is constant at I_a, continues to flow through $V_{dc} - C - T_A - v_o$ and back to V_{dc}.

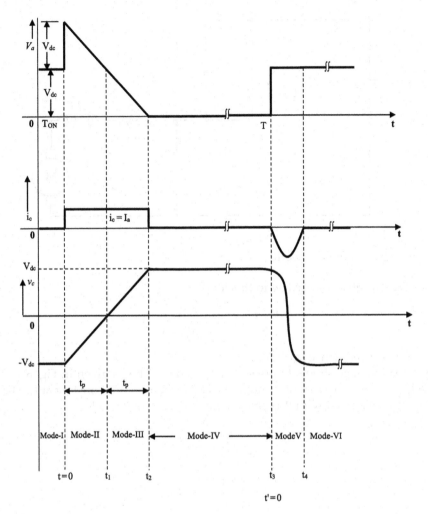

FIGURE 6.5
Variation of V_a, i_c, and V_c during commutation.

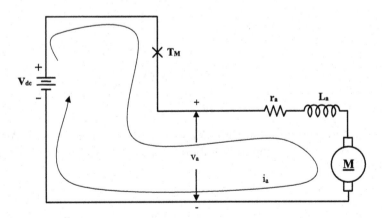

FIGURE 6.6
Equivalent circuit for mode-I.

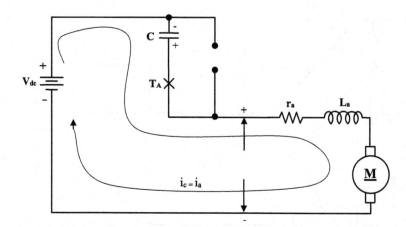

FIGURE 6.7
Equivalent circuit for mode II and mode III.

Applying Kirchhoff's voltage law to this loop,

$$V_{dc} + v_c - v_a = 0$$

$$v_a = V_{dc} + v_c$$

At $t = 0$, $v_c = V_{dc}$ and hence $v_a = 2V_{dc}$.

Thus the load voltage shoots to twice its rated value. This is given in Fig. 6.5. The load current, which is also the current through the capacitor, remains constant at I_a. The capacitor voltage can be derived with this charging current as shown below:

At $0 < t < t_1$

$$i_c = C \frac{dv_c}{dt}$$

$$v_c = \frac{1}{C} \int i_c \, dt + K_1$$

$$v_c = \frac{1}{C} \int I_a \, dt + K_1$$

$$v_c = \frac{I_a}{C} t + K_1$$

At $t = 0, v_c = -V_{dc}$

$$V_{dc} = K_1 \text{ and hence}$$

$$v_c = \frac{I_a}{C} t - V_{dc} \tag{6.9}$$

Thus, the capacitance is charging from $-V_{dc}$ to a positive voltage linearly. As v_c becomes zero at $t = t_p$, $v_c = V_{dc}$. Substituting in Equation (6.9), at $t = t_p$, $v_c = 0$, this time can be computed as follows:

$$V_{dc} = \frac{I_a}{C} t_p$$

$$t_p = \frac{V_{dc}}{I_a} C$$

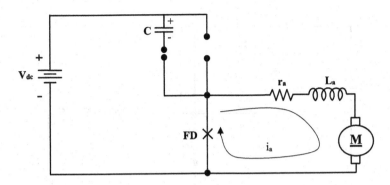

FIGURE 6.8
Equivalent circuit for mode IV.

Mode III: As the capacitance voltage increases from the zero value, the capacitor polarity is now interchanged, with the top plate becoming positive and bottom plate negative. At $t = 2t_p$, $v_c = V_{dc}$ and v_a goes to zero. This makes the FD takeover the load current, forcing T_A to switch OFF. Thus T_A is turned OFF and load current starts flowing through the FD. An equivalent circuit is shown in Fig. 6.8.

Mode IV: During this period, no SCR is conducting and the load current freewheels through the FD. This circuit is same as that in Fig. 6.8. Depending upon the nature of load torque as well as the length of the OFF period of the chopper, the current may be either continuous or discontinuous. If the current is continuous, the output voltage remains at zero value; otherwise it will be motor back-emf during current discontinuity.

Mode V: At the end of the OFF period of the chopper, the main SCR is again triggered to start the new cycle. The FD is turned OFF and load current starts from source V_{dc} and reaches the load through T_M, thus making $v_a = V_{dc}$. When T_M is triggered, due to the polarity of the capacitance voltage, diode D is forward-biased and makes a closed path for discharging of C through $T_M - D - L$ and back to C. This is shown in Fig. 6.9. The equation governing the capacitor current i_c is derived below:

$$L\frac{di_c}{dt'} + \frac{1}{C}\int i_c\, dt' + V_{dc} = 0, \quad \text{where} \quad t' = t - (T - T_{on})$$

$$L\frac{d^2 i_c}{dt'^2} + \frac{i_c}{C} = 0$$

$$\frac{d^2 i_c}{dt'^2} + \frac{i_c}{LC} = 0$$

$$D^2 = -\frac{1}{LC}$$

$$D = \pm\sqrt{\frac{-1}{LC}} = \pm\frac{j}{\sqrt{LC}} = \pm j\omega_{LC}$$

where

$$\omega_{LC} = \frac{1}{\sqrt{LC}}$$

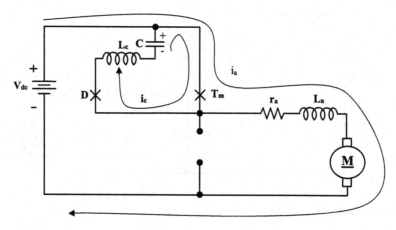

FIGURE 6.9
Equivalent circuit for mode-V.

Thus

$$i_c = K_1 \cos \omega_{LC} t' + K_2 \sin \omega_{LC}$$

at $t' = 0$, $i_c = 0$.
Hence $K_1 = 0$ and

$$i_c = K_2 \sin \omega_{LC} \qquad (6.10)$$

Differentiating Equation (6.9),

$$\frac{di_c}{dt} = K_2 \omega_{LC} \cos \omega_{LC} t'$$

Multiply L_a on both sides

$$L \frac{di_c}{dt} = L K_2 \omega_{LC} \cos \omega_{LC} t'$$

At $t' = 0$, $v_c = L \dfrac{di_c}{dt} = V_{dc}$. Substituting,

$$L K_2 \omega_{LC} \cos \omega_{LC} t' |_{t'=0} = V_{dc}$$

$$K_2 = \frac{V_{dc}}{\omega_{LC} L}$$

Thus Equation (6.10) becomes

$$i_c = \frac{V_{dc}}{\omega_{LC} L} \sin \omega_{LC} t'$$

$$i_c = V_{dc} \frac{\sqrt{LC}}{L} \sin \omega_{LC} t' \qquad (6.11)$$

$$i_c = V_{dc} \sqrt{\frac{C}{L}} \sin \omega_{LC} t'$$

The voltage across capacitance is

$$v_c = \frac{1}{C} \int i_c \, dt' + K_1$$

$$= \frac{1}{C} \int (-i_c) \, dt' + K_1$$

$$= \frac{1}{C} \int -\left[V_{dc} \sqrt{\frac{C}{L}} \sin \omega_{LC} t' \right] dt' + K_1$$

$$= \frac{1}{C} V_{dc} \sqrt{\frac{C}{L}} \frac{\cos \omega_{LC} t'}{\omega_{LC}} + K_1$$

$$v_c = V_{dc} \cos \omega_{LC} t' + K_1$$

At $t' = 0$, $V_c = V_{dc}$

$$V_{dc} = V_{dc} + K_1$$
$$K_1 = 0 \tag{6.12}$$
$$v_c = V_{dc} \cos \omega_{LC} t'$$

It is evident that the capacitor current is sinusoidal, and hence the voltage across the capacitance is a cosine wave as shown in Fig. 6.5. When the capacitor current passes through the natural zero value, diode D is turned OFF and the capacitance voltage is clamped at $-V_{dc}$. This voltage can be utilized for turning OFF of T_M again.

Mode VI: This is similar to mode I and is shown in Fig. 6.10.

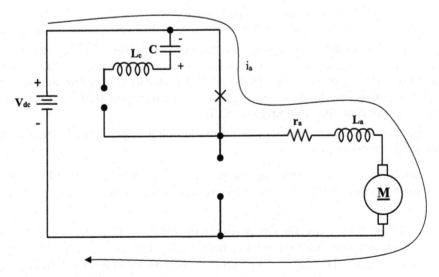

FIGURE 6.10
Equivalent circuit for mode VI.

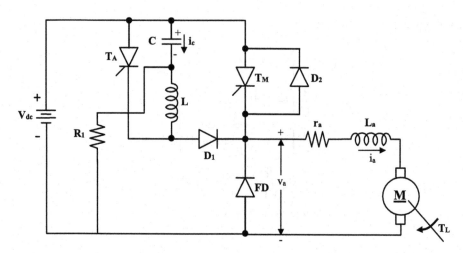

FIGURE 6.11
Type-A chopper using current commutation.

6.4.2 Current-Commutated First-Quadrant Chopper

It is also possible to turn OFF the main thyristor in the first-quadrant chopper using another commutation procedure called current commutation. The practical circuit employing this commutation is shown in Fig. 6.11. The labels of various components are explained below:

T_M: The main SCR carrying the load current

FD: Freewheeling diode

r_a, L_a: Resistance and self-inductance of dc motor

T_A: The auxiliary thyristor; commutation of T_M is initiated by triggering this SCR

D_1: Diode employed in the commutation network

D_2: Diode connected across main SCRT_M in such a way that the anode of one is connected to the cathode of other and vice versa; known as anti-parallel connection

R_1: Resistance that enables the capacitance voltage to be clamped at V_{dc} after the commutation process is over. The value of this resistor is selected such that the time constant of R_1C network is much larger than the duration of commutation (which is of the order of few microseconds); however, the time constant of R_1C is selected much smaller than the chopping period T.

C,L: Capacitance and inductance of commutation circuit; generally termed as commutating components.

The variations of v_c and i_c for the entire commutation interval is plotted in Fig. 6.12. The entire commutation process is divided into various modes, and each mode is elaborated below.

Mode I: This mode starts when the ON period of the chopper, T_{on}, is reached. The commutation is initiated by triggering the auxiliary thyristor T_A. The charged capacitor discharges through the inductor L. The circuit representation of this mode is shown in Fig. 6.13. The capacitor current, i_c, and voltage, v_c, are the same as those derived in Equations (6.11) and (6.12), respectively.

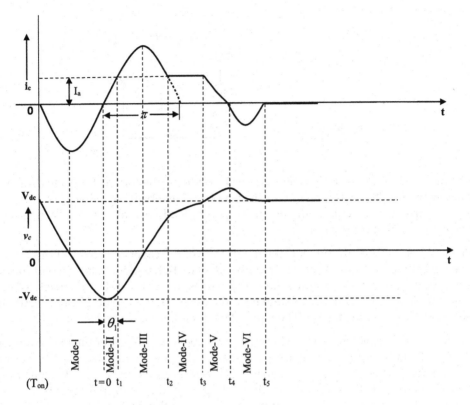

FIGURE 6.12
Capacitor current and voltage during commutation.

It is evident that the current is sinusoidal and that the capacitance voltage is a cosine wave. This is shown in Fig. 6.12. At the end of the half-cycle, the capacitor current becomes zero, and T_A is then turned OFF.

Mode II: From the figure, it is seen that v_c has become $-V_{dc}$ at the end of Mode I. Because T_A is turned OFF, the capacitor current now finds an alternate path through L-D_1-T_M-C. This is illustrated in Fig. 6.14. It may be noted that the capacitor is discharging again through the inductor, and hence the current will continue to be sinusoidal in nature. A new time axis is marked from Mode I onward such that t = 0 starts from this point onward. From the variation of i_c shown in Fig. 6.14, it is seen that, at $\omega_{LC}t_1 = \theta_1$, the capacitor current equals the load current, and thereby the net current flowing through T_M is reduced to zero.

Mode III: From Fig. 6.12, it is seen that the peak value of i_c is larger than the load current, and hence the excess current now passes through diode D_2. The circuit corresponding to

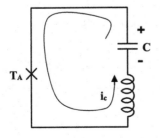

FIGURE 6.13
Circuit during mode I.

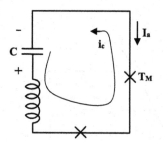

FIGURE 6.14
Circuit during mode II.

this mode is shown in Fig. 6.15. When diode D_2 conducts, the forward voltage drop across this device makes T_M in the reverse-biased mode, thus ensuring reliable commutation. When the capacitor current again becomes equal to load current at $t = t_2$, D_2 gets turned OFF. The duration of conduction of D_2 is $t_2 - t_1$.

Mode IV: When D_2 gets turned OFF, the magnitude of the capacitor voltage is less than V_{dc} and hence FD will remain in the reverse-biased state. Because the load current has to be continuous, it starts from the source now and passes through C-L-D_1-load back to source. Because the capacitor current is constant now, the capacitor voltage builds up linearly from the interval t_3 until it reaches V_{dc}. The equivalent circuit is shown in Fig. 6.16.

Mode V: Once the capacitor voltage shoots slightly above V_{dc}, consider the loop comprising source-C-L-D_1-FD. Because the current is constant, the voltage drop across L is zero and, neglecting the drop across D_1, FD gets forward-biased after mode IV. Now the load current freewheels through FD and may decay depending upon the motor parameters.

However, the capacitor current cannot drop to zero at this instant because L is in series with C. Hence, the capacitor discharges through the path starting from source-C-L-FD (FD is short-circuited now). Thus two currents exist in this mode, namely load current and capacitor current, which are independent of each other. After the duration of $(t_4 - t_3)$, the capacitor current becomes zero and D_1 is turned OFF. At the end of mode V, it is seen that the capacitor is charged to a potential slightly above source voltage V_{dc}. The circuit operation in modeV is represented in Fig. 6.17.

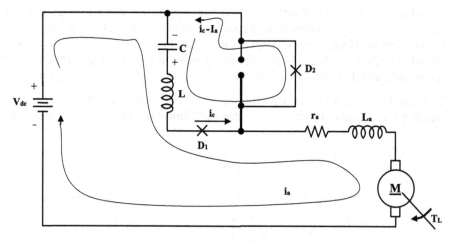

FIGURE 6.15
Circuit during mode III.

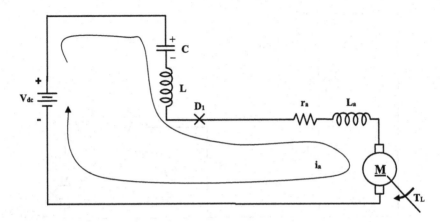

FIGURE 6.16
Circuit during mode IV.

Mode VI: Once D_1 is turned OFF at t_4, now consider the loop consisting of source voltage-C-R_1. The overcharged capacitor now discharges to the source through R_1. At the end of discharge, the capacitor voltage is clamped at V_{dc}. This mode is shown in Fig. 6.18. Load current freewheels through FD and may be continuous or discontinuous depending on the values of T_L and $T - T_{on}$.

6.5 Design of Commutation Elements

Referring to Fig. 6.12, it is evident that the peak value of capacitor current I_m should be larger than load current I_a. for reliable commutation. Let

$$I_m = xI_a, \quad \text{where} \quad x > 1. \tag{6.13}$$

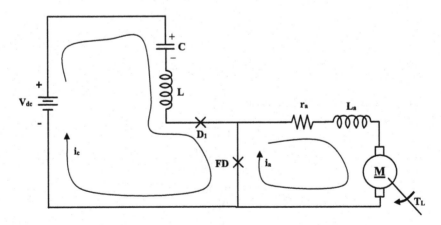

FIGURE 6.17
Circuit during mode V.

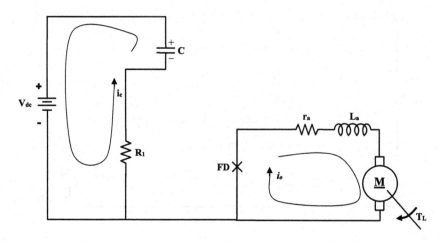

FIGURE 6.18
Circuit during mode VI.

Consider mode II. Here at $t = t_1$, i_c equals I_a:

$$I_a = I_m \sin\theta_1, \quad \text{where} \quad \theta_1 = \omega_{LC} t_1$$

$$\sin\theta_1 = \frac{I_a}{I_m}$$

$$\theta_1 = \sin^{-1}\frac{I_a}{I_m}$$

$$\theta_1 = \sin^{-1}\frac{1}{x}$$

$$\omega_{LC} t_1 = \pi - 2\theta_1$$

$$\omega_{LC} t_1 = \pi - 2\sin^{-1}\left(\frac{1}{x}\right)$$

The right side is a function of x—let it be written as F(x). Hence

$$\omega_{LC} t_1 = F(x)$$

$$t_1 = \frac{F(x)}{\omega_{LC}} = F(x)\sqrt{LC} \tag{6.14}$$

$$xI_a = V_{dc}\sqrt{\frac{C}{L}} \tag{6.15}$$

Dividing Equation (6.14) with Equation (6.15):

$$\frac{t_1}{xI_a} = \frac{F(x)\sqrt{LC}}{V_{dc}\sqrt{\dfrac{C}{L}}}$$

$$\frac{t_1}{xI_a} = \frac{F(x)L}{V_{dc}}$$

$$L = \frac{t_1}{F(x)}\frac{V_{dc}}{xI_a} \tag{6.16}$$

Further, it can be derived that

$$C = \frac{t_1}{F(x)} \frac{xI_a}{V_{dc}}$$ (6.17)

EXAMPLE 6.7

A first-quadrant chopper is feeding a separately excited motor rated 200 V, 10 A, 1,500 rpm. The chopper is supplied from a constant bus bar voltage of 300 V. If the SCR turn-OFF time is 38 μs, compute the value of commutation components.

$$L = \frac{t_1}{F(x)} \frac{V_{dc}}{xI_a}$$

where t_1 = circuit turn-OFF time (and must be greater than SCR turn-OFF time) and V_{dc} = source voltage:

$$\omega_{LC}t_1 = \pi - 2\sin^{-1}\left(\frac{1}{x}\right) = F(x)$$

I_a: Maximum load current
I_m: Peak capacitor current
V_{dc} = 300 V
$t_1 = 38 + 38 = 76$ μs
$x = 1.5$
$I_a = 10$ A

$$L = \frac{76 \times 10^{-6} \times 300}{\left(\pi - 2\sin^{-1}\left(\frac{1}{1.5}\right)\right) \times 1.5 \times 10}$$

$$L = 903.61 \ \mu H$$

$$C = \frac{t_1}{F(x)} \frac{xI_a}{V_{dc}}$$

$$C = \frac{76 \times 10^{-6} \times 1.5 \times 10}{\left(\pi - 2\sin^{-1}\left(\frac{1}{1.5}\right)\right) \times 300}$$

$$C = 2.259 \ \mu F$$

EXAMPLE 6.8

First-quadrant chopper using current-commutated SCRs is used to drive a 230-V, 25-A, 1,450-rpm separately excited dc motor. If the motor is supplied from a 110-V busbar and the turn-OFF time of the SCR is 25 μs, design the commutation circuit.

SOLUTION:

$$L = \frac{t_1}{F(x)} \frac{V_{dc}}{xI_a}$$

$$\omega_{LC} t_1 = \pi - 2\sin^{-1}\left(\frac{1}{x}\right) = F(x)$$

$V_{dc} = 110$ V
$t_1 = 25 + 25 = 50$ μs
$x = 1.5$
$I_a = 25$ A

$$L = \frac{50 \times 10^{-6} \times 110}{\left(\pi - 2\sin^{-1}\left(\frac{1}{1.5}\right)\right) \times 1.5 \times 25}$$

$$L = 87.19 \text{ μH}$$

$$C = \frac{t_1}{F(x)} \frac{xI_a}{V_{dc}}$$

$$C = \frac{50 \times 10^{-6} \times 1.5 \times 25}{\left(\pi - 2\sin^{-1}\left(\frac{1}{1.5}\right)\right) \times 110}$$

$$C = 10.13 \text{ μF}$$

EXAMPLE 6.9

The speed of a 230-V, 25-A, 150-rpm separately excited dc motor is controlled by connecting a current-commutated first-quadrant chopper in the armature. The input voltage to the chopper is 300 V. The circuit turn-OFF time is fixed at 25 μs and the peak capacitor current is twice the load current. Compute the values of commutating components.

SOLUTION:

$V_{dc} = 300$ V
$t_1 = 25$ ms
$x = 2$
$I_a = 25$ A

$$L = \frac{t_1}{F(x)} \frac{V_{dc}}{xI_a}$$

$$L = \frac{25 \times 10^{-6} \times 300}{\left(\pi - 2\sin^{-1}\left(\frac{1}{2}\right)\right) \times 2 \times 25}$$

$$L = 71.62 \text{ μH}$$

$$C = \frac{t_1}{F(x)} \frac{xI_a}{V_{dc}}$$

$$C = \frac{25 \times 10^{-6} \times 2 \times 25}{\left(\pi - 2\sin^{-1}\left(\frac{1}{2}\right)\right) \times 300}$$

$$C = 1.989 \text{ μF}$$

6.6 dc/dc Converters as Switching Mode Regulators

Several portable electronic devices, namely laptops and mobile phones, work on dc power suppliers powered from batteries. These electronic gadgets have many electronic circuits within them, and each one operates at different voltage levels. These different voltage levels are provided by dc/dc converters supplied from the dc source/battery. These converters operate at high frequencies of the range 10 KHz to 1 MHz, leading to the small size of inductors and capacitors. This section introduces operational features of three basic types of dc/dc converters.

6.6.1 Buck Converter

This is a step-down circuit, where the average output voltage is less than the input dc voltage. The circuit is shown in Fig. 6.19(a). In the absence of L and C, v_0 is the same as that shown in Fig. 6.1(a), which is undesirable for powering integrated circuits, biasing transistors, and similar applications. Hence, L and C are added to deliver a ripple-free dc voltage across the load.

When the switch is ON, the FD is reverse-biased; current flows from V_{dc} to load through inductor. The value of capacitor C is large so that v_0 is assumed constant throughout. During ON period, inductor current builds up as shown in Fig. 6.19(b), where it is assumed that i_L is continuous. Applying Kirchhoff's voltage law to the loop of V_{dc}, v_L, and v_o,

$$V_{dc} - v_L - v_0 = 0$$
$$v_L = V_{dc} - v_0$$
$$L\frac{di_L}{dt} = V_{dc} - v_o$$

(6.18)

For large C, $v_0 = V_0$ (constant). Hence,

$$\frac{di_L}{dt} = \frac{V_{dc} - V_0}{L}$$
$$i_L = \left(\frac{V_{dc} - V_0}{L}\right)t + K$$

(6.19)

where K is the constant of integration. At $t = 0$, $i_L = I_1$, and hence $K = I_1$.

$$i_L = \left(\frac{V_{dc} - V_0}{L}\right)t + I_1$$

(6.20)

When the switch is turned OFF, inductor current continues to flow through FD and load. Using the voltage law, it is seen that $v_L = -V_0$. Hence

$$L_a\frac{di_L}{dt'} = -V_0$$
$$i_L = \frac{-V_0}{L}t' + K_1$$

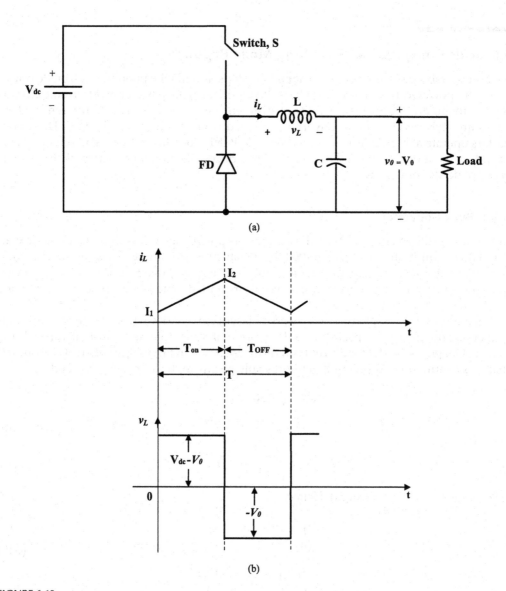

FIGURE 6.19

(a) Buck converter circuit. (b) Inductor current and voltage.

where $t' = t - T_{on}$ and K_1 is a constant. At $t' = 0$, $i_L = I_2$, and hence $K_1 = I_2$.

$$\therefore i_L = \frac{-V_0}{L} t' + I_2 \tag{6.21}$$

It is obvious that inductor current decays linearly as shown in Fig. 6.19(b).

Assuming an ideal inductor, the energy stored in the inductor during the ON period is transferred to the load when the switch is OFF. Because the average power in an ideal inductor is zero, and in the present case the average inductor current is non-zero, it implies that the time integral of the inductor voltage over one cycle of output voltage should be zero.

$$\int_{0}^{T} v_L dt = 0$$

$$\int_{0}^{T_{on}} v_L dt + \int_{T_{on}}^{T} v_L dt = 0$$

$$\int_{0}^{T_{on}} (V_{dc} - V_0) dt + \int_{T_{on}}^{T} (-V_0) dt = 0$$

$$(V_{dc} - V_0) T_{on} + (-V_0)(T - T_{on}) = 0$$

Solving this, we get

$$\frac{V_o}{V_{dc}} = \frac{T_{on}}{T} = D \tag{6.22}$$

Equation (6.21) shows that, with continuous inductor current, the output voltage varies linearly with the duty ratio, D, and its circuit parameters.

6.6.2 Boost Converter

In a boost regulator, the output voltage is higher than the input dc voltage. The circuit of the boost converter is presented in Fig. 6.20(a). The value of C is high enough to maintain a constant output voltage. Typical waveforms of inductor voltage and current are shown in Fig. 6.20(b).

When the switch is closed, V_{dc} appears across the inductor and i_L increases linearly. Thus,

$$v_L = V_{dc}, \ 0 < t < T_{on} \tag{6.23}$$

During the OFF period, the diode conducts and i_L flows through load. Due to the presence of V_0, i_L decays and the polarity of v_L reverses as shown in Fig. 6.20(b). Thus,

$$V_{dc} - v_L - v_0 = 0$$
$$v_L = V_{dc} - v_0 \tag{6.24}$$

It may be noted that $v_0 = V_0$ and $v_L = V_{dc} - V_0$.

Furthermore, i_L is unidirectional, and hence the average inductor voltage is zero. Thus,

$$\int v_L dt = 0$$

$$\int_{0}^{T_{on}} v_L dt + \int_{T_{on}}^{T} v_L dt = 0$$

$$\int_{0}^{T_{on}} V_{dc} dt + \int_{T_{on}}^{T} (V_{dc} - V_0) dt = 0 \tag{6.25}$$

$$V_{dc} T_{on} + (V_{dc} - V_0)(T - T_{on}) = 0$$

$$\frac{V_o}{V_{dc}} = \frac{1}{1 - D}$$

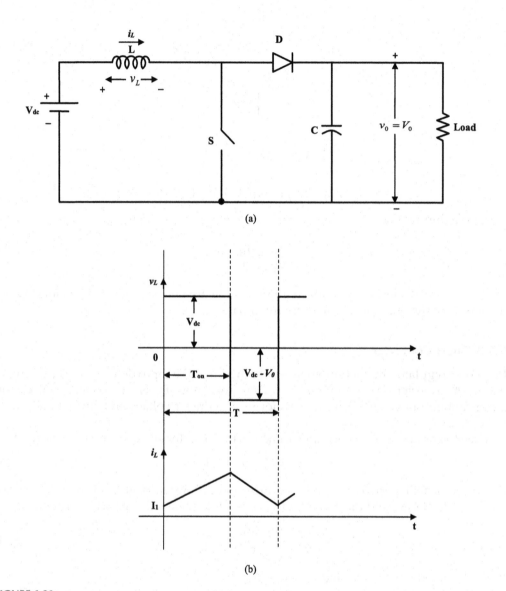

FIGURE 6.20
(a) Boost regulator. (b) Inductor voltage and current waveform.

Thus, for any value of $D = \dfrac{T_{on}}{T}$ between 0 and unity, the output voltage is larger than the input V_{dc}.

6.6.3 Buck-Boost Converter

This circuit is an integrated version of the previously described circuits and is employed when a negative-polarity output dc voltage is required. The amplitude of output voltage can be stepped up or stepped down as required. The buck-boost dc/dc converter is shown in Fig. 6.21(a). During the ON period, L is connected across V_{dc}; the diode is reverse-biased, which thereby disconnects the load from L. The equivalent circuit is shown in Fig. 6.21(c).

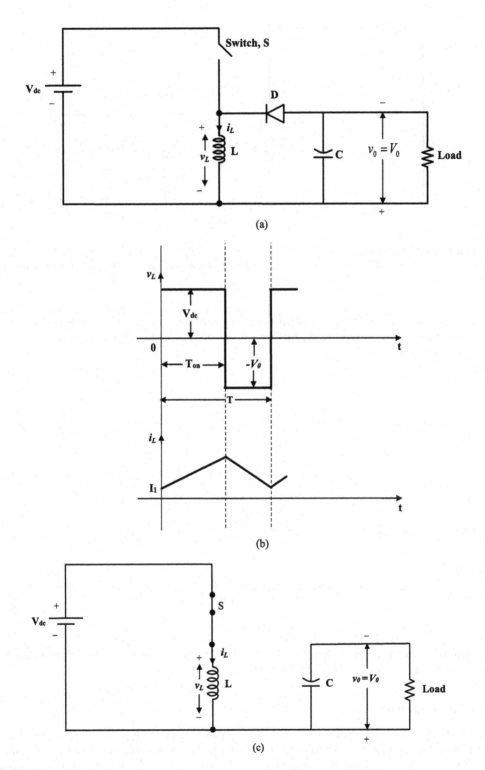

FIGURE 6.21
(a) Buck-boost converter. (b) Inductor voltage and current. (c) Circuit during the ON period. (d) Circuit when the switch is OFF.

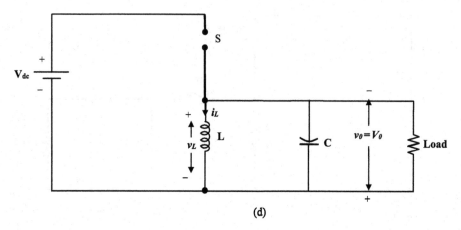

(d)

FIGURE 6.21 (*CONTINUED*)
(a) Buck-boost converter. (b) Inductor voltage and current. (c) Circuit during the ON period. (d) Circuit when the switch is OFF.

The inductor voltage is V_{dc} and i_L builds up linearly. During the OFF period, the circuit appears to the one shown in Fig. 6.21(d), and $v_L = -V_0$. As in the earlier section, the average inductor voltage is zero.

$$\int_0^T v_L\, dt = 0$$

$$\int_0^{T_{on}} v_L\, dt + \int_{T_{on}}^T v_L\, dt = 0$$

$$\int_0^{T_{on}} V_{dc}\, dt + \int_{T_{on}}^T (-V_0)\, dt = 0$$

$$V_{dc} T_{on} + (-V_0)(T - T_{on}) = 0$$

Hence,

$$\frac{V_o}{V_{dc}} = \frac{D}{1-D} \tag{6.26}$$

Equation (6.25) suggests that, up to $D = 0.5$, the circuit operates as a buck converter, and above this value it works in the boost mode.

6.6.4 Ćuk Converter

This dc/dc converter provides an adjustable-amplitude dc voltage of negative polarity at the output. The magnitude of the output dc voltage can be lower or higher than the input dc voltage depending upon the duty ratio, D. The circuit of a Ćuk converter is shown in Fig. 6.22(a). There are additional L and C filters in this configuration compared to the previous one, which ensures continuous inductor currents.

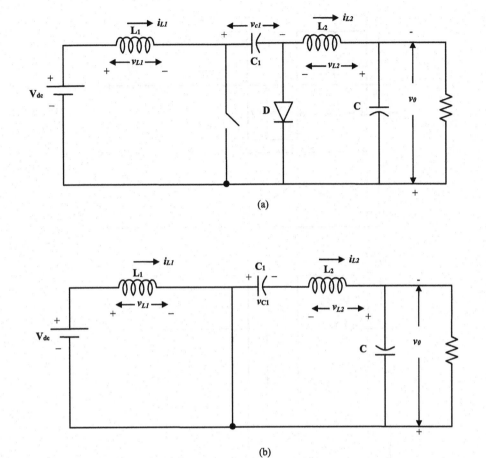

(a)

(b)

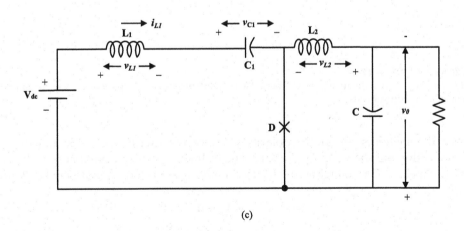

(c)

FIGURE 6.22
(a) Ćuk converter. (b) Circuit during the ON period. (c) Circuit during the OFF period. (d) Waveforms of voltage and currents of inductor coils.

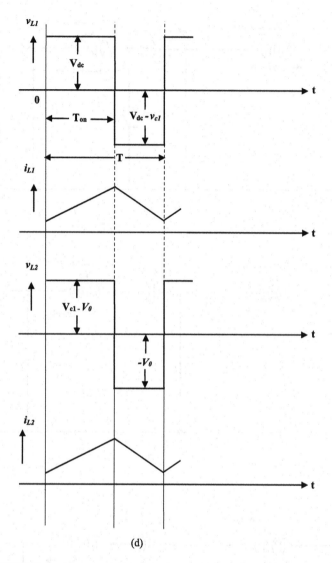

(d)

FIGURE 6.22 (*CONTINUED*)
(a) Ćuk converter. (b) Circuit during the ON period. (c) Circuit during the OFF period. (d) Waveforms of voltage and currents of inductor coils.

When the switch is closed, the polarity of V_{c1} reverse-biases the diode and the circuit appears as the one shown in Fig. 6.22(b). The inductor L_1 is connected across V_{dc}, and i_{L1} increases linearly. The voltage across L_1 is V_{dc}. Looking at the loop comprising L_2, C_1, and V_0,

$$-v_{c1} + v_{L2} + V_0 = 0$$
$$V_{L2} = v_{c1} - V_0 \tag{6.27}$$

During the OFF period, V_{L1} had been charged to a larger voltage than V_0, and hence v_{L2} is positive, indicating the linear rise of i_{L2}. The plots of inductor voltages and currents are shown in Fig. 6.22(d).

When the switch is turned OFF, the inductor currents force their path through the diode, D, and the circuit equivalent is represented as shown in Fig. 6.22(c). The drop across L_1 can be written as

$$V_{dc} - v_{L1} - v_{c1} = 0$$
$$v_{L1} = V_{dc} - v_{c1}$$

Thus, i_{L1} falls linearly.

When KVL is applied to the loop of L_2 and V_0, we get

$$-v_{L2} - V_0 = 0$$
$$v_{L2} = -V_0$$

This causes i_{L2} also to decay linearly.

It should be noted that during the ON period, inductor L_1 stores energy, and C_1 transfers its stored energy to both output and L_2. When the switch is turned OFF, C_1 gets charged from both L_1 and V_{dc}. The energy stored in L_2 appears at output.

The voltage transfer ratio of a Ćuk converter is obtained by equating each inductor voltage to zero.

For inductor L_1,

$$\int_0^T v_L \, dt = 0$$

$$\int_0^{T_{on}} V_{dc} \, dt + \int_{T_{on}}^T (V_{dc} - v_{c1}) dt = 0$$

$$V_{dc} T_{on} + (V_{dc} - v_{c1})(T - T_{on}) = 0 \qquad (6.28)$$

$$v_{c1} = \left(\frac{1}{1-D}\right) V_{dc}$$

For inductor L_2,

$$\int_0^{T_{on}} (v_{c1} - V_0) dt + \int_{T_{on}}^T (-V_0) dt = 0$$

$$(v_{c1} - V_0) T_{on} + (-V_0)(T - T_{on}) = 0 \qquad (6.29)$$

$$v_{c1} = \left(\frac{1}{D}\right) V_0$$

Equating Equation (6.28) to Equation (6.29) yields

$$\frac{V_0}{V_{dc}} = \frac{D}{1-D} \qquad (6.30)$$

This circuit can buck or boost the output voltage depending on the value of D.

Questions

1. What is dc/dc converter and principle of operation of dc/dc converter?
2. Analyze the steady state operation of first quadrant chopper fed DC motor drive.
3. Describe the different types of commutation process in brief.
4. Explain the different types of dc/dc converters.

Unsolved Problems

1. A first-quadrant chopper employing current commutation feeds a dc motor at half the rated load. The chopper is supplied from a 220-Vdc supply. The motor is rated at 220 V, 20 A, and 1,440 rpm. The armature resistance is 0.5 Ω. If commutation capacitance is 100 micro-farads, compute the minimum value of commutating inductance at all loads. For reliable commutation, capacitor current during commutation should be twice that of the current in the SCR to be turned off.

2. A first-quadrant chopper is feeding a dc motor load with the following specifications: supply voltage 220 V, $T_{on} = 3$ ms, $T = 10$ ms, motor resistance = 1 Ω, inductance = 45 mH, back-emf = 45 V. Compute the minimum and maximum values of the motor current.

3. A first-quadrant chopper is feeding a dc motor load with the following specifications: supply voltage 220 V, $T_{on} = 3$ ms, $T = 10$ ms, motor resistance = 2.3 Ω, inductance = 45 mH, back-emf = 150 V. Compute the minimum and maximum values of the motor current.

4. A dc motor with rated terminal voltage of 140 V is to be fed from a 220-V dc mains with the help of a dc/dc converter. Find the duty ratio required for the converter.

5. A separately excited dc motor fed from a 230-V mains through a dc/dc converter is running at 1,500 rpm and produces a back-emf of 120 V. Given an armature resistance of 2 Ω and a duty ratio of 65%, calculate the armature current.

Answers

1. 3.025 mH
2. $I_{min} = 15.9$ A, $I_{max} = 26.21$ A
3. I_{min} is negative, discontinuous conduction mode, $I_{max} = 4.326$ A
4. 0.636
5. 14.75 A

7

dc Motor Speed Control Employing dc/dc Converters

7.1 Introduction

When the existing power source is a dc grid or battery, a dc chopper or dc/dc converter can be used for either armature voltage control or field excitation control of a dc motor drive for variable-speed operation. Chopper circuits with a dc motor drive system are conveniently developed and analyzed for either first-, second-, or four-quadrant operation. In addition to regenerative braking, dynamic braking can also performed with dc motors using chopper circuits.

7.2 Classification of dc/dc Converters

Considering the average armature voltage, V_a, and average armature current, I_a, as the two deciding factors, the operation of a separately excited dc motor can be represented in the V-I diagram. Based on the V-I diagram, the dc/dc converters driving separately excited dc motors can be classified into 1-quadrant, 2-quadrant, and 4-quadrant chopper circuits. The characteristics of these dc/dc converters are explained in the following sections.

7.2.1 1-Quadrant Chopper

There are several low-power applications where speed reversal is not necessary and faster braking is also not required. In such cases, 1-quadrant choppers consisting of a single switching element and a freewheeling diode perform the operation. The switching element employed in the past was a silicon-controlled rectifier; however, due to the availability of high-power insulated-gate bipolar transistors (IGBTs) and reduced costs, these devices are now largely employed in dc/dc converters. Fig. 7.1(a) shows the circuit diagram of a 1-quadrant chopper. Here, when T_1 is turned ON, the armature voltage, v_a, is supply voltage, V_{dc}, and when this device is turned OFF, the armature current freewheels through diode D_1. The motor terminals are labeled as A and AA. The typical motor terminal voltage and current waveforms are depicted in Fig. 7.1(b), where armature current, i_a, is assumed to be continuous.

Depending on the value of the armature inductance as well as the load on the shaft, the armature current will be either continuous or discontinuous. If the armature current is discontinuous, torque and hence speed fluctuations may occur at the load side. Motor terminal voltage and armature current waveform for this case is shown in Fig. 7.1(c). Therefore,

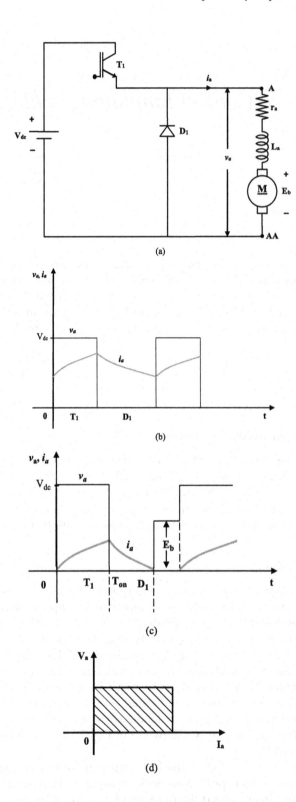

FIGURE 7.1
(a) A first-quadrant dc/dc converter-fed dc motor drive circuit. (b) Motor voltage and continuous armature current. (c) Motor voltage with discontinuous armature current. (d) V-I diagram.

sensitive loads may demand a continuous armature current, which must be ripple-free. However, this requires additional inductance to be inserted between the converter and the armature. It is evident that armature voltage and current are always positive, leading to first-quadrant operation as indicated in Fig. 7.1(d).

EXAMPLE 7.1

A 230-V, 6-A, 1,500-rpm separately excited dc motor has an armature resistance of 5.1 Ω. A 1-quadrant chopper supplied from a 300-V dc bus is operating at a duty ratio of 60% and supplies power to the motor armature at rated current. Compute the motor speed.

SOLUTION:

At rated condition,

$$E_{b\,rated} = V_a - I_{a\,rated} \times r_a$$
$$= 230 - (5.1 \times 6)$$
$$= 199.4 \text{ V}$$

$$E_{b\,rated} \propto 1,500 \text{ rpm} (N_{rated})$$

At 60% duty ratio, $E_{b1} \propto N_1$.

$$\text{Armature voltage} = 300 \times (60/100)$$
$$= 180 \text{ V}$$

$$\text{Back-emf}, E_{b1} = 180 - (5.1 \times 6)$$
$$= 149.4 \text{ V}$$

Therefore,

$$N_1 = \frac{E_{b1}}{E_{b\,rated}} N_{rated}$$
$$= \frac{149.4}{199.4} \cdot 1500 \approx 1124 \text{ rpm}$$

EXAMPLE 7.2

A first-quadrant dc/dc converter is fed from a 300-V dc bus. When the converter supplies power to a separately excited dc motor at 40% duty ratio, the average armature current is 5 A at 1,560 rpm. What is the duty ratio required to reduce the speed to 1,300 rpm for the same armature current? The armature resistance is 5.1 Ω.

SOLUTION:

At 40% duty ratio, armature voltage is

$$V_a = 300 \times \frac{40}{100}$$
$$= 120 \text{ V}$$

$$\text{Back-emf at 40\% duty ratio, } E_{b1} = 120 - (I_a r_a)$$
$$= 120 - (5 \times 5.1)$$
$$= 94.5 \text{ V}$$

$$E_{b1} \propto 1,560 \text{ rpm}$$

Now back-emf, E_{b2}, at 1,300 rpm can be related as $E_{b2} \propto 1,300$ rpm.

$$E_{b2} = \frac{N_2}{N_1} E_{b1} = 78.75 \text{ V}$$

$$\text{Armature voltage, } V_a = E_{b2} + I_a r_a$$

$$V_a = 300 \times D$$

Therefore,

$$D = 34.75\%$$

EXAMPLE 7.3

A separately excited dc motor is fed from a 440-V dc source through a single-quadrant chopper, $r_a = 0.2\ \Omega$, and armature current is 175 A. The voltage and torque constants are equal at 1.2 V/rad/s. The field current is 1.5 A. The duty cycle of chopper is 0.5. Find (a) speed and (b) torque.

SOLUTION:

a.
$$E_b = 220 - 175 \times 0.2 = 185 \text{ V}$$
$$= 1.2 \times \omega_r \times I_f$$
$$\omega_r = 185/(1.2 \times 1.5)$$
$$= 102.77 \text{ rad/s}$$
$$= 981.38 \text{ rpm}$$

b.
$$\text{Torque} = 1.2 \times 1.5 \times 175 = 315 \text{ N-m}$$

7.2.2 2-Quadrant Chopper

The circuit diagram of a 2-quadrant chopper is given in Fig. 7.2(a). Here, T_1-D_1 is used for the motoring mode, and T_2-D_2 is used for regenerative braking. Thus the converter is capable of forward rotation and regenerative braking. The voltage and current waveforms during motoring mode when T_1 and D_1 are conducting are sketched in Fig. 7.2(b). Here both voltage and current are positive and hence the drive is in the motoring mode.

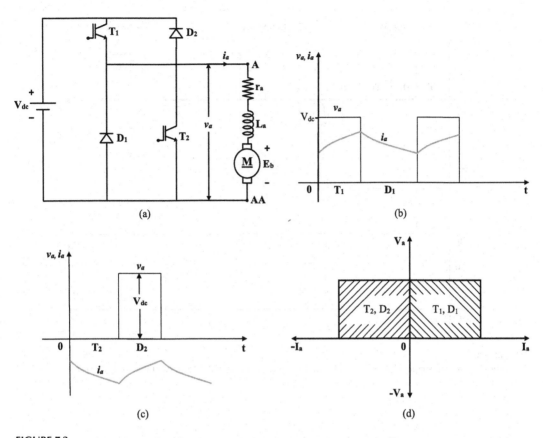

FIGURE 7.2
(a) A 2-quadrant chopper-fed dc motor circuit. (b) Motor voltage and armature current during motoring. (c) Motor voltage and armature current during regenerative braking. (d) V-I diagram.

Consider a case where the motor is in the running condition and the back-emf is higher, either due to increased field excitation or to downward movement of the motor. To apply regenerative braking, T_2 is now turned ON such that the armature current builds up in the opposite direction. Current i_a leaves A and enters AA and is negative. If T_2 is turned OFF, the armature current passes through D_2 and enters V_{dc}. Now the armature voltage is V_{dc} and the current is in the opposite direction, indicating regenerative braking. Typical motor voltage and current waveforms during regenerative braking are given in Fig. 7.2(c). Accordingly, the V-I diagram is given in Fig. 7.2(d).

7.2.3 Forward-Running and Regenerative Braking of dc Motor with a 2-Quadrant Chopper

Figure 7.2(a) shows a 2-quadrant chopper used for motoring the drive as well as regenerative braking. During the motoring mode, T_1 and D_1 are operated such that v_a and i_a are positive, thus the average output power is positive. When regenerative braking is intended, T_1 and D_1 are kept OFF and the braking is carried out through T_2 and D_2. Figure 7.3 explains the transition process from motoring to regenerative braking.

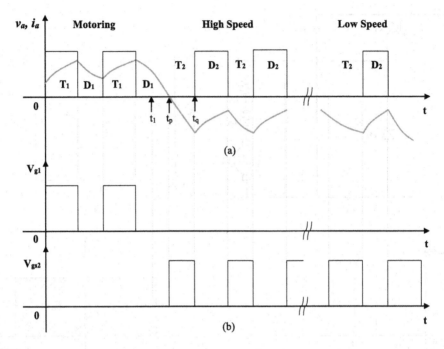

FIGURE 7.3
Transition from motoring mode to regenerative braking mode. (a) Motor voltage and current waveforms. (b) Gating signals.

Consider Fig. 7.3, where first-quadrant operation is in progress. At the end of first-quadrant operation, the controller (not shown in Fig. 7.3) starts the process of regenerative braking. Let the regenerative braking start at $t = t_1$. As a first step toward regenerative braking, the power switch is T_2 gated. At $t = t_1$, because D_1 is conducting, the voltage drop across D_1 does not allow T_2 to turn ON. At $t = t_p$, armature current i_a falls to zero and, because T_2 is already ON, i_a reverses its direction and passes from E_b-L_a-r_a-T_2 and back to E_b. The exponential increase of current is obtained by solving Equation (7.1).

$$L_a \frac{di_a}{dt} + i_a r_a = E_b,   \tag{7.1}$$

When armature current reaches sufficient amplitude at $t = t_q$, T_2 is turned OFF and i_a continues in the reverse direction from E_b-L_a-r_a-D_2-V_{dc} back to E_b. During this interval, $v_a = V_{dc}$ and i_a is negative, indicating regenerative braking. During the interval of regenerative braking, i_a falls and T_2 is again turned ON to boost the amplitude of i_a. During the process of regenerative braking, speed falls and, to maintain the reverse direction of i_a, E_b needs to be increased. This naturally happens in the case of downward movement; otherwise field excitation needs to be increased so that the amplitude of the armature current is retained at its rated value for faster braking. During regenerative braking, the voltage pulse width is suitably varied so that the average value of the motor current is retained at rated value. In other words, at the start of braking, the motor speed is higher and hence T_2 is gated for shorter durations. As braking proceeds, speed falls and T_2 is gated for longer durations. This is shown in Fig. 7.3.

EXAMPLE 7.4

A separately excited dc motor has the following name plate data: 220 V, 100 A, 2,200 rpm. The armature resistance is 0.1 Ω, and inductance is 5 mH. The motor is fed by a chopper that is operating from a dc supply of 250 V. Due to restrictions in the power circuit, the chopper can be operated over a duty cycle ranging from 30% to 70%. Determine the range of speeds over which the motor can be operated at rated torque.

SOLUTION:

Because e torque is constant, i_a is the same for all the values of D.

$$V_{o(av)} = DV_{dc}$$

At D = 0.3,

$$V_{o(av)} = 0.3 \times 250$$
$$= 75 \text{ V}$$
$$E_{b(0.3)} = V_{o(av)} - I_a r_a$$
$$= 75 - (100 \times 0.1)$$
$$= 65 \text{ V}$$

At D = 0.7,

$$V_{o(av)} = 0.7 \times 250$$
$$= 175 \text{ V}$$
$$E_{b(0.7)} = 175 - 10$$
$$= 165 \text{ V}$$

Under rated conditions, V_a = 220 V, I_a = 100 A, r_a = 0.1 Ω

$$E_{b(rated)} = 220 - (100 \times 0.1)$$
$$= 210 \text{ V}$$

$$N_r = 2200 \text{ rpm}$$

$$\frac{N_{0.7}}{N_r} = \frac{E_{b(0.7)}}{E_{b(rated)}}$$

$$N_{(0.7)} = \frac{165}{210} \times 2200$$
$$= 1,728.5714 \text{ rpm}$$

$$N_{(0.3)} = \frac{65}{210} \times 2200$$
$$= 680.95 \text{ rpm}$$

Hence speed can be varied in the range $680.95 \leq N \leq 1728.5714$.

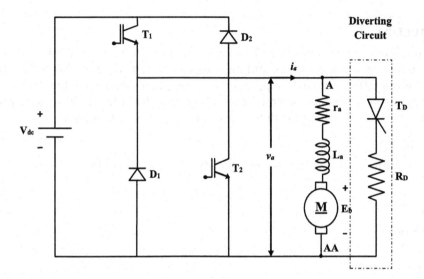

FIGURE 7.4
Diverter circuit.

7.3 Nonreceptive Lines

The regenerative braking causes the kinetic energy of the motor to be fed back to the dc mains. This power will be utilized by the other traction motors, which are either accelerating or running at constant speed. It is also possible that several of the trains are stopped by regenerative braking, and the power thus regenerated may not find takers. If the regenerated power is less utilized, then the dc bus bar voltage increases. Generally, an increase up to 20% is permitted, beyond which the safety relays will operate. Thus regenerative braking needs to be carried out by sensing the dc bus bar voltage. If the dc bus bar voltage is found to increase during regenerative braking, then the regenerative process must be stopped. Because the locomotive has to be stopped at the correct platform, the regenerative braking is suspended and dynamic braking is initiated. The thyristor T_D and resistor R_D as shown in Fig. 7.4 are employed for dynamic braking. This circuit is called a diverter circuit because the power flow is diverted to R_D through this process.

7.4 4-Quadrant Chopper

The circuit diagram to facilitate 4-quadrant operation is given in Fig. 7.5(a). This circuit contains four controlled switches and four diodes. Here, v_a and i_a are indicated with reference to motor terminals A and AA. The controlled switches are realized using IGBTs in this chapter. This circuit is capable of providing motoring, regenerative braking, operation of the motor in the reverse speed and regenerative braking in that direction corresponding to the four quadrants of V-I diagram, which is shown in Fig. 7.5(b). The following section explains the converter fed drive characteristics in four different quadrants.

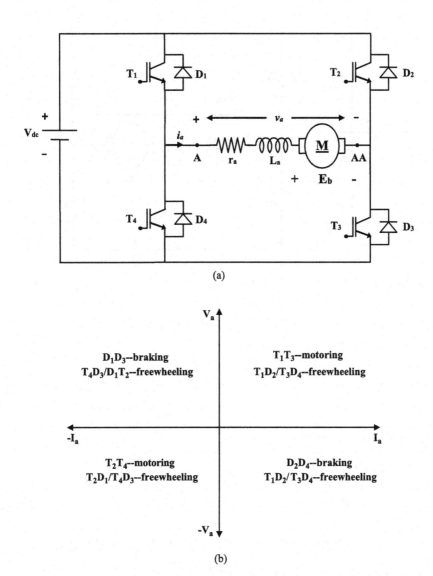

(a)

(b)

FIGURE 7.5
(a) 4-quadrant chopper circuit. (b) V-I diagram.

7.4.1 Motoring in the Forward Direction

Fig. 7.6(a) shows the devices conducting during the motoring of the drive in the forward direction. During the period T_{on} of the converter, the IGBTs T_1 and T_3 are simultaneously gated so that $v_a = V_{dc}$ and the armature current i_a is positive. During the OFF period of the dc/dc converter, T_1 alone is switched OFF, such that the armature current i_a now freewheels through T_3 and D_4, making the motor terminal voltage zero. The motor terminal voltage and current waveforms for continuous mode are shown in Fig. 7.6(b). As seen in this figure, the average values of voltage and current are positive, thus the average output power is always positive, leading to first-quadrant operation as indicated in Fig. 7.6(c). It may be noted that freewheeling of the armature current is also possible through D_2 and T_1, which is indicated by a dotted line.

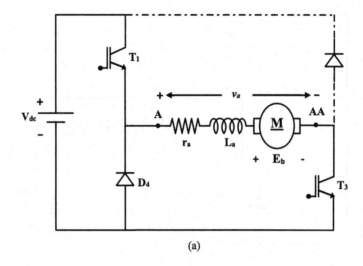

(a)

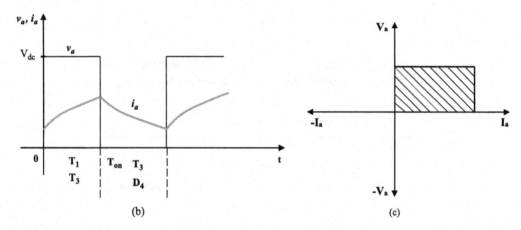

(b)   (c)

FIGURE 7.6
First-quadrant circuit. (a) Equivalent circuit. (b) Motor voltage and current waveforms. (c) V-I diagram.

7.4.2 Regenerative Braking after Forward Rotation

When regenerative braking is required, the IGBT T_1 is switched OFF. For the regenerative braking to take place, the electromagnetic torque (T_e) must be made negative. Because $T_e \propto (\Phi_m I_a)$, reversal of polarity of T_e is possible by changing the polarity of either Φ_m or I_a. We consider the case of reversal of I_a alone because reversal of Φ_m requires more time due to the increased field time constant.

Consider Fig. 7.7(a). Assume that freewheeling was taking place through T_3 and D_4 and that the regenerative braking command has come during the freewheeling action. To initiate the armature current reversal process during regenerative braking, T_4 is triggered and the gating signal to T_3 is continued till armature current goes to zero. Although T_4 is gated, the forward voltage drop across D_4 prevents T_4 from conducting and, as such, T_3-D_4 continues to cause freewheeling armature current. When this current drops to zero, the back-emf E_b causes reversal of armature current through T_4 and D_3. Armature current, i_a now flows from A to AA and is in the negative direction as shown in Fig. 7.7(b). This current rises exponentially, and when T_4 is turned OFF, the armature current maintains its direction

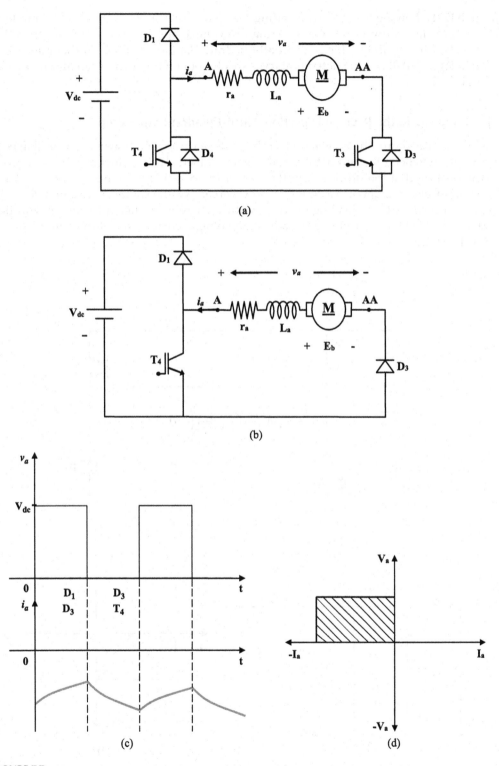

FIGURE 7.7
Second-quadrant operation. (a) Circuit during transition. (b) Circuit during regenerative braking. (c) Motor voltage and current waveforms. (d) V-I diagram.

through D_3-D_1 flowing to the source leading to regenerative braking. Pulse-width modulation of T_4 results in uniform braking. Armature voltage and current during regeneration are sketched in Fig. 7.7(c). This is a second-quadrant operation in the V-I diagram and is given in Fig. 7.7(d). It may be noted that free-wheeling of motor current can also take place through D_1-T_2.

7.4.3 Motoring in the Reverse Direction/Third-Quadrant Operation

Now the motor should be accelerated in the reverse direction. To achieve this, IGBTs T_2 and T_4 are switched ON, and both v_a and i_a get reversed. The product of v_a and i_a is positive, indicating that the process is in the motoring operation. Because i_a is reversed, the direction of electromagnetic torque is also reversed, so the motor is accelerated in the opposite direction. Freewheeling of the armature current can take place either through T_4-D_3 or D_1-T_2. The power circuit, steady-state voltage, current waveforms, and V-I diagram are given in Fig. 7.8.

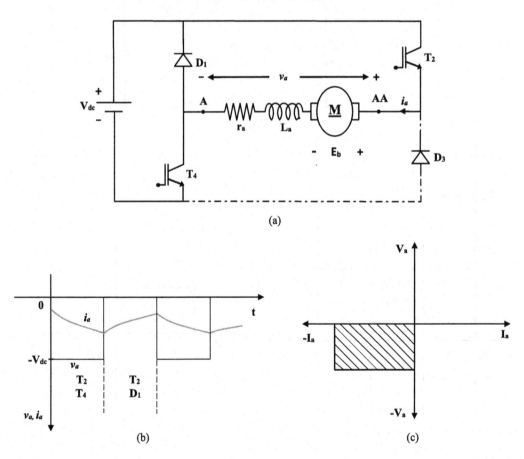

(a)

(b)

(c)

FIGURE 7.8
Motoring in reverse-braking. (a) Equivalent circuit. (b) Motor voltage and current waveforms. (c) V-I diagram.

7.4.4 Regenerative Braking after Speed Reversal

A process similar to that for second-quadrant operation takes place for the regenerative braking mode. Referring to Fig. 7.9(a), D_1-T_2 is the path of freewheeling, and when T_1 is triggered for regenerative braking, T_1 is prevented from conducting because of the forward voltage drop across D_1. The devices D_1-T_2 stop conducting when the armature current becomes zero, and the back-emf now drives the armature current through D_2 and T_1. Now armature current i_a flows from A to AA and hence is positive. When T_1 is turned OFF,

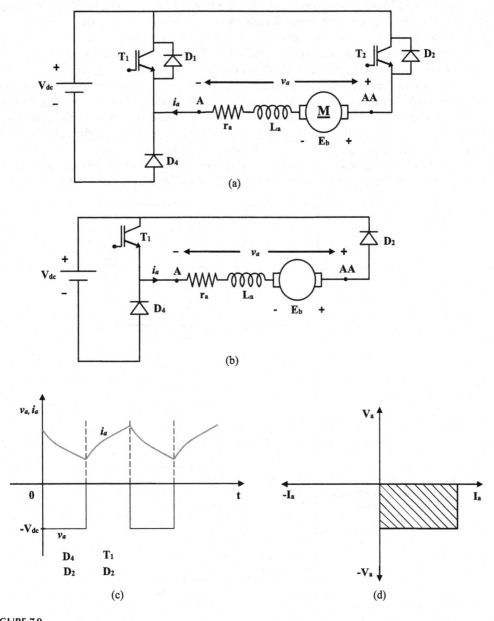

FIGURE 7.9
Fourth-quadrant operation. (a) Circuit during transition. (b) Circuit during regenerative braking. (c) Motor voltage and current waveforms. (d) V-I diagram.

regeneration takes place through D_4-D_2 and leads to fourth-quadrant operation. The power circuit, typical steady-state waveforms, and V-I diagram are given in Fig. 7.9.

7.5 Dynamic Braking

Traditionally, a motor is stopped by employing friction brakes, and the magnitude of braking torque is controlled through pneumatic air pressure. While frictional air braking is highly reliable and efficient, the major drawback lies in frequent repair and replacement of braking shoes because brakes are subjected to wear and tear during braking.

An alternative is to use electric braking, which may be dynamic braking or regenerative braking. In dynamic braking, the armature is disconnected from the supply and reconnected across a variable resistance. This resistance can be starter resistance with suitable modifications in connection. A simple schematic is shown in Fig. 7.10(a). Here R_b is braking resistance and can vary. The braking torque is $T_b \propto \Phi_m I_a$ and depends on field excitation and armature current. Ideally, the variation of braking torque (T_b) with the speed must be constant as indicated in Fig. 7.10(b). At high speeds, R_b is kept constant and reduction in I_a is compensated by increasing the field flux up to the saturation level. To stop, R_b is decremented and, when the speed becomes very low, mechanical brakes are applied.

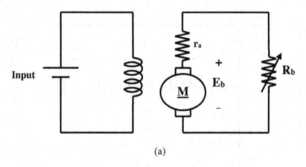

(a)

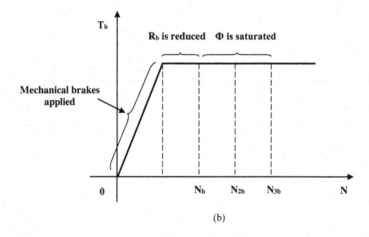

(b)

FIGURE 7.10
(a) Dynamic braking circuit. (b) Ideal characteristic.

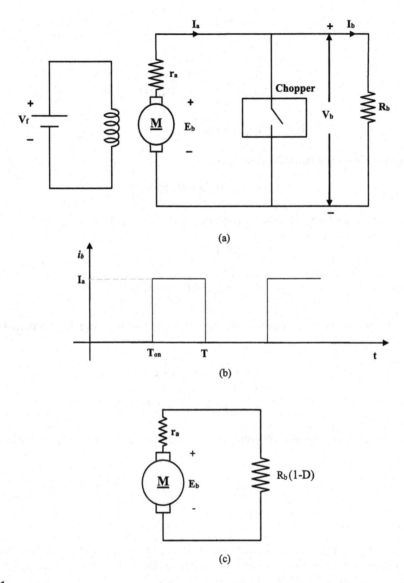

(a)

(b)

(c)

FIGURE 7.11
(a) Dynamic (rheostatic) braking circuit. (b) Current waveforms in R_b. (c) Equivalent circuit.

To achieve the characteristics shown in Fig. 7.10(b), a simple schematic of a chopper-controlled rheostat braking is given in Fig. 7.11(a), and the current through R_b is given in Fig. 7.11(b). The rms value of current can be written as

$$I_{b(rms)} = \sqrt{\frac{1}{T} \int_0^T (i_a)^2 \, dt} \tag{7.2}$$

$$= \sqrt{\frac{1}{T} \int_{T_{on}}^T (i_b)^2 \, dt}$$

$$I_{b(rms)} = \sqrt{\frac{I_a^2}{T}[t]_{T_{on}}^T}$$

$$I_{b(rms)} = \sqrt{\frac{I_a^2}{T}(T - T_{on})} = I_a \sqrt{\left(1 - \frac{T_{on}}{T}\right)}$$

$$I_{b(rms)} = I_a \sqrt{(1-D)} \tag{7.3}$$

Power dissipated in the braking resistance,

$$P_b = I_{b(rms)}^2 R_b = I_a^2 R_b (1-D) \tag{7.4}$$

Let $R_{b(eff)}$ be the effective value of R_b with dc/dc converter. Then

$$I_a^2 R_{b(eff)} = P_b$$

$$\therefore R_{b(eff)} = \frac{P_b}{I_a^2} = R_b (1-D) \tag{7.5}$$

Accordingly, the equivalent circuit during regenerative braking is given in Fig. 7.11(c). Referring to this figure,

$$E_b = I_a r_a + I_a R_b (1-D)$$
$$= I_a (r_a + R_b (1-D)) \tag{7.6}$$

Similarly, the average values of voltage and current across braking resistance can be expressed as

$$V_{b(avg)} = V_a (1-D) \tag{7.7}$$

$$I_{b(avg)} = I_a (1-D) \tag{7.8}$$

While the circuit in Fig. 7.11 works satisfactorily, it may be noted that during braking the armature voltage is higher than the rated value; to reduce the voltage across the chopper, the braking resistance is divided into R_{b1} and R_{b2} as shown in Fig. 7.12. Here there are two

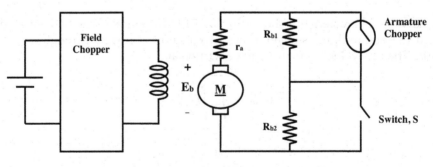

FIGURE 7.12
Dynamic braking with two braking resistors.

choppers, namely the field chopper and the armature. A mechanical switch S is also used. At the start of braking, the armature chopper and switch S are kept open so that armature current is limited by $r_a + R_{b1} + R_{b2}$. During this period, field chopper is controlled to maintain constant braking torque. Once the field flux is saturated, the armature chopper is operated with a variable duty ratio so that braking resistance is $r_a + R_{b1} + R_{b2}$ or $r_a + R_{b2}$. At low speeds, the mechanical switch S is permanently closed and the armature chopper controls the effective value of R_{b1} to provide constant braking torque.

EXAMPLE 7.5

A separately excited dc motor has an armature resistance 2.3 Ω, and armature current is 100 A. (a) Find the voltage across the braking resistance for a duty ratio of 25%. (b) Find the power dissipated in braking resistance.

SOLUTION:

$$\text{Average current} = I_a(1-D) = 100(1-0.25) = 75 \text{ A}$$

$$\text{Average Voltage} = I_{b(avg)} \times R_b = 75 \times 2.3 = 172.5 \text{ V}$$

$$P_b = I_a^2 R_b (1-D)$$

$$P_b = 100^2 \times 2.3 \times (1-0.25) = 17250 \text{ W}$$

EXAMPLE 7.6

A separately excited dc motor has the following name plate data: 200 V, 75 A, and 1,500 rpm. The armature resistance is 0.2 Ω. If dynamic baking takes place at 600 rpm at rated torque, compute the duty ratio. The braking resistance is 5 Ω.

SOLUTION:

E_b under rated condition,

$$E_{b(rated)} = V - I_a r_a$$
$$= 200 - (75 \times 0.2)$$
$$= 185 \text{ V}$$

$$\frac{E_{b(600)}}{E_{b(rated)}} = \frac{N}{N_{rated}}$$

$$E_{b(600)} = \left(\frac{N}{N_{rated}}\right) E_{b\,rated}$$

$$E_{b(600)} = \left(\frac{600}{1500}\right) \times 185 = 74 \text{ V}$$

Now,

$$E_{b(600)} = I_a \left(r_a + R_b(1-D)\right)$$

Because braking takes place at rated torque, $I_a = I_{arated} = 75$ A,

$$\text{i.e., } E_{b(600)} = 75\big(0.2 + 5(1 - D)\big)$$

$$74 = 75 \times 0.2 + 75 \times 5(1 - D)$$

$$\therefore D = 0.84$$

7.6 Dual Input dc/dc Converter

In the case of a propulsion drive, two or more dc sources power the vehicle. One source can be a renewable energy source such as a photovoltaic system, and the other source can be a reliable one, like a battery. The two sources supply power to the vehicle depending on the availability. A switching pattern of two dc/dc converters can be modulated in such a way that the renewable power source supplies power as long as it is available, and the balance is met from the battery. A simple diagram of a dual-input buck converter supplying a dc motor load is given in Fig. 7.13(a). Typical load voltage is given in Fig. 7.13(b). Here, D_1 is the duty ratio of switch S_1 and D_2 is that of S_2.

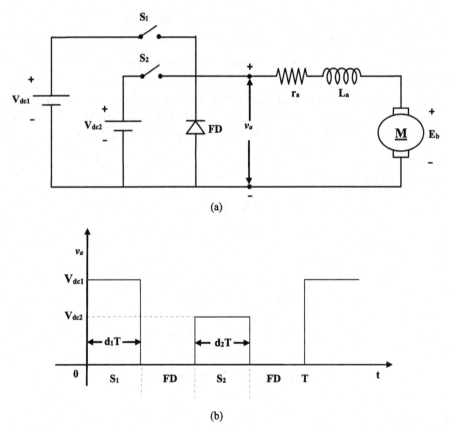

(a)

(b)

FIGURE 7.13
Dual-input dc/dc converter (a) circuit and (b) motor terminal voltage.

EXAMPLE 7.7

A dual-input dc/dc converter is supplied from two dc sources: 12-V and 24-V batteries. The duty ratio of the power switch connected to the first source is 40%, while that of the second source is 25%. Compute the average output voltage. The load consists of large inductance and resistance.

SOLUTION:

$$V_{o(av)} = 12 \times \frac{40}{100} + 24 \times \frac{25}{100}$$
$$= 4.8 + 6$$
$$= 10.8 \text{ V}$$

Questions

1. Explain the operation of different quadrant chopper-fed dc motor drives.
2. Describe the forward-running and regenerative braking of a dc motor.
3. Explain the different modes of operation of a four-quadrant chopper-fed dc motor drive.
4. Explain the concept of dynamic braking.
5. Why do we need a dual-input dc/dc converter.

Unsolved Problems

1. A first-quadrant chopper is feeding a constant, ripple-free current to the armature of a separately excited dc motor. Derive an expression for rms value of a chopper current, a ripple current of chopper current, and determine the duty ratio at which the chopper current is maximum.
2. A separately excited dc motor rated at 220 V, 10 A, and 1,500 rpm has an armature resistance of 1.00 Ω. The motor terminal voltage is controlled using a first-quadrant chopper, which is supplied from a 230-V source. If the motor speed is to be controlled from 500 rpm to 1,000 rpm at rated load, compute the desired ON period of the chopper for the two speeds at a chopping frequency of 1.0 KHz.
3. A separately excited dc motor has the following name plate data: 230 V, 50 A, 1,400 rpm. The armature resistance is 0.1 Ω. If dynamic baking takes place at 500 rpm at 1.5 times the rated torque, compute the duty ratio. The braking resistance is 6 Ω. (Hint: $I_a = 1.5 \times I_{a \text{ rated}}$)
4. A 230-V, 8-A, 1,500-rpm separately excited dc motor has an armature resistance of 2.5 Ω. A single-quadrant chopper supplied from a 300-V dc bus is operating at a duty ratio of 55% and supplies power to the motor armature at rated current. Compute the motor speed.

5. A first-quadrant dc/dc converter is fed from a 280-V dc bus. When the converter supplies power to a separately excited dc motor at 45% duty ratio, the average armature current is 4 A at 1,560 rpm. What is the duty ratio required to reduce the speed to 1,400 rpm for the same armature current? The armature resistance is 4 Ω.

6. A separately excited dc motor is fed from a 440-V dc source through a single-quadrant chopper with $r_a = 1.2 \ \Omega$ and armature current of 180 A. The voltage and torque constants are equal at 1.25 V/rad/s. The field current is 2 A. The duty cycle of chopper is 0.65. Find (a) speed and (b) torque.

7. A dual-input dc/dc converter is supplied from two dc batteries (12 V and 24 V). The duty ratio of the power switch connected to the first source is 45%, while that of the second is 30%. Compute the average output voltage. Load consists of large inductance and resistance.

8. A separately excited dc motor has an armature resistance of 2.3 Ω and an armature current of 100 A. (a) Find the voltage across the braking resistance for a duty ratio of 25%. (b) Find the power dissipated in braking resistance.

Answers

1. rms chopper current $= \left[\dfrac{1}{T} \displaystyle\int_0^{T_{on}} I_a^2 \, dt \right]^2 = I_a \left[\dfrac{1}{T} \displaystyle\int_0^{T_{on}} dt \right]^2 = I_a \left(\dfrac{T_{on}}{T} \right)^2 = I_a D^{1/2}$

   Average chopper current $= D I_a$

   Chopper ripple current $= \left[\left(I_a D^{1/2} \right)^2 - \left(D I_a \right)^2 \right]^{1/2} = I_a \left(D - D^2 \right)^{1/2}$

   For maximum value of chopper ripple current, $\dfrac{d}{dt} \left(I_a \left(D - D^2 \right)^{1/2} \right) = 0$

   $D = 0.5$

2. 0.348 ms, 0.652 ms
3. 0.838
4. 1,036 rpm
5. 0.41
6. 1092.44 rpm, 450 N-m
7. 12.6 V
8. 210 V, 25.2 kW

8

---

ac/ac Converters

8.1 Introduction

Single-stage ac/ac power conversion is possible by connecting two silicon-controlled recti-fiers (SCRs) in anti-parallel manner or a single triac between source and load and by vary-ing the SCR firing angle. The root mean square (rms) value of output voltage is controlled to deliver variable-output power. This type of power conversion finds application in indus-trial heating, illumination control, fan speed control, and transformer tap changing.

ac/ac power converters can be put into three categories:

1. ac voltage regulators
2. Cycloconverters
3. ON–OFF controls

All of the above converters are supplied from constant frequency and constant voltage supply lines. With ac voltage regulators, the amplitude of the output voltage can be var-ied, but its frequency remains at line frequency. In the case of a cycloconverter/ON–OFF control scheme, both the amplitude and frequency of the output voltage can be varied.

8.2 ac Voltage Regulators

In ac voltage converters, the output voltage amplitude can be conveniently varied without any change in frequency. Depending on the control scheme, there are two configurations:

1. Phase angle control
2. Pulse-width modulation

These methods differ in operation and type of devices employed, and each scheme has some advantages over the other.

8.2.1 Phase Angle Control

The principle of phase angle control in a single-phase circuit is represented in Fig. 8.1. As shown in the figure, there are two SCRs connected in an anti-parallel manner between the source and load. To start with, a resistive load is considered. During the positive half-cycle,

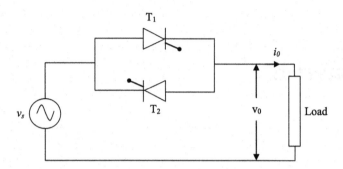

FIGURE 8.1
Circuit diagram of an ac voltage controller with R load.

T_1 is triggered at α, making v_0 the same as v_s. At $\omega_s t = \pi$, both v_0 and i_0 go to zero, and T_1 is turned OFF by natural commutation. Similarly, T_2 is fired at $\pi + \alpha$ and is naturally commutated at 2π. The waveforms are given in Fig. 8.2. Referring to the output voltage wave shape, the rms value of load voltage can be derived as follows:

$$
\begin{aligned}
V_{orms}^2 &= \frac{1}{2\pi} \int_{\alpha}^{\pi} (V_m \sin \omega_s t)^2 \, d(\omega_s t) + \int_{\pi+\alpha}^{2\pi} (V_m \sin \omega_s t)^2 \, d(\omega_s t) \\
&= \frac{V_m^2}{2\pi} \int_{\alpha}^{\pi} \frac{1 - \cos 2\omega_s t}{2} \, d(\omega_s t) + \int_{\pi+\alpha}^{2\pi} \frac{1 - \cos 2\omega_s t}{2} \, d(\omega_s t) \\
&= \frac{V_m^2}{4\pi} \left[\left(\omega_s t - \frac{\sin \omega_s t}{2} \right)_{\alpha}^{\pi} + \left(\omega_s t - \frac{\sin \omega_s t}{2} \right)_{\pi+\alpha}^{2\pi} \right] \\
&= \frac{V_m^2}{4\pi} \left[\pi - 0 - \alpha + 2\pi - 0 - \pi - \alpha + \frac{1}{2} (\sin 2\alpha + \sin 2(\pi + \alpha)) \right] \\
&= \frac{V_m^2}{4\pi} [2\pi - 2\alpha + \sin 2\alpha]
\end{aligned}
\tag{8.1}
$$

$$
V_{orms} = \frac{V_m}{\sqrt{2\pi}} \left[\pi - \alpha + \frac{\sin 2\alpha}{2} \right]^{\frac{1}{2}}
$$

Load is now assumed to be inductive in nature. Let R and L represent the load:

$$
L \frac{di_o}{dt} + R i_o = V_m \sin(\omega_s t), \alpha \le \omega_s t \le \beta
$$

where β is the angle at which current falls to zero. The solution is given in Equation (2.16) and is reproduced below:

$$
i_o = \frac{V_m}{Z} \left(\sin(\omega_s t - \phi) + \sin \phi \, e^{\frac{-R}{L} t} \right)
\tag{8.2}
$$

In the Equation 8.2, Z is the load impedance and is $\sqrt{R^2 + (\omega_s L)^2}$, and ϕ is the load power factor angle and is $\tan^{-1} \left(\frac{\omega_s L}{R} \right)$. The output current is plotted in Fig. 8.3.

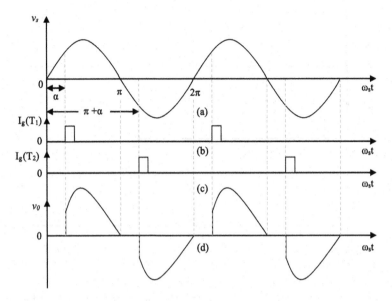

FIGURE 8.2
An ac voltage controller with R load. (a) Source voltage. (b) Gate pulses to T_1 and T_2. (c) Load voltage.

The angle β can be evaluated by substituting at $\omega_s t = \beta$ and $i_0(t) = 0$ in Equation (8.2). The value of β depends upon the relative value of R and $\omega_s L$. With increasing $\dfrac{\omega_s L}{R}$ values, β reaches $\pi + \alpha$, and there is no voltage control possible at the output. Hence, it should be mentioned that in the phase angle control scheme, voltage control is effective only up to $\alpha < \phi$, thereby imposing a stringent constraint. In other words, the load power factor is an important parameter that decides the amplitude of load voltage (i.e., the value of β) and the voltage control range. This is demonstrated in Fig. 8.3(e).

For low-power applications, a single triac can replace the two thyristors. Then the circuit becomes very compact and economical. Such a configuration is commonly employed for speed control of domestic fans, commercially known as electronic fan regulators. Stator voltage control of three-phase induction motors is discussed in Section 10.4.

EXAMPLE 8.1

A single-phase ac regulator has a resistive load R = 20 Ω and input voltage of 250 V and rms = 50 Hz. The firing angles of both thyristors are same at 90°. Determine

    a. the rms output voltage;

    b. the power dissipated in the resistor;

    c. the supply power factor;

    d. the average current of the thyristor; and

    e. the rms current of the thyristor.

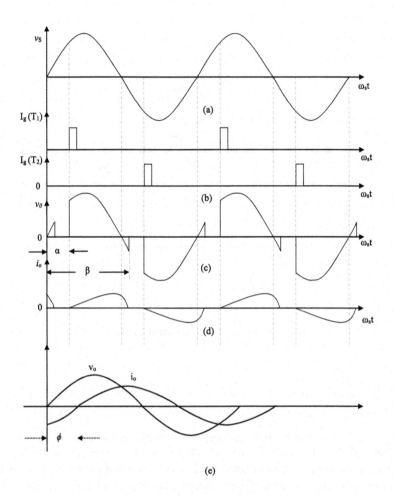

FIGURE 8.3
Waveforms of an ac voltage controller with R-L load. (a) Source voltage. (b) Gating signals to T_1 and T_2. (c) Load voltage. (d) Load current. (e) Effect of power factor angle, ϕ, on firing angle, α.

SOLUTION:

a.

$$V_{orms} = \frac{V_m}{\sqrt{2\pi}} \left[\pi - \alpha + \frac{\sin 2\alpha}{2} \right]^{\frac{1}{2}}$$

$$V_{orms} = \frac{250 \times \sqrt{2}}{\sqrt{2\pi}} \left[\pi - 1.57 + \frac{\sin(2 \times 90)}{2} \right]^{\frac{1}{2}}$$

$$V_{o(rms)} = 176.82 \text{ V}$$

b.

$$\text{Power } P_L = \frac{((V_o)_{rms})^2}{R_2}$$

$$= \frac{(176.82)^2}{20} = 1,563.26 \text{ W}$$

c.

$$\text{Power Factor}(\text{PF}) = \text{P.F} = \frac{\dfrac{(V_{orms})^2}{R}}{V_{rms}I_{rms}}$$

$$= \frac{V_{o(rms)}I_{rms}}{V_{rms}I_{rms}}$$

$$= \frac{176.82}{250}$$

$$\cos\phi = 0.707 \text{ Lag}$$

d. Average current of thyristor

$$I_{T(avg)} = \frac{V_m}{2\pi R}(1 + \cos 90) = 2.81 \text{ A}$$

e. rms value—refer to Equation (2.31B):

$$i_{o(rms)} = \frac{V_m}{2R}\left[\frac{1}{\pi}\left[\pi - \alpha + \frac{\sin 2\alpha}{2}\right]\right]^{\frac{1}{2}}$$

where

$$i_{o(rms)} = i_{T(rms)}$$

$$\therefore i_{T(rms)} = \frac{250 \times \sqrt{2}}{2 \times 20}\left[\frac{1}{\pi}\left[\pi - 90° + \frac{\sin 2 \times 90°}{2}\right]\right]^{\frac{1}{2}}$$

$$= 6.25 \text{ A}$$

EXAMPLE 8.2

A single-phase supply of $200\sqrt{2}\sin(314t)$ is connected to the R-L load through the triac. The load resistance is 150 Ω. It is observed that for a triac firing angle of $\alpha \le 30°$, the load voltage always remains the same as the supply voltage. Compute the value of the load inductance.

SOLUTION:

$\alpha \le 30°$ indicates that load power factor angle ϕ is 30°.

$$\tan\phi = \frac{X_L}{R}$$

$$\tan(30) = \frac{2\pi fL}{150} = \frac{314\,L}{150}$$

$$L = 0.276 \text{ H}$$

EXAMPLE 8.3

A triac is connected between $220\sqrt{2}\sin(314t)$ and R-L load of $R = 100\ \Omega$ and $L = 116$ mH. Compute the minimum value of the triac firing angle below which the load voltage remains the same as the supply voltage.

SOLUTION:

The voltage and current waveforms of R-L load with ac supply is given in Fig. 8.3(e). Here ϕ is the power factor angle and for the triac, with a firing angle $\alpha \leq \phi$, the load voltage becomes the supply voltage:

$$\phi = \tan^{-1}\frac{L\omega_s}{R}$$

$$\phi = \tan^{-1}\frac{116\times10^{-3}\times314}{100} = 20°$$

Here, $\alpha \leq 20°$, for which the load voltage is the supply voltage.

8.2.2 Pulse-Width Modulation

The pulse-width modulation (PWM) technique is performed with an ac chopper, which is another ac/ac converter that converts fixed voltage line frequency into variable ac voltage without a change in frequency. The simple schematic is shown in Fig. 8.4, where two switches are employed; one switch, S_1, is in series with load and the other switch, S_2, shunts the load. The switches are realized using self-commutating devices such as insulated-gate bipolar transistors (IGBTs) or power metal-oxide semiconductor field-effect transistors (MOSFETs). When switch S_1 is closed, $v_0 = v_s$, and once S_2 is switched ON, $v_0 = 0$. It may be noted that S_2 can be dispensed with resistance load; further simultaneous closing of S_1 and S_2 must be avoided to prevent short-circuiting of the source voltage.

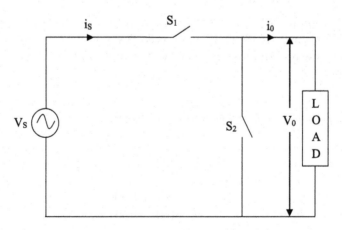

FIGURE 8.4
An ac chopper circuit.

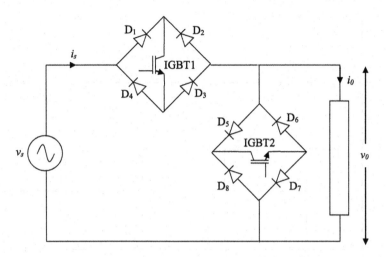

FIGURE 8.5
An ac chopper using insulated-gate bipolar transistors (IGBTs).

The practical implementation of switches S_1 and S_2 is shown in Fig. 8.5. The use of a diode rectifier along with each IGBT permits bi-directional conduction of IGBTs. The operation is explained with the help of Fig. 8.6. Here, a resistance load is considered, and hence S_2 does not have a freewheeling function. The IGBT-1 is gated between α and $\pi - \alpha$ during the positive half-cycle, making $v_0 = v_s$ during this period. This device is gated again in the negative half-cycle from $\pi + \alpha$ to $\pi - \alpha$. This produces one voltage pulse per half-cycle, which possesses half- and quarter-wave symmetries and leads to fewer harmonics. The load current waveform resembles that of load voltage.

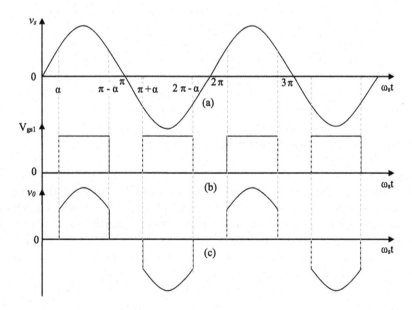

FIGURE 8.6
Waveforms of a pulse-width modulated ac chopper with R load. (a) Source voltage. (b) Gating signal. (c) Output voltage.

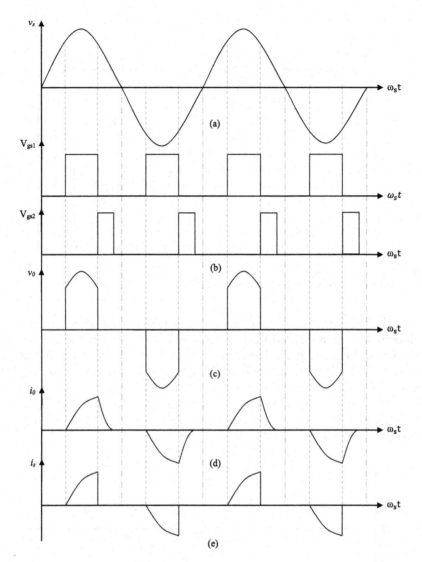

FIGURE 8.7
Waveforms of pulse-width modulation in an ac chopper with R-L load. (a) Source voltage. (b) Gating pulses. (c) Load voltage. (d) Load current. (e) Source current.

Figure 8.7 shows ac chopper waveforms with R-L load. In this case, the triggering pattern of IGBT-1 is retained, as was seen with R load, but at $\omega t = \pi - \alpha$, when IGBT-1 is turned OFF, IGBT-2 is gated for freewheeling of the load current. The rms value of output voltage can be derived as follows:

$$V_{\text{orms}}^2 = \frac{1}{2\pi}\left[\int_{\alpha}^{\pi-\alpha}\left(V_m \sin \omega_s t\right)^2 d\omega_s t + \int_{\pi+\alpha}^{2\pi-\alpha}\left(V_m \sin \omega_s t\right)^2 d\omega_s t\right]$$

$$V_{\text{orms}}^2 = \frac{V_m^2}{2\pi}\left[\int_{\alpha}^{\pi-\alpha}\left(\frac{1-\cos(2\omega_s t)}{2}\right) d\omega_s t + \int_{\pi+\alpha}^{2\pi-\alpha}\left(\frac{1-\cos(2\omega_s t)}{2}\right) d\omega_s t\right]$$

$$V_{orms}^2 = \frac{V_m^2}{4\pi}\left[\int_\alpha^{\pi-\alpha}(1-\cos(2\omega_s t))\,d\omega_s t + \int_{\pi+\alpha}^{2\pi-\alpha}(1-2\cos(2\omega_s t))\,d\omega_s t\right]$$

$$V_{orms}^2 = \frac{V_m^2}{4\pi}\left[\left(\omega_s t - \frac{\sin(2\omega_s t)}{2}\right)_\alpha^{\pi-\alpha} + \left(\omega_s t - \frac{\sin(2\omega_s t)}{2}\right)_{\pi+\alpha}^{2\pi-\alpha}\right]$$

$$V_{orms}^2 = \frac{V_m^2}{4\pi}\left[\left(\pi - 2\alpha - \frac{\sin 2(\pi-\alpha)}{2} + \frac{\sin 2(\alpha)}{2}\right) + \left(\pi - 2\alpha - \frac{\sin 2(2\pi-\alpha)}{2} - \frac{\sin 2(\pi+\alpha)}{2}\right)\right]$$

$$V_{orms}^2 = \frac{V_m^2}{4\pi}[2(\pi - 2\alpha) + 2\sin 2(\alpha)]$$

$$V_{orms} = \frac{V_m}{2}\left[\frac{2(\pi - 2\alpha) + 2\sin(2\alpha)}{\pi}\right]^{\frac{1}{2}}$$

The PWM technique can be used with an ac chopper in which the two IGBTs are switched ON/OFF sequentially so that the output voltage comprises several voltage pulses per cycle. This is illustrated in Fig. 8.8, where there are seven voltage pulses per half-cycle. The gating signal V_{gs1} alone is shown, and V_{gs2} is the complemented version of V_{gs1}. The output voltage and current waveforms in Fig. 8.8 show that there are far fewer harmonic components. This is the major advantage that an ac chopper has over the phase angle control scheme.

EXAMPLE 8.4

A single-phase chopper supplied from $230\sqrt{2}\sin(314t)$ produces a single-output voltage pulse per half-cycle. If the width of the pulse is 120°, compute the rms value of the output voltage.

$$\text{Width of the pulse} = \pi - 2\alpha$$

Therefore,

$$\alpha = 30° = \frac{\pi}{6}$$

From Equation (8.1),

$$V_{orms} = \frac{V_m}{2}\left[\frac{2(\pi - 2\alpha) + \sin(\alpha)}{\pi}\right]^{\frac{1}{2}}$$

$$V_{orms} = \frac{230\sqrt{2}}{2}\left[\frac{2\left(\pi - 2\times\frac{\pi}{6}\right) + \sin\left(\frac{\pi}{6}\right)}{\pi}\right]^{\frac{1}{2}} = 198.68 \text{ V}$$

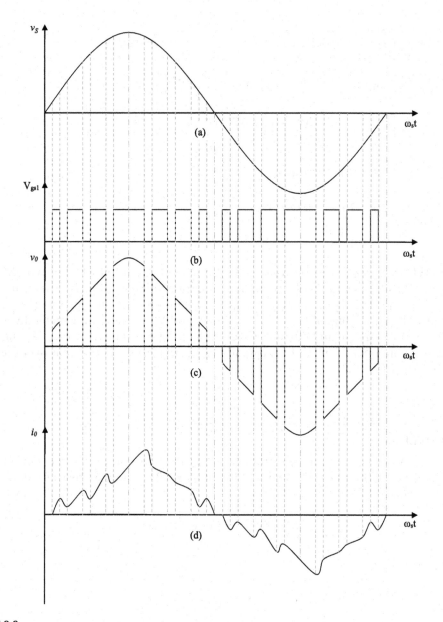

FIGURE 8.8
Waveforms of the multiple pulse-width modulation technique in an ac chopper. (a) Source voltage. (b) Gating pulses. (c) Load voltage. (d) Load current.

8.3 Cycloconverters

A cycloconverter is an ac/ac power converter supplied from fixed-frequency, fixed-voltage ac mains and provides a lower frequency ac voltage at the output. A bridge-type cyclo-converter is shown in Fig. 8.9. Here, there are two ac/dc converters labeled as C-1 and C-2. The operation of the circuit can be explained with the help of waveforms depicted

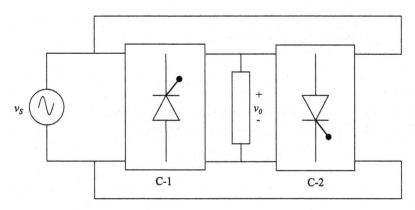

FIGURE 8.9
Circuit of a bridge-type cycloconverter.

in Fig. 8.10. Here Fig. 8.10(a) shows the supply voltage waveform, and C-1 is gated for the one cycle of v_s, and C-2 is made to conduct for the second cycle; if this sequence continues, the output voltage appears as shown in Fig. 8.10(b) and the output frequency is now $f_1/2$. The waveforms in Fig. 8.10(c) shows the output frequency at $f_1/3$. Output voltage control can be achieved by implementing phase angle control on C-1 and C-2, leading to

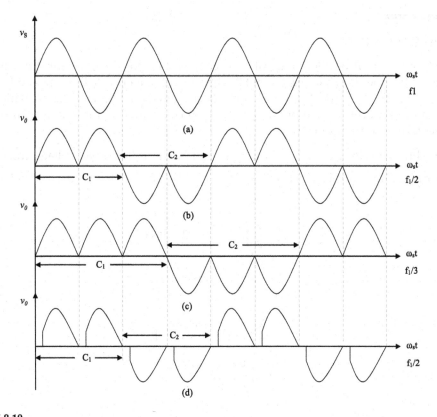

FIGURE 8.10
Waveforms of a cycloconverter with R-load. (a) Source voltage. (b) Output voltage at $f_1/2$. (c) Load voltage $f_1/3$. (d) Phase voltage control of output voltage at $f_1/2$.

variable-voltage, low-frequency output voltage as given in Fig. 8.10(d). Cycloconverters are used for low-speed operation of ac motor drives.

EXAMPLE 8.5

A two-pole, three-phase induction motor fed with a cycloconverter runs at 475 rpm at a slip of 5%. Calculate the input frequency of the motor.

$$\omega_r = \omega_s (1-S)$$

$$\omega_s = \frac{\omega_r}{(1-S)} = 500 \text{ rpm}$$

Because $\omega_s = \frac{120f}{P}$,

$$f = \frac{\omega_s P}{120} = \frac{500 \times 2}{120} = \frac{25}{3} \text{Hz}$$

8.4 On–Off Control

ON–OFF control is also called integral cycle control. The power circuit for this scheme is the same as the one shown in Fig. 8.1, but the triggering sequence is different and is shown in Fig. 8.11. The SCRs T_1 and T_2 are continuously triggered for a finite duration, which is an integral multiple of the period of input supply. Thus, during T_{on}, the load is connected to the source, and it is disconnected during $T - T_{on}$.

By suitably controlling $\frac{T_{on}}{T}$, the load voltage can be controlled. Load voltage and current waveforms are shown in Fig. 8.11. The rms value of the load voltage is derived below:

$$V_{orms}^2 = \frac{1}{T} \int_0^{T_{on}} (V_m \sin \omega_s t)^2 d\omega_s t$$

$$= \frac{V_m^2}{T} \int_0^{T_{on}} \frac{1 - \cos 2\omega_s t}{2} d\omega_s t$$

$$= \frac{V_m^2}{2T} \left[\omega_s t - \frac{\sin 2\omega_s t}{2} \right]_0^{T_{on}}$$

$$= \frac{V_m^2}{2T} \left[T_{on} - \frac{\sin 2T_{on}}{2} \right]$$

The interval T_{on} corresponds to multiple of π and hence, $\sin(\omega_s T_{on}) = 0$.

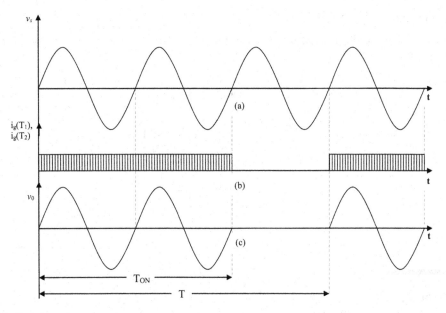

FIGURE 8.11
Waveforms of an integral cycle control scheme. (a) Source voltage. (b) Gating signal. (c) Output voltage.

Hence,

$$V_{orms}^2 = \frac{V_m^2}{2T}[T_{on} - 0]$$

$$= \frac{V_m^2}{2T}T_{on}$$

(8.3)

$$V_{orms} = \frac{V_m}{\sqrt{2T}}\sqrt{T_{on}}$$

$$= \frac{V_m}{\sqrt{2}}\sqrt{\frac{T_{on}}{T}}$$

While this scheme can be effectively employed for temperature or illumination control, it is not suitable for induction motor control because considerable speed fluctuations can hamper many industrial applications.

EXAMPLE 8.6

A room heater rated 1.6 kW, 220 V, and 50 Hz is controlled with an integral cycle controller. Calculate the duty ratio of the gate pulse such that the heater delivers 25% of the rated power if it is supplied from a 220-V, 50-Hz supply.

SOLUTION:

To have 25% of rated power, voltage should be reduced to 50% because the load is resistive. From Equation (8.3),

$$V_{orms} = \frac{V_m}{\sqrt{2}} \sqrt{\frac{T_{on}}{T}}$$

$$\therefore \frac{220}{2} = \frac{220\sqrt{2}}{\sqrt{2}} \sqrt{\frac{T_{on}}{T}}$$

$$\text{Duty ratio, } D = \frac{T_{on}}{T} = \frac{1}{4}$$

Questions

1. Derive an expression for the rms value of the output voltage of an integral cycle control scheme.
2. Explain the operation of single-phase multiple-pulse PWM ac chopper.
3. Derive an expression for the rms value of the output voltage of an ac voltage regulator.
4. With neat circuits and waveforms, explain the operation of a single-phase ac voltage regulator feeding a resistive load.
5. The output voltage of a PWM ac chopper has a single pulse width of β degrees. The output voltage pulse has quarter-wave and half-wave symmetries. If the load is highly inductive, explain the operation of the circuit with the help of load voltage and load current.
6. A PWM ac chopper is feeding an inductive load. If there are five pulses per half-cycle, derive an expression for the load current. Also sketch the load voltage, load current, and source current.

Unsolved Problems

1. A single-phase ac voltage regulator supplied from $220\sqrt{2} \sin(314t)$ is feeding a resistive load of 100 Ω. If the SCR firing angle is 45°, compute the rms values of the load voltage and current.
2. A single-phase supply of $200\sqrt{2} \sin(314t)$ is connected to the R-L load through a triac. The load resistance is 50 Ω. It is observed that for a triac firing angle of $\alpha \leq 25°$, the load voltage always remains the same as the supply voltage. Compute the value of the load inductance.

   (Hint: L = 10 mH)

3. A single-phase supply of $200\sqrt{2}\sin(314t)$ is connected to the R-L load through a triac. The load resistance is 100 Ω. It is observed that for a triac firing angle of $\alpha \le 30°$, the load voltage always remains the same as the supply voltage. Compute the value of the load inductance.

4. A room heater rated 2 kW, 220 V, 50 Hz is controlled with an integral cycle controller. Calculate the duty ratio of the gate pulse such that the heater delivers 40% of the rated power if it is supplied from a 220-V, 50-Hz supply.

5. A two-pole, three-phase induction motor fed with a cycloconverter runs at 800 rpm at a slip of 4%. Calculate the input frequency of the motor.

Answers

1. 136.80 V, 1.368 A
2. L = 10 mH
3. 183.80 mH
4. 0.633
5. 13.88

9

Induction Motor Fundamentals

9.1 Introduction

The theory of operation of three-phase induction motors is well articulated in numerous text books. When the three-phase stator winding is excited by a three-phase balanced supply, a rotating magnetic field is created on the stator. This rotating field causes rotation of the rotor and the conversion of electrical energy to mechanical energy. Induction-motor performance study includes motor efficiency, power factor, etc., during steady state and transient response of speed and torque. Computation of steady state and transient characteristics of an induction motor require a suitable motor model. This chapter introduces the development of a two-axis induction motor model, which is subsequently reduced to the commonly available per-phase equivalent circuit.

9.2 Derivation of a Two-Axis Model of an Induction Motor

As a first step toward the development of a two-axis model of an induction motor, consider the three-phase stator windings excited from voltages v_R, v_Y, v_B, with the rotor rotating at the angular speed ω_r. This is given in Fig. 9.1. Under normal running conditions, rotor windings are generally shorted. Now consider an imaginary two-winding stator and rotor equivalence. One stator winding is aligned to that of one winding of the three-phase motor; this is called the quadrature axis, and the voltage on this winding is indicated by v_{qs}. The other stator winding is vertically placed and is called the direct axis, and the voltage in this winding is named v_{ds}. In a similar manner, rotor windings can also be represented in quadrature given as v_{qr} and v_{dr}. The angular displacement between the stator and rotor axis is θ_r given by $\theta_r = \omega_r t$, where ω_r = rotor speed in rad/s at time t.

The voltages of the direct and quadrature axes on the stator can be resolved in terms of the three-phase stator voltages v_R, v_Y, and v_B. Note that the axis of v_R is made to coincide with that of v_{qs} so that the following equations can be obtained:

$$v_{qs} = v_R \tag{9.1}$$

$$v_{ds} = \frac{1}{\sqrt{3}}(v_B - v_Y) \tag{9.2}$$

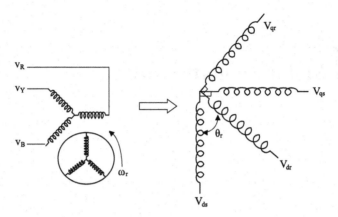

FIGURE 9.1
Two winding models of a three-phase induction motor.

9.3 Derivation on an Arbitrary Reference Frame

Consider an arbitrary reference frame rotating at an angle, ω_a, such that it makes the angle $\theta_a = \omega_a t$ at time t. Two-axis voltages are indicated as v_{qs}^a and v_{ds}^a as shown in Fig. 9.2. The relationship between voltages on fixed frame and arbitrary frame can be written as

$$\begin{bmatrix} v_{qs} \\ v_{ds} \end{bmatrix} = \begin{bmatrix} \cos\theta_a & \sin\theta_a \\ -\sin\theta_a & \cos\theta_a \end{bmatrix} \begin{bmatrix} v_{qs}^a \\ v_{ds}^a \end{bmatrix} \tag{9.3}$$

The voltage in a coil is the sum of the resistive drop and rate of change of flux linkage; accordingly, let ψ stand for flux linkage and R_s and R_r denote stator and rotor resistances,

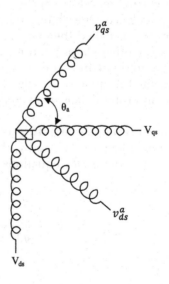

FIGURE 9.2
Arbitrary reference frame.

respectively. The voltage equations of the arbitrary reference frame can be written as follows:

$$v_{qs}^a = R_s i_{qs}^a + \frac{d}{dt}\left(\psi_{qs}^a\right) \tag{9.4}$$

$$v_{ds}^a = R_s i_{ds}^a + \frac{d}{dt}\left(\psi_{ds}^a\right) \tag{9.5}$$

$$v_{qr}^a = R_r i_{qr}^a + \frac{d}{dt}\left(\psi_{qr}^a\right) \tag{9.6}$$

$$v_{dr}^a = R_r i_{dr}^a + \frac{d}{dt}\left(\psi_{dr}^a\right) \tag{9.7}$$

The above equations can be summarized as a matrix differential equation:

$$
\begin{bmatrix} v_{ds}^a \\ v_{ds}^a \\ v_{qr}^a \\ v_{dr}^a \end{bmatrix} =
\begin{bmatrix}
R_s + L_s p & \omega_a L_s & L_m p & \omega_a L_m \\
-\omega_a L_s & R_s + L_s p & -\omega_a L_m & L_m p \\
L_m p & (\omega_a - \omega_r) L_m & R_r + L_r p & (\omega_a - \omega_r) L_m \\
-(\omega_a - \omega_r) L_m & L_m p & -(\omega_a - \omega_r) L_m & R_r + L_r p
\end{bmatrix}
\begin{bmatrix} i_{qs}^a \\ i_{ds}^a \\ i_{qr}^a \\ i_{dr}^a \end{bmatrix} \tag{9.8}
$$

where p denotes d/dt.

9.4 Stator Reference Frame Model

Putting $\omega_a = 0$ in the above matrix differential equation, the stator reference model of a three-phase induction motor can be obtained:

$$
\begin{bmatrix} v_{qs} \\ v_{ds} \\ v_{qr} \\ v_{dr} \end{bmatrix} =
\begin{bmatrix}
R_s + L_s p & 0 & L_m p & 0 \\
0 & R_s + L_s p & 0 & L_m p \\
L_m p & (\omega_a - \omega_r) L_m & R_r + L_r p & (\omega_a - \omega_r) L_m \\
-(\omega_a - \omega_r) L_m & L_m p & -(\omega_a - \omega_r) L_m & R_r + L_r p
\end{bmatrix}
\begin{bmatrix} i_{qs} \\ i_{ds} \\ i_{qr} \\ i_{dr} \end{bmatrix} \tag{9.9}
$$

9.5 Derivation of a Per-Phase Equivalent Circuit of a Three-Phase Induction Motor

Consider the three-phase induction motor that is supplied from balanced three-phase voltages as given by:

$$v_R = V_m \sin \omega_s t \tag{9.10}$$

$$v_Y = V_m \sin\left(\omega_s t - \frac{2\pi}{3}\right) \tag{9.11}$$

$$v_B = V_m \sin\left(\omega_s t + \frac{2\pi}{3}\right) \tag{9.12}$$

The d-q axis voltages are now derived:

$$v_{qs} = v_R = V_m \sin(\omega_s t) \tag{9.13}$$

$$
\begin{aligned}
v_{ds} &= \frac{1}{\sqrt{3}}(v_B - v_Y) = \frac{1}{\sqrt{3}}\left(V_m \sin\left(\omega_s t + \frac{2\pi}{3}\right) - v_{bs} = V_m \sin\left(\omega_s t - \frac{2\pi}{3}\right)\right) \\
&= \frac{1}{\sqrt{3}}V_m\left(\sin\omega_s t\cos\frac{2\pi}{3} + \cos\omega_s t\sin\frac{2\pi}{3} - \left(\sin\omega_s t\cos\frac{2\pi}{3} - \cos\omega_s t\sin\frac{2\pi}{3}\right)\right) \\
&= V_m \cos\omega_s t
\end{aligned}
\tag{9.14}
$$

Under normal running conditions, the rotor windings are short circuited, such that $v_{qr} = 0$ and $v_{dr} = 0$. Because the stator voltages are at 90° from each other, respective currents are also phase shifted by 90°. Hence,

$$i_{ds} = ji_{qs}$$

and

$$i_{dr} = ji_{qr}$$

Thus, the matrix equation given by Equation (9.9) can be modified:

$$
\begin{bmatrix} v_m \sin\omega_s t \\ v_m \cos\omega_s t \\ 0 \\ 0 \end{bmatrix} = \begin{bmatrix} R_s + j\omega_s L_s & 0 & \omega_s L_m & 0 \\ 0 & R_s + j\omega_s L_s & 0 & j\omega_s L_m \\ j\omega_s L_m & -\omega_r L_m & R_r + j\omega_s L_r & -\omega_r L_r \\ \omega_r L_m & j\omega_s L_m & \omega_r L_r & R_r + j\omega_s L_r \end{bmatrix} \begin{bmatrix} i_{qs} \\ i_{ds} \\ i_{qr} \\ i_{dr} \end{bmatrix} \tag{9.15}
$$

The above equation corresponds to a two-axis model in which there are two stator and two rotor windings. For the per-phase equivalent circuit, we need to consider only one stator winding and one rotor winding. Let V_s, I_s, and I_r denote the rms values of stator voltage, stator current, and rotor current, respectively. Then $V_s = V_m \sin\omega_s t$, and I_s and I_r are the rms values of the stator and rotor currents (i.e., $i_{ds} = I_s$ and $i_{qr} = I_r$). Now the first row of Equation (9.15) can be simplified:

For the first row in Equation (9.15),

$$V_s = (R_s + j\omega_s L_s)I_s + j\omega_s L_m I_r \tag{9.16}$$

By substituting $L_S = L_{ls} + L_m$ and $L_r = L_{lr} + L_m$, the above equation becomes

$$V_s = R_s + j\omega_s(L_{ls} + L_m)I_s + j\omega_s L_m I_r \tag{9.17}$$

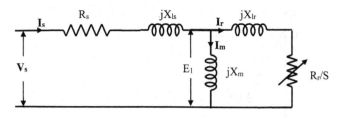

FIGURE 9.3
Induction motor per-phase equivalent circuit.

The rotor side equation is obtained by considering the third row of Equation (9.15):

$$0 = \left(j\omega_s L_m - j\omega_r L_m\right)I_s + \left(R_r + j\omega_s L_r - \omega_r L_r\right)I_r$$
$$= R_s + j\omega_s L_{ls}I_s + j\omega_s L_m\left(I_s + I_r\right)$$
$$0 = j\left(\omega_s - \omega_r\right)L_m I_s + R_r I_s + \left(R_r + j\left(\omega_s - \omega_r\right)\right)L_r I_r$$
$$0 = jS\omega_s L_m I_s + \left(R_r I_r + jS\omega_s L_{lr} + jS\omega_s L_m\right)I_r \tag{9.18}$$
$$0 = jS\omega_s L_m\left(I_s + I_r\right) + \frac{R_r}{S}I_r + j\omega_s L_{lr}$$
$$0 = I_r\frac{R_r}{S} + j\omega_s L_{lr} + jS\omega_s L_m\left(I_s + I_r\right)$$

Equations (9.17) and (9.18) define the per-phase equivalent circuit model of the motor.

It is important to mention that the above equivalent circuit is valid only if the following conditions are satisfied:

1. The supply voltage is sinusoidal in nature with angular frequency ω_s.
2. This circuit is valid for steady-state conditions alone, i.e., there are no changes in amplitude and frequency of supply voltage and further speed or slip is constant.

It is conventional that the rotor current, I_r, flows into the rotor. Accordingly, the per-phase equivalent circuit is shown in Fig. 9.3, where E_1 is the stator-induced emf.

Considering that the magnetization current, I_m, establishes a flux linkage of Φ_m in the airgap, the phasor diagram can be sketched as given in Fig. 9.4. It is evident that the stator voltage, V_s, is higher than E_1 by stator impedance drop.

9.6 Expression for Electromagnetic Torque

Let m be the number of phases of the motor. Then the airgap power, P_{ag}, is

$$P_{ag} = m_1 I_r^2 \frac{R_r}{S} \tag{9.19}$$

Rotor copper loss, P_{cu} is

$$P_{cu} = m_1 I_r^2 R$$
$$= S P_{ag} \tag{9.20}$$

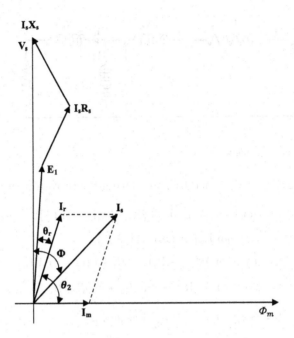

FIGURE 9.4
Induction motor phasor diagram.

But

$$P_{ag} = P_{cu} + P_{mech}$$

where P_{mech} is the mechanical power developed:

$$= m_1 I_r^2 R_r \left(\frac{1-S}{S} \right)$$

But

$$P_{mech} = \omega_r T_e$$

$$\therefore \omega_r T_e = m_1 I_r^2 R_r \left(\frac{1-S}{S} \right) \tag{9.21}$$

But

$$\omega_r = \omega_s (1-S) \quad \text{and} \quad \text{since } \omega_s = \frac{2\pi f_1}{p_1}$$

where f_1 = stator frequency in Hz and p_1 = the number of pairs of poles for which the rotor is wound.

$$T_e = \frac{p_1 m_1}{2\pi f_1} \left(I_r^2 \frac{R_r}{S} \right) \tag{9.22}$$

For the purpose of simplifying the calculations, the stator voltage drop is neglected and so the per-phase equivalent circuit in Fig. 9.3 can be approximated as shown in Fig. 9.5.

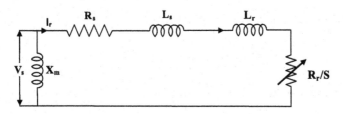

FIGURE 9.5
Simplified induction motor per-phase equivalent circuit.

Using the approximate equivalent circuit in Fig. 9.5, rotor current can be expressed as

$$I_r = \frac{V_s}{\sqrt{\left(R_s + \dfrac{R_r}{S}\right)^2 + \omega_s^2\left(L_s + L_r\right)^2}} \tag{9.23}$$

where $\omega_s = 2\pi f_1$.

Substituting Equation (9.23) in Equation (9.22),

$$T_e = \frac{p_1 m_1}{2\pi f_1} \times \frac{R_r}{S} \times \frac{V_s^2}{\left(R_s + \dfrac{R_r}{S}\right)^2 + \omega_s^2\left(L_s + L_r\right)^2} \tag{9.24}$$

Equation (9.24) suggests that the electromagnetic torque is a function of slip when stator frequency and voltage are kept constant. This equation is now used to sketch the torque-slip curve as shown in Fig. 9.6. This curve shows starting torque, maximum torque (also known as breakdown torque), and motoring, generating, and braking modes.

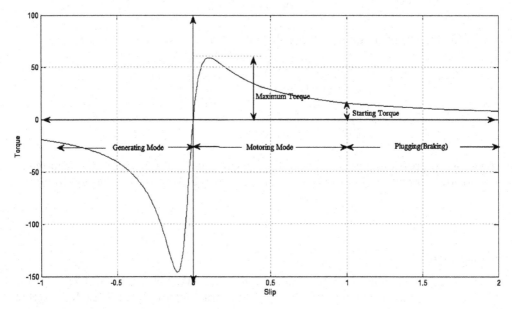

FIGURE 9.6
Induction motor torque-slip curve.

Starting torque (T_s), can be found by substituting $S = 1$ in Equation (9.24):

$$T_s = \frac{p_1 m_1 R_r}{2\pi f_1} \times \frac{V_s^2}{\left(R_s + R_r\right)^2 + \omega_s^2 \left(L_s + L_r\right)^2} \tag{9.25}$$

The above equation can be written as a function of slip as follows:

$$T_e = \frac{N(s)}{D(s)}$$

where $N(s) = \dfrac{p_1 m_1 R_r V_s^2}{2\pi f_1}$ and $D(s) = S\left(R_s + \dfrac{R_r}{S}\right)^2 + S(L_s + L_r)^2$

Torque, T_e, will be maximum when the denominator is minimum. The minimum can be found by differentiating $D(s)$ and equating to zero:

$$D(s) = S\left(R_s^2 + \frac{R_r^2}{S^2} + 2 \cdot R_s \cdot \frac{R_r}{S}\right) + S\omega_s^2 (L_s + L_r)^2 \tag{9.26}$$

$$\frac{dD(s)}{dS} = 0$$

$$R_s^2 - \frac{R_r^2}{S^2} + \omega_s^2 (L_s + L_r)^2 = 0$$

$$\frac{R_s^2 + \omega_s^2 (L_s + L_r)^2}{R_r^2} = \frac{1}{S^2}$$

$$S = \pm \frac{R_r}{\sqrt{R_s^2 + \omega_s^2 (L_s + L_r)^2}}$$

Therefore, the electromagnetic torque, T_e, will be maximum for slip $S = S_{max}$:

$$S_{max} = \pm \frac{R_r}{\sqrt{R_s^2 + \omega_s^2 (L_s + L_r)^2}} = \pm \frac{R_r}{\sqrt{R_s^2 + (\omega_s L_s + \omega_s L_r)^2}}$$

$$= \pm \frac{R_r}{\sqrt{R_s^2 + (X_{Ls} + X_{Lr})^2}} \tag{9.27}$$

In the above expression, a positive sign indicates the motoring mode, and a negative sign stands for the generating mode. The slip-torque characteristic curve is plotted in Fig. 9.6.

EXAMPLE 9.1

A three-phase four-pole induction motor rated 3.73 kW, 1,500 rpm, and 7.5 A is fed from a rated 415-V, 50-Hz supply. The motor parameters are $R_s = 2.76\ \Omega$, $R_r = 0.68\ \Omega$, and $X_{01} = 5.9\ \Omega$. Find the electromagnetic torque developed if the machine is running at 1,440 rpm.

SOLUTION:

$$\text{Slip S} = \frac{1500-1440}{1500} \times 100 = 4\%$$

Electromagnetic torque developed is given by

$$T_e = \frac{p_1 m_1}{2\pi f_1} \times \frac{R_r}{S} \times \frac{V_s^2}{\left(R_s + \dfrac{R_r}{S}\right)^2 + \omega_s^2 \left(L_s + L_r\right)^2}$$

$$T_e = \frac{2 \times 3}{2\pi \times 50} \times \frac{0.68}{0.04} \times \frac{(415/\sqrt{3})^2}{\left(2.76 + \dfrac{0.68}{0.04}\right)^2 + 5.9^2} = 43.83 \text{ N-m}$$

EXAMPLE 9.2

A three-phase four-pole induction motor rated 3.73 kW, 1,500 rpm, and 7.5 A is fed from a rated 415-V, 50-Hz supply. The motor parameters are $R_s = 2.76\ \Omega$, $R_r = 0.68\ \Omega$, and $X_{01} = 5.9\ \Omega$. Find the ratio of maximum torque to starting torque.

SOLUTION:

Starting torque is given by

$$T_s = \frac{p_1 m_1 R_r}{2\pi f_1} \times \frac{V_s^2}{\left(R_s + R_r\right)^2 + \omega_s^2 \left(L_s + L_r\right)^2}$$

$$T_s = \frac{2 \times 3 \times 0.68}{2\pi \times 50} \times \frac{\left(415/\sqrt{3}\right)^2}{\left(2.76 + 0.68\right)^2 + 5.9^2} = 15.98 \text{ Nm}$$

$$S_{max} = \frac{R_r}{\sqrt{R_s^2 + \omega_s^2 \left(L_s + L_r\right)^2}}$$

$$S_{max} = \frac{0.68}{\sqrt{2.76^2 + 5.9^2}} = 0.104$$

$$T_{max} = \frac{p_1 m_1}{2\pi f_1} \times \frac{R_r}{S_{max}} \times \frac{V_s^2}{\left(R_s + \dfrac{R_r}{S_{max}}\right)^2 + \omega_s^2 \left(L_s + L_r\right)^2}$$

$$T_{max} = \frac{2 \times 3}{2\pi \times 50} \times \frac{0.68}{0.104} \times \frac{(415/\sqrt{3})^2}{\left(2.76 + \dfrac{0.68}{0.104}\right)^2 + 5.9^2} = 59.11 \text{ Nm}$$

$$\frac{T_{max}}{T_s} = \frac{59.11}{15.98} = 3.69$$

EXAMPLE 9.3

A 415-V, 50-Hz, 1,435-rpm, three-phase, four-pole induction motor has the following parameters: $R_s = 8.0\ \Omega$, $R_r = 8.3\ \Omega$, $L_s = 0.0449$ H, $L_r = 0.0449$ H, and $L_m = 1.01795$. Compute the stator and rotor currents, the input power factor, and the electromagnetic torque developed at rated speed.

SOLUTION:

$$\text{Slip } S = \frac{1,500-1,435}{1,500} \times 100 = 4.33\%g$$

$$X_s = X_r = 2\pi \times 50 \times 0.0449 = 14.10\ \Omega$$

$$X_m = 2\pi \times 50 \times 1.01795 = 319.79\ \Omega$$

$$I_s = \frac{415/\sqrt{3}}{(8+14.10j)+\left(\dfrac{8.3}{0.043}+14.10j\right)\|(319j)}$$

$$= \frac{415/\sqrt{3}}{(8+14.10j)+(132.52+90.29j)} = 1.09-0.82j = 1.36\angle-36.6°$$

$$\text{Power Factor} = \cos(36.6) = 0.80 \text{ lagging}$$

$$I_r = I_s \times \frac{319.79j}{319.79j+\left(14.10j+\dfrac{8.3}{0.043}\right)} = 1.122-0.136j = 1.13\angle-6.92°$$

$$T_e = \frac{p_1 m_1}{2\pi f_1}\left(I_r^2 R_r/S\right)$$

$$T_e = \frac{2\times3}{2\pi\times50}\left(1.13^2\times8.3/0.043\right) = 3.71 \text{ N-m}$$

EXAMPLE 9.4

A three-phase induction motor rated 7 kW, 400 V, 14.7 A, 50 Hz, and 1,440 rpm is supplied with 330 V and speed is found to be 1,400 rpm. Calculate the load torque, neglecting the losses, with the following machine parameters: $R_s = 2.745\ \Omega$, $X_{01} = 33.45$, and $R_r = 0.385\ \Omega$.

SOLUTION:

$$\text{Slip } S = \frac{1500-1400}{1500} = 0.066$$

$$T_e = \frac{p_1 m_1}{2\pi f_1} \times \frac{R_r}{S} \times \frac{V_s^2}{\left(R_s+\dfrac{R_r}{S}\right)^2 + \omega_s^2\left(L_s+L_r\right)^2}$$

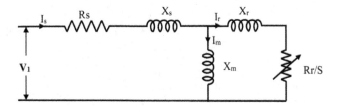

FIGURE 9.7
Induction motor per-phase equivalent circuit.

$$p_1 = \text{No. of pair of poles} = \frac{4}{2} = 2$$

$$T_e = \frac{2 \times 3}{2\pi \times 50} \times \frac{0.385}{0.066} \times \frac{\left(330/\sqrt{3}\right)^2}{\left(2.745 + \frac{0.385}{0.066}\right)^2 + 33.45^2} = 3.39 \text{ N-m}$$

9.7 Derivation of Harmonic Equivalent Circuit

Assume that the supply voltage is non-sinusoidal and, to get the stator current under steady-state conditions, the stator voltage is split into fundamental components and harmonics:

$$\text{i.e., } v_s = v_1 + \sum_{n=2}^{n} v_n$$

where v_1 represents the fundamental component and v_n is the harmonic voltage. The rms value of v_1 can be applied to the per-phase equivalent circuit, and the stator current I_1 can be computed. For the harmonic voltage, v_n, the per-phase equivalent circuit is modified as shown in Fig. 9.7.

The stator and rotor resistances remain the same at higher frequencies, neglecting the skin effect. The leakage and magnetic reactance increases as per the order of harmonics:

$$S_n = \frac{n\omega_s - \omega_r}{n\omega_s} \approx 1$$

In comparison with the harmonic order of the reactances, the stator and the rotor resistances can be neglected, and the per-phase equivalent circuit simplifies to the one shown in Fig. 9.8.

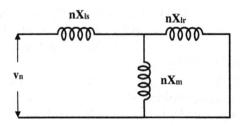

FIGURE 9.8
Harmonic equivalent circuit of induction motor neglecting stator and rotor resistances.

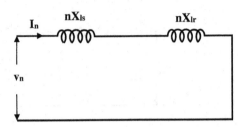

FIGURE 9.9
Harmonic equivalent circuit of induction
motor neglecting stator and rotor resistances
and magnetizing reactance.

Because the magnetizing reactance is very high, the above circuit can be further simplified into the circuit shown in Fig. 9.9.

$$\text{Thus,} \quad I_n = \frac{V_n}{nj(X_{ls} + X_{lr})} \quad (9.28)$$

$$i_s = i_1 + \sum_{n=2}^{\infty} i_n \quad (9.29)$$

The above analysis shows that harmonic currents do not flow through the magnetizing circuit. Hence it is important to mention that the airgap flux is always sinusoidal, even when a non-sinusoidal voltage is applied to the stator. This aspect is further examined in section 12.4.

Unsolved Questions

1. A three-phase, four-pole induction motor rated 7 kW, 1,500 rpm, and 14.7 A is fed from a rated 400-V, 50-Hz supply. The motor parameters are $R_s = 2.56\ \Omega$, $R_r = 0.385\ \Omega$, are $X_{01} = 33.45\ \Omega$. Find the electromagnetic torques developed for slips of 10%, 15%, and 100%.

2. Calculate the maximum/starting torque for a three-phase, four-pole induction motor rated 7 kW, 1,500 rpm, 14.7 A when fed from a rated 400-V, 50-Hz supply. The motor parameters are $R_s = 2.56\ \Omega$, $R_r = 0.385\ \Omega$, and $X_{01} = 33.45\ \Omega$.

3. A 415-V, 50-Hz, 1,435-rpm three-phase, four-pole induction motor has the following parameters: $R_s = 3\ \Omega$, $R_r = 3.5\ \Omega$, $X_{ls} = 16.72\ \Omega$, $X_{lr} = 16.72\ \Omega$, and $X_m = 400\ \Omega$. Compute the stator and rotor currents, the input power factor, and the electromagnetic torque developed at rated speed.

4. A three-phase, four-pole induction motor rated 2.2 kW, 415 V, 2.7 A, 1,435 rpm, and 50 Hz has the following per-phase equivalent circuit parameters: $R_s = 8\ \Omega$, $R_r = 8.3\ \Omega$, $L_s = 0.0449\ H$, $L_r = 0.0449\ H$, moment of inertia(J) $= 0.0060\ kg\text{-}m^2$. Calculate the peak current if the machine is accelerating from 0 to 1,000 rpm in 10 s at no load.

5. A three-phase induction motor rated 7 kW, 400 V, 14.7 A, 50 Hz, 1,440 rpm is supplied with 300 V and delivers torque of 2 N-m. Calculate the speed of the machine, neglecting the losses. Machine parameters are: $R_s = 2.745\ \Omega$, $X_{01} = 33.45\ \Omega$, $R_r = 0.385\ \Omega$.

Answers

1. At 10% slip $T_e = 3.45$ N-m, at 15% $T_e = 2.31$ N-m, at 100% $T_e = 0.35$ N-m

2. $T_s = 0.35$ N-m, $T_{max} = 13.6$ N-m

3. $I_s = 2.70\angle - 31.92$, pf $= 0.84$ lagging, $I_2 = 2.55\angle$-22.4, T $= 8.66$ N-m

4. $T_e = J\dfrac{d\omega}{dt}$, $T_e = 3I_r^2\dfrac{R_r}{S\omega_s}$, $I_{peak} = 3.92$ A

5. $S = 0.09$, speed $= 1,365$ rpm

10

Stator Voltage Control of Induction Motors

10.1 Introduction

The synchronous speed and hence the rotor speed of a given induction motor depends solely on the stator voltage and stator frequency for specific load torque. When an ac voltage regulator is introduced between an ac source and motor, the motor terminal voltage can be conveniently changed to allow variable-speed operation. This scheme is simple and economical, and it is used to control the speed of fan and pump drives and to allow smooth starting of induction motors. This chapter describes theoretical concepts behind stator voltage-controlled induction motors and circuits for ac/ac conversion, as well as a discussion on speed control range with constant load torque and fan/pump drives. This chapter also features rotor resistance control and slip power recovery schemes.

10.2 Stator Voltage Control: Theoretical Investigations

Stator voltage control is one of the cheapest and most reliable methods of induction motor speed control. While this scheme has a few disadvantages, such as a limited range of speed control, reduced efficiency, and increased line current harmonics, this scheme is widely used for fan and pump drives and the starting of large induction motors.

It was shown in Chapter 9 that the input power to the rotor is divided into rotor copper loss and mechanical power, as noted in Equation (10.1):

$$P_{ag} = P_{cu} + P_{mech} \tag{10.1}$$

We also know the following:

$$P_{cu} = SP_{ag} \tag{10.2}$$

Thus,

$$P_{mech} = (1 - S)P_{ag} \tag{10.3}$$

Therefore, neglecting stator copper loss, iron loss, and friction and windage losses, motor efficiency can be written as

$$\frac{P_{mech}}{P_{ag}} = 1 - S \tag{10.4}$$

Equation (10.4) shows that, for fixed frequency stator excitation, motor efficiency decreases as slip, S, increases. Thus, reduced speed operation at rated frequency operation results in poor operating efficiency; furthermore, at low speed, S is high, making rotor copper loss higher as indicated by Equation (10.2).

The electromagnetic torque of an induction motor is

$$T_e = \frac{p_1 m_1}{2\pi f_1} I_r^2 \frac{R_r}{S} \tag{10.5}$$

Neglecting magnetizing current, $I_s \approx I_r$. Thus at a given slip,

$$T_e \propto I_s^2$$

$$T_e \propto V_s^2 \tag{10.6}$$

Equation (10.6) is used to construct torque-slip curves of an induction motor as shown in Fig. 10.1. This figure shows that the range of speed control appears to be small.

Consider that a constant load torque of magnitude T_{L1} is applied to the shaft. It is seen from Fig. 10.1 that T_{L1} crosses 100% and 80% of rated torque speed curves. In other words, a reduction from 100% to 80% of rated voltage gives a limited change in speed for the load torque T_{L1}. Next consider a lighter load condition in which the load torque is T_{L2}. This torque line intersects several torque-speed characteristic curves for 100% to 50% of the rated voltage. Comparing this to the previous case of T_{L1}, even though there are several intersecting points, the speed range is also limited with T_{L2}. From these two case studies, it can be concluded that, for constant load torques, stator voltage control results in limited speed range.

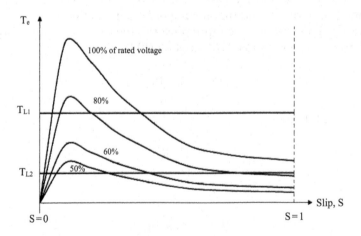

FIGURE 10.1
Torque-slip curves of a variable-speed induction motor.

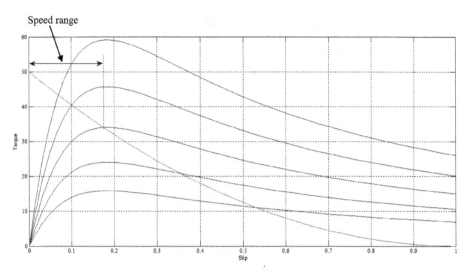

FIGURE 10.2
Speed control of fan/pump loads by stator voltage control.

Now consider a fan or pump load where the load torque, which is proportional to the square of the speed, is connected to the induction motor. The motor torque-speed curve along with the load torque curve are given in Fig. 10.2.

As indicated in Fig. 10.2, even with a fan-type load, stator voltage control allows a limited speed range. Because fan and pump drives require large variations in speed, the characteristic curve in Fig. 10.2 primarily indicates that stator voltage control is also unsuitable as a method for speed control of this kind.

To make stator voltage control suitable for the speed control of fan or pump drives, the rotor resistance is now increased so that the stable region of the speed torque speed characteristic curve of the motor is wider now. The new torque-speed characteristics with a fan-type load is shown in Fig. 10.3. It is apparent that the speed control range is now

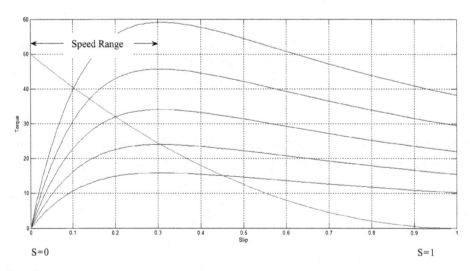

FIGURE 10.3
Speed control of fan/pump loads by stator voltage control with increased rotor resistance.

increased. Thus, to achieve a wide range of speeds for the fan or pump loads employing stator voltage control, larger rotor resistance is required. This is now analytically probed.

Equation (10.5) can be rewritten as

$$T_e \propto I_r^2 \frac{R_r}{S}$$

$$\propto I_r^2 \frac{R_r}{S} \tag{10.7}$$

For a fan-type load,

$$T_L = K_1 \omega_r^2 \tag{10.8}$$

Or

$$T_L \propto \omega_r^2$$

$$\propto \left[\omega_s (1-S)\right]^2 \tag{10.9}$$

As stator frequency is constant, ω_s is constant. Thus,

$$T_L \propto (1-S)^2 \tag{10.10}$$

Because $T_e = T_L$, we can write

$$I_s^2 \frac{R_r}{S} = (1-S)^2$$

$$I_s = (1-S)\sqrt{S/R_r} \tag{10.11}$$

Stator current is maximum at $\dfrac{dI_s}{dS} = 0$.

$$\frac{d}{dS}\left((1-S)S^{\frac{1}{2}}\right) = 0$$

$$(1-S)\frac{1}{2} \cdot S^{-\frac{1}{2}} + S^{\frac{1}{2}}(-1) = 0 \tag{10.12}$$

$$(1-S) - 2S = 0$$

$$S = \frac{1}{3}$$

Let $I_{s\,max}$ represent this current at $S = \frac{1}{3}$. If S_{FL} and I_{sFL} denote full load slip and full load current, respectively, then we obtain the following from Equation (10.11):

$$I_{sFL} = (1-S_{FL})\sqrt{\frac{S_{FL}}{R_r}} \tag{10.13}$$

and

$$I_{s\,max} = \frac{2}{3\sqrt{3R_r}} \text{ (by putting } S = \frac{1}{3} \text{ in Equation 10.11).} \tag{10.14}$$

TABLE 10.1

Relationship between s_{FL} and $I_{smax}\big/I_{sFL}$.

s_{FL}	I_{smax}/I_{sFL}
1%	3.8879
5%	1.8119
10%	1.3524
20%	1.0758
30%	1.0039
50%	1.0887

Now,

$$\frac{I_{smax}}{I_{sFL}} = \frac{2}{3\sqrt{3}\ (1-S_{FL})\sqrt{S_{FL}}}. \tag{10.15}$$

Equation (10.15) helps us evaluate $\dfrac{I_{smax}}{I_{sFL}}$ as a function of S_{FL}. Consider a full load slip of 16% with fan loads, then the ratio $\dfrac{I_{smax}}{I_{sFL}}$ is 1.14. This means that the maximum stator current is only 1.14 times that of the full load current. Table 10.1 illustrates the relationship between s_{FL} and $I_{smax}\big/I_{sFL}$.

For a fan-type load, a slip of 30–50% is generally preferred because values of I_{smax} and I_{sFL} are almost equal. The findings Table 10.1 clearly suggest that a larger full load slip makes the ratio $I_{smax}\big/I_{sFL}$ near to unity. This shows that under such situations, the full load and the maximum currents are equal. Because the stator is designed for the full load current, now this winding can safely carry the full load current because $I_{smax} = I_{sFL}$. To get s_{FL} around 30–50%, the rotor resistance of the motor must be large enough to produce flatter torque-speed curves. Thus, it can be established that stator voltage control is quite suitable for induction motor drives, provided the rotor resistance is larger and the load is a fan- or pump-type.

EXAMPLE 10.1

Calculate the ratio of the maximum current to the full load current for a three-phase induction motor driving a fan/pump load. The machine's specifications are as follows: 2.2 kW, 415 V, 50 Hz, 1,435 rpm.

SOLUTION:

Because the full load speed is given to be 1,435 rpm, we can assume the synchronous speed is 1,500 rpm.

$$\text{Full load slip } S_{FL} = \frac{1500 - 1435}{1500} = 0.043$$

$$\frac{I_{smax}}{I_{sFL}} = \frac{2}{3\sqrt{3}(1 - s_{FL})\sqrt{S_{FL}}} = 1.93$$

EXAMPLE 10.2

A start-connected, three-phase induction motor, rated 2.2 kW, 415 V, 50 Hz, 1,435 rpm with parameters $R_s = 2.745\ \Omega$, $R_r = 5\ \Omega$, $X_s = X_r' = 11.225\ \Omega$, is initially driving a fan-type load at rated conditions. Find the voltage to be applied across the motor terminals to drive the fan at 1,200 rpm. Also calculate the current and power factor at 1,200 rpm.

SOLUTION:

$$\text{Full load torque} = \frac{2,200}{1,435 \times \dfrac{2\pi}{60}} = 14.64\ \text{Nm}$$

$$\text{Full load Slip} = \frac{1,500 - 1,435}{1,500} = 0.043$$

For a fan-type load, Torque ∞ Speed^2

$$\text{Torque} = K(1-S)^2$$

$$14.64 = K(1 - 0.043)^2$$

$$K = 15.98$$

$$\text{Slip at } 1,200\ \text{rpm} = \frac{1,500 - 1,200}{1,500} = 0.2$$

Torque required to drive the fan at 1200 rpm is given by,

$$\text{Torque} = 15.98(1 - 0.2)^2 = 10.22\ \text{Nm}$$

We have

$$T_e = \frac{p_1 m_1}{2\pi f_1} \times \frac{R_r}{S} \times \frac{V_s^2}{\left(R_s + \dfrac{R_r}{S}\right)^2 + \omega^2 (L_s + L_r)^2}$$

$$10.22 = \frac{2 \times 3}{2\pi \times 50} \times \frac{5}{0.2} \times \frac{V_s^2}{\left(2.745 + \dfrac{5}{0.2}\right)^2 + 22.45^2}$$

On solving we get,

$$V_s = 165.15\ \text{V}$$

Terminal voltage to be applied = 286.04 V
Because X_m is not mentioned, it can be neglected, so the current is given by

$$I_s = \frac{165.15}{(2.745 + 11.225j) + \left(\dfrac{5}{0.2} + 11.225j\right)} = 4.627\ \angle{-38.97}$$

Power factor = cos 38.97 = 0.77 (lagging)

EXAMPLE 10.3

A three-phase induction motor rated 2.2 kW, 415 V, 50 Hz, 4.7 A, 1,435 rpm is supplied with 300 V to drive a constant torque load of 5 N-m. The parameters of the machine are: $R_s = 8.0\ \Omega$, $R_r = 8.3\ \Omega$, $X_s = X_r' = 14.11\ \Omega$. Compute the speed of the drive.

SOLUTION:

$$\text{Full load torque} = \frac{2{,}200}{1{,}435 \times \dfrac{2\pi}{60}} = 14.64\ \text{Nm}$$

From Equation (9.26),

$$T_e = \frac{p_1 m_1}{2\pi f_1} \times \frac{R_r}{S} \times \frac{V_s^2}{\left(R_s + \dfrac{R_r}{S}\right)^2 + \omega^2 (L_s + L_r)^2}$$

$$5 = \frac{2 \times 3}{2\pi \times 50} \times \frac{8.3}{S} \times \frac{\left(300/\sqrt{3}\right)^2}{\left(8 + \dfrac{8.3}{S}\right)^2 + 14.11^2}$$

$$5 = \frac{1}{S^2} \times \frac{S \times 4{,}755.54}{64 + \dfrac{68.89}{S^2} + \dfrac{132.8}{S} + 199.01}$$

$$263 S^2 - 818.2 S + 68.89 = 0$$

On solving the above equation, we get S = 3 and S = 0.087. Here we choose 0.087 because S = 3 corresponds to the braking operation.

$$\text{Speed} = \omega_s (1 - S) = 1{,}500(1 - 0.0866) = 1{,}369.5\ \text{rpm}$$

EXAMPLE 10.4

A three-phase four-pole, 50- Hz, 440-V, 1,460- rpm induction motor has the following parameters: stator resistance 1.68 Ω, rotor resistance 2.3 Ω, rotor and stator reactance 5.8 Ω at 50 Hz. If the motor is driven from a three-phase ac voltage controller, compute the minimum possible firing angles of a silicon-controlled rectifier (SCR) at no load and at full load. The motor runs at 1,480 rpm in the no-load condition.

SOLUTION:

$$\text{No load slip } S = \frac{1{,}500 - 1{,}480}{1{,}500} = 0.013$$

$$\text{No load current} = \frac{\dfrac{440}{\sqrt{3}}}{(1.98 + 5.8j) + \left(\dfrac{2.3}{0.013} + 5.8j\right)} = 1.608\angle -4.2°\text{A}$$

The firing angle should be greater than the power factor angle, so the minimum α at no-load is 4.2°.

$$\text{Full load slip S} = \frac{1{,}500 - 1{,}460}{1{,}500} = 0.02$$

$$\text{Full load current} = \frac{\frac{440}{\sqrt{3}}}{(1.98 + 5.8j) + \left(\frac{2.3}{0.02} + 5.8j\right)} = 2.16 \angle{-5.66°}\text{A}$$

Minimum possible α at no-load is 5.66°

10.3 Efficiency of Voltage-Controlled Induction Motor Drives

Equation (10.4) is reproduced below.

$$\frac{P_{mech}}{P_{ag}} = 1 - S \tag{10.16}$$

The above expression is valid for all induction motors, provided that the stator frequency is kept constant at the rated value. Thus, for the induction motor to achieve higher efficiency, the rated speed is always kept nearer to the synchronous speed. As an example, for an induction motor, if the synchronous speed is 1,500 rpm, the rated speed should generally be 1,450–1,500 rpm so that at the rated speed the motor possesses higher efficiency. Now consider the stator voltage control of an induction motor with a fan-type load. It was shown that it is possible to achieve larger speed variation with increased rotor resistance. However, when the speed is lower, it can be concluded that the efficiency of the motor is poor. Thus the major drawback associated with variable-speed operation of fan/pump motors with stator voltage control is poor efficiency. It may be noted that, in spite of this drawback, stator voltage control is still preferred for fan/pump-type applications of low- to medium-power range due to the reduced cost of the stator voltage control scheme and smaller size.

EXAMPLE 10.5

A three-phase, 400-V, four-pole, 50-Hz induction motor is to be run at 1,000 rpm, (a) from a 50-Hz supply or (b) from a 40-Hz supply. Compare the approximate efficiencies. Compute the voltage to be applied to the drive when it is operated with 40 Hz such that torque remains constant at rated value.

SOLUTION

Neglecting rotational losses, efficiency of an induction motor is (1-S).

$$\text{Slip at 50 Hz} = (1{,}500 - 1{,}000)/1{,}500 = 0.33$$

$$\text{Efficiency at 50 Hz} = (1 - 0.33) \times 100 = 66.67\%$$

$$\text{Slip at 40 Hz} = (1,200 - 1,000)/1,200 = 0.167$$

$$\text{Efficiency at 40 Hz} = (1 - 0.167) \times 100 = 83.33\%$$

This clearly indicates that a variable-frequency drive is more efficient than a voltage-controlled drive.

To maintain the torque constant at rated value V/f ratio should be constant. Therefore,

$$\frac{V_{rated}}{f_{rated}} = \frac{400}{50}$$

$$\frac{V_{50Hz}}{f_{50Hz}} = \frac{V_{40Hz}}{f_{40Hz}}$$

$$V_{40Hz} = 400 \times 40/50 = 320 \text{ V}$$

10.4 Power Converters for Stator Voltage Control

Sometimes called ac voltage regulators, ac voltage controllers employing triacs or SCRs are generally used for stator voltage control of induction motor drives with fan- or pump-type loads. For low-power applications such as domestic fan speed control, triacs are used, whereas SCRs are used for medium- to high-power fan or pump loads. The ac voltage controller-fed domestic fan motor uses a single triac with a simple R-C firing circuit. Commercially it is known as an electronic fan regulator, shown in Fig. 10.4. Here the variable resistance or the potentiometer is suitably projected outside for convenient variation of resistance, which results in a change of the triac firing angle. Thus, R can be conveniently varied, changing the stator voltage and hence resulting in fan speed control.

A three-phase ac voltage controller feeding a three-phase induction motor is shown in Fig. 10.5(a). Here, there are three pairs of SCRs connected back-to-back, namely $T_1 - T_1^1$, $T_2 - T_2^1$, and $T_3 - T_3^1$. The firing angle of T_1 is fixed at α such that T_1^1 is triggered at $180° + \alpha$. The pair $T_2 - T_2^1$ receives firing pulses at $120° + \alpha$ and $120° + 180° + \alpha$, respectively. For the pair $T_3 - T_3^1$, the firing pulses are generated at $240° + \alpha$ and $240° + 180° + \alpha$, respectively.

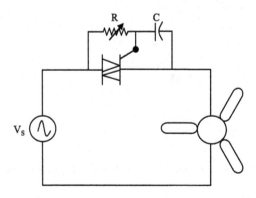

FIGURE 10.4
Triac-based domestic fan speed control.

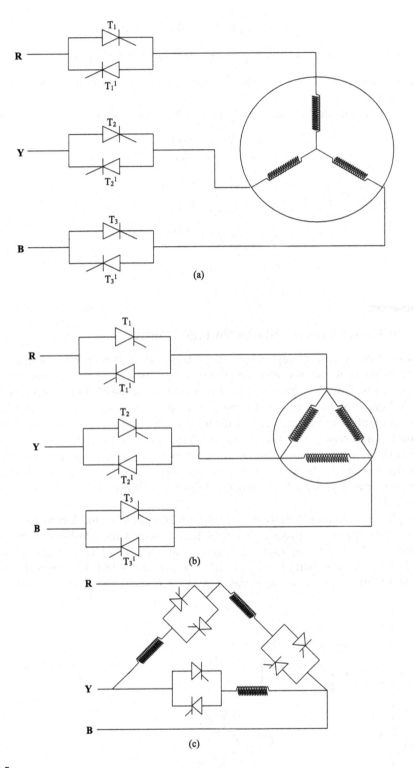

FIGURE 10.5
Three-phase SCR-based voltage regulator for (a) a star-connected induction motor and (b) and (c) delta-connected induction motors.

With a delta-connected induction motor, there are two thyristor-fed voltage-control schemes as shown in Fig. 10.5(b) and 10.5(c). The circuit in Fig. 10.5(b) is the same as that in Fig. 10.5(a). The firing pulses to SCRs in Fig. 10.4 and Fig. 10.5(b) can be generated with reference to phase voltages. However, for the circuit in Fig. 10.5(c), line voltage is taken as the reference for firing SCRs. Furthermore, in this topology, it is mandatory to have all six terminals of the motor available externally.

10.5 Energy-Efficient Operation of an ac Voltage Controller-Fed Induction Motor Drive

While an ac voltage controller is primarily suited for the speed control of induction motors with fan- or pump-type loads, this controller can also be used as an energy saver while driving three-phase induction motors. When a three-phase induction motor is supplied directly from three phase lines, the stator voltage, V_s, and frequency, f, remain at the rated values. Hence, the air gap flux, which is proportional to V/f, remains constant at its rated value regardless of the load on the shaft. If the load on the shaft varies largely, say 10% to rated value at different intervals, the motor efficiency will also change depending on the load condition. To be precise, the motor efficiency is poor under light load conditions, and the efficiency increases as the load on the shaft increases. The motor power factor is also poor at low load conditions, but this increases to higher values with increase in load. Thus, with a variable load on the shaft, the motor efficiency and the power factor are poor under light load conditions. The poor performance of the motor under light load conditions can be attributed to the fact that the air gap flux, and hence the magnetization current and the flux related core losses, remain the same at all load conditions. In other words, when the load on the shaft is less, it demands only less air gap flux, although it is maintained at the rated value because the motor is permanently connected to the supply lines.

The concept of enhancing the energy efficiency of an induction motor drive is based on the fact that the stator voltage can be adjusted depending on the load on the shaft; i.e., the stator voltage can be low for a light load, medium for medium loads, and at rated voltage for rated load conditions. When the load is less on the shaft and a low voltage is applied to the stator with the frequency remaining the same, the V/f ratio is less than the rated value. This reduces the air gap flux and the amplitude of magnetization current. This lowers stator copper loss as well as flux-related core losses, leading to improved motor efficiency and power factor. With an ac voltage controller now supplying power to the induction motor, the converter firing angle can be suitably adjusted for the required stator voltage to meet the load torque demand. In a closed-loop mode, the load is continuously sensed and the stator voltage is controlled accordingly by adjusting the converter firing angle to obtain higher efficiency and power factor.

10.6 ac Voltage Controller for Starting Induction Motor Drives

Another application of an ac voltage controller is to start three-phase induction motor drives. The thyristorized voltage regulator is called a soft starter or solid-state starter. Compared to the conventional methods of soft starting, such as a star-delta switch, series

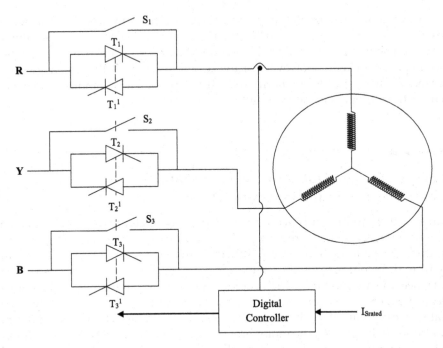

FIGURE 10.6
Induction motor soft starting.

resistance starting, and autotransformer-based starting, the solid-state starter possesses several advantages such as neatness, compactness, ease of implementation of the closed-loop control, lower cost, and near-zero maintenance cost. The schematic of a soft starter–fed induction motor drive is given in Fig. 10.6. Here the ac voltage controller is used to adjust the stator voltage during starting. The motor current is sensed with a current trans-former and fed to a digital controller. The digital processor finds the difference between the measured current and the rated current (I_{srated}), and the error is processed by the digital controller. Subsequently, the SCR firing angle α is varied to achieve smooth starting with rated current. Once the starting process is over, the thyristor units are bypassed through mechanical switches indicated as S_1, S_2, and S_3 as shown in the Fig. 10.6.

10.7 Starting an Induction Motor Using a Thyristorized Star-Delta Switch

In this method, the conventional star-delta switch is replaced with a SCR-based circuit configuration as shown in Fig. 10.7. When continuously triggered, a pair of anti-parallel connected SCRs will act as a closed switch, and this principle is used to construct the solid-state star-delta switch. There are six pairs of SCRs connected back-to-back; three pairs constitute the star group, and the other three pairs comprise the delta group. The operation of the circuit emulates conventional electromechanical star-delta starting. Here, SCRs in the star group are fired first, and as motor speed picks up, firing pulses to the star group SCR are withdrawn, at which point the SCRs in the delta group are fired. However, it is evident from the circuit that simultaneous triggering of SCR pairs in both groups

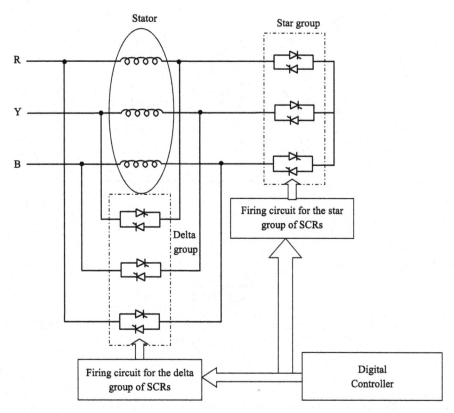

FIGURE 10.7
Speed control of an induction motor using thyristorized star-delta switch.

causes a short-circuit of the supply lines and should be avoided. Hence, a reasonable delay time, typically taken to be three cycles of supply voltage, should be allowed during the star-delta transition. The firing pulses to the SCRs, on both the star and the delta groups, are generated using a digital controller.

10.8 Speed Control of a Wound-Type Induction Motor through the Rotor Side

In the case of a wound-type induction motor (i.e., slip-ring type), it is possible to control the rotor resistance such that a variable-speed operation is achieved. The analytical equations that govern the speed control are given here.

The electromagnetic torque is given by

$$T_e = \frac{p_1 m_1}{2\pi f_1} I_r^2 \frac{R_r}{S} \tag{10.17}$$

Assuming the rotor current I_r is approximately equal to stator current I_s,

$$T_e = \frac{p_1 m_1}{2\pi f_1} I_s^2 \frac{R_r}{S} \tag{10.18}$$

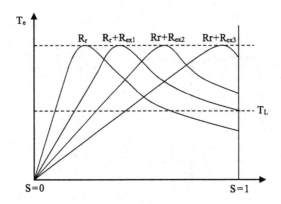

FIGURE 10.8
Torque-speed curves with rotor resistance control.

Because $I_s \propto V_s$,

$$T_e = \frac{p_1 m_1}{2\pi f_1} V_s^2 \frac{R_r}{S} \tag{10.19}$$

In the case of a wound-type induction motor with rotor-side control, stator voltage, V_s, and frequency, f_1, are constants. Thus, the above equation reduces to

$$T_e \propto \frac{R_r}{S} \tag{10.20}$$

Equation (10.20) clearly suggests that, for a given slip S, the electromagnetic torque is proportional to rotor resistance R_r:

$$T_e \propto R_r \tag{10.21}$$

If it is possible to connect external resistances in the rotor circuit, it is evident that different torque values can be obtained for a given slip or speed. If we assume that R_{ex1}, R_{ex2}, R_{ex3} are three values of externally connected rotor resistances, where $R_{ex1} < R_{ex2} < R_{ex3}$, it is possible to get a family of torque-slip characteristic curves, as given in Fig. 10.8. Considering a constant load torque of T_L, Fig. 10.8 suggests a wide range of speeds with variable rotor resistance. While the wide speed control range as indicated in Fig. 10.8 is very attractive, one must remember that the increased speed range is achieved at the cost of increased rotor copper loss or poor efficiency.

The implementation of rotor resistance variation is now discussed. Assuming that a balanced three-phase variable resistance is available, this can be directly connected to the slip rings as indicated in Fig. 10.9, and manual control of the balanced three-phase resistance will yield different speeds for a given torque. However, this requires

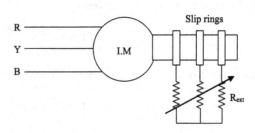

FIGURE 10.9
Balanced rotor resistance control.

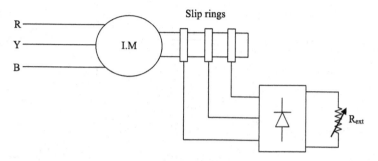

FIGURE 10.10
Rotor resistance control with a SCR diode rectifier and variable resistance.

human intervention for speed control. There is also a possibility of unbalanced rotor resistance while controlling.

To circumvent the problem of unbalance resistance, a diode rectifier together with a variable external resistance is now connected to the slip rings of the motor as shown in Fig. 10.10.

In Fig. 10.10, a human operator is mandatory, making the scheme infeasible for closed-loop speed control. To address this, the diode bridge rectifier in Fig. 10.10 is replaced with a SCR-based bridge rectifier, such that the SCR firing angle can be varied to achieve a variable speed. The closed-loop scheme is shown in Fig. 10.11.

The SCRs in the rectifier shown in Fig. 10.11 are triggered based on rotor voltage, and because the rotor frequency is slip × stator frequency, for each change in speed, the rotor frequency must be predetermined to generate the firing pulses to the SCRs. Hence a rotor chopper-based scheme is now proposed in Fig. 10.12. Here the rotor voltage is first rectified into dc using a diode bridge rectifier, and a static chopper circuit is connected across the external resistance R_{ext}. The change in the duty ratio of the static chopper will result in a variable rotor resistance and speed control.

All of the schemes described in Fig. 10.9 through Fig. 10.12 have the problem of increased rotor copper loss and poor efficiency. When power semiconductor devices are employed for closed-loop speed control as shown in Fig. 10.11 and Fig. 10.12, the rotor current is interrupted, and this is reflected on the stator side also. In other words, with solid-state switches on the rotor side, the stator current is polluted, yielding to poor efficiency and torque pulsations. To summarize the methods described in Fig. 10.9 through Fig. 10.12, they are less efficient, they cause poor input power factor, and they may also lead to torque pulsations. Therefore, these schemes are not generally recommended.

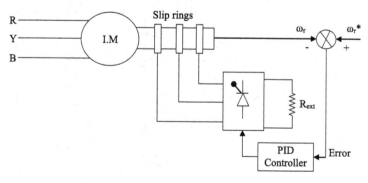

FIGURE 10.11
Rotor resistance control with SCR converter.

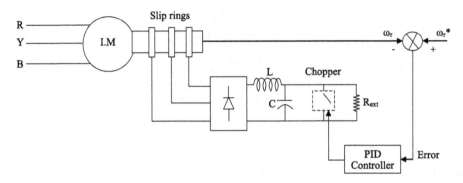

FIGURE 10.12
Rotor resistance control using chopper.

EXAMPLE 10.6

A three-phase wound-rotor induction motor is connected directly to a three-phase supply of constant voltage and frequency. Pairs of inverse parallel SCRs are incorporated onto the secondary windings using the connection seen in Fig. 10.13. In values referred to as primary turns, $R_A = 6 R_B$ and R_A is two times the secondary winding resistance, R_r. Sketch and explain the approximate torque speed characteristic for two conditions: (a) $\alpha = 0°$, and (b) $\alpha = 180°$.

SOLUTION:

a. $\alpha = 0°$

R_A and R_B are connected in parallel, and R_r comes in series to this combination. Overall rotor resistance is given by

$$R = R_r + (R_A \| R_B)$$
$$R = R_r + (12R_r \| 2R_r) = 2.71R_r$$

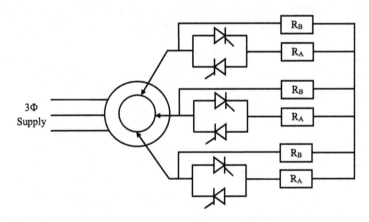

FIGURE 10.13
Circuit for Example 10.6.

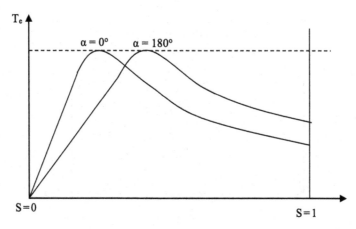

FIGURE 10.14
Figure for Example 10.6.

b. $\alpha = 180°$

R_A will be removed from the circuit, and R_B comes in series to R_s. Overall rotor resistance is given by

$$R = R_r + R_B$$
$$R = R_r + 2R_r = 3R_r$$

Approximate torque slip characteristics are given in Fig. 10.14.

EXAMPLE 10.7

A three-phase wound-rotor induction motor is connected directly to a three-phase supply of constant voltage and frequency. The rotor circuit consists of a chopper followed by a rectifier. A fixed resistance of 100 Ω is connected in the chopper output. Draw the approximate torque-slip characteristic curves for the following duty ratios: $D = 100\%$, $D = 60\%$, $D = 40\%$. Comment about the nature of the curves with variations in the duty ratio.

SOLUTION:

At $D = 100\%$, the external resistance is shorted and rotor resistance is R_r only. With the decreased duty ratios, R_{ext} is connected to the rotor circuit, leading to increased values of effective rotor resistance. The characteristic curves are given in Fig. 10.15.

EXAMPLE 10.8

A three-phase wound-type induction motor is rated 440 V, 25 A, 50 Hz, and 1,440 rpm. It uses a SCR-controlled three-phase semi-converter at the rotor side for rotor resistance variation. For 50% of rated speed, the SCR firing angle is computed as 90°. Find the necessary time delay of firing pulses to the SCRs.

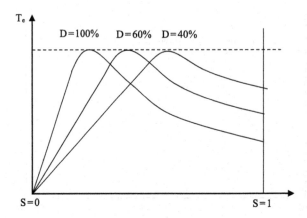

FIGURE 10.15
Figure for Example 10.7.

SOLUTION:

Synchronous speed, $N_s = 1,500$ rpm
50% of the rated speed $= 1,440/2 = 720$ rpm

$$\text{Slip S} = \frac{1,500 - 720}{1,500} = 0.52$$

Rotor frequency $= S$, $f_1 = 0.52 \times 50 = 26$ Hz
For the 90° firing angle, the time delay $= \dfrac{1}{26} \times \dfrac{90}{360} = 9.6$ ms.

EXAMPLE 10.9

A three-phase SCR-based ac/dc converter is connected to the slip rings of a three-phase induction motor. A resistance of 100 Ω is connected at the output of the converter. On a single diagram, sketch the torque-speed characteristics (approximately) under the following conditions: (a) slip ring short-circuited; (b) SCRs fired at 0°; (c) SCRs fired at 60°.

SOLUTION:

When the slip rings are shorted, only the rotor resistance will come into the circuit. As the firing angle increases, the effective resistance is increased. The characteristic curves are shown in Fig. 10.16.

10.9 Slip Power Recovery Scheme

To overcome the increased copper loss and lower efficiency for the schemes described earlier, a noble scheme called the slip power recovery scheme is introduced and is shown in Fig. 10.17. Here there are two converters labeled as conv-1 and conv-2. Conv-1 is a diode bridge rectifier, which converts the rotor voltage and current into dc and filters them

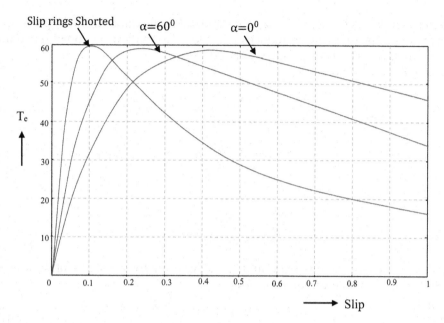

FIGURE 10.16
Figure for Example 10.9.

through the inductance, L. Thus $V_1 \times I_{dc}$ as indicated in figure is the quantum of rotor power, generally called slip power. Conv-2 is a thyristorized bridge rectifier triggered at firing angle α. Depending on the amplitude of α, V_2 changes and $V_2 \times I_{dc}$ is the power fed back to the ac mains through a transformer. Conv-2 is operating in the inverting mode such that α is in the range $\pi/2 \le \alpha \le \pi$, which makes V_2 negative. The rotor voltage is of low amplitude, as are the magnitudes of both V_1 and V_2. Hence, to cope with the amplitude of line voltage, a transformer of transformation ratio, N_T, is introduced in the scheme as shown in Fig. 10.17.

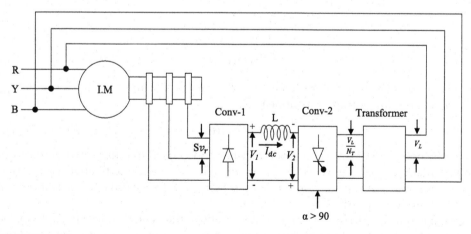

FIGURE 10.17
Slip power recovery scheme.

Let V_r be the rotor-induced line voltage at standstill of the rotor. As motor picks up speed at a slip, S, the rotor line voltage reduces to $S \times V_r$. The analytical expression connecting motor slip and converter firing angle is derived below.

Putting $\alpha = 0$ in Equation (3.26):

$$V_1 = \frac{3V_{LM}}{\pi} = \frac{3\sqrt{2}V_L}{\pi} \tag{10.22}$$

$$V_1 = \frac{3\sqrt{2}}{\pi} SV_r \tag{10.23}$$

Similarly,

$$V_2 = \frac{3\sqrt{2}}{\pi} \frac{V_L}{N_T} \cos\alpha \tag{10.24}$$

Considering the loop comprising v_1 and v_2 and ignoring the resistance of the filter inductance, the following equation can be written:

$$V_1 + V_2 = 0 \tag{10.25}$$

$$\frac{3\sqrt{2}}{\pi} SV_r + \frac{3\sqrt{2}}{\pi} \frac{V_L}{N_T} \cos\alpha = 0 \tag{10.26}$$

$$SV_r = -\frac{V_L}{N_T} \cos\alpha \tag{10.27}$$

$$S = -\frac{V_L}{V_r} \frac{1}{N_T} \cos\alpha \tag{10.28}$$

In the above expression, $\dfrac{V_L}{V_r}$ indicates the ratio of stator turns to rotor turns of the motor; let this ratio be indicated by N. Hence, we can write

$$S = -\frac{N}{N_T} \cos\alpha \tag{10.29}$$

$$S \propto -\cos\alpha \tag{10.30}$$

Remember that the value of α is constrained between $\pi/2$ and π, so the negative sign in the above expression naturally disappears.

$$S \propto \cos\alpha \tag{10.31}$$

It may be noted that conv-1 is a diode rectifier and is used to take away the rotor power from the motor. Hence, the speed of the motor will be constrained between zero to near synchronous speed in the above-mentioned scheme. In other words, the operation will be below synchronous speed.

In case it is required to have the motor speed above the synchronous speed, then conv-1 must be replaced with a thyristorized bridge circuit. In such a case, by suitably adjusting the firing angles of the two converters, it is possible either to take away the rotor power or feed power to the rotor circuit. In the second case, because air gap power and the rotor power injected are added together, the motor speed increases above the synchronous speed. This mode is called super-synchronous speed. In other words, when both the converters are thyristor converters, it is possible to adjust the firing angles of both convertors such that either sub-synchronous speed or super-synchronous speed can be achieved. When the machine works in the super-synchronous speed, it is no longer a conventional induction motor; instead it is called a doubly-fed induction motor.

EXAMPLE 10.10

A three-phase, four-pole, 400-V, 50-Hz induction motor employs a slip power recovery scheme. The turns ratio of the induction motor is 2. A transformer of turns ratio 5:1 is used between the supply lines and the SCR converter. Find the firing angle of the SCR converter to run the motor at a speed of 1,200 rpm.

SOLUTION:

We have

$$\text{Synchronous speed, } N_s = \frac{120f}{P}$$
$$N_s = 120 \times 50/4$$
$$= 1,500 \text{ rpm}$$

$$\text{Slip, } S = (N_s - N_r)/N_s$$
$$= (1,500 - 1,200)/1,500$$
$$= 0.2$$

In the slip recovery scheme,

$$S = -(N_a/N_T)\cos\alpha$$
$$\cos\alpha = -0.2 \times 5/2$$
$$= -0.5$$

$$\alpha = \cos^{-1}(-0.5)$$
$$= 120°$$

Questions

1. Write the necessary equations to show the effect of stator voltage control on the speed of an induction motor with suitable graphs.

2. Justify this statement: A wide speed control range for stator voltage-controlled fan/pump-type loads needs increased rotor resistance.

3. For a variable-voltage induction motor-driven fan, derive the relationship between slip and stator current at maximum and full loads.

4. Sketch and explain a triac-based speed control circuit for domestic fan motors.

5. Explain the principle of efficiency enhancement of lightly loaded induction motors using stator voltage control.

6. Draw and explain the SCR-based star-delta starter.

7. Elucidate different power electronics employed for rotor resistance control.

8. Describe the theory of slip-power recovery scheme.

Unsolved Problems

1. The slip-power recovery scheme (sub-synchronous speed mode) is employed for variable-speed operation (0.5–0.9 times synchronous speed) of a three-phase slip ring induction motor drive. If the firing angle of the inverter is limited to 170°, find the useful range of inverter firing angles in the entire operating range with and without a transformer between the inverter and ac supply. (Turns ratio IM = 2; Turns ratio T/F = 3.5)

2. A three-phase, 400-V, four-pole, 50-Hz induction motor rated 1,440 rpm is to be run at 1,000 rpm from a 40-Hz supply. Calculate the efficiency of the machine.

3. What is the minimum firing angle that can be employed for an ac/ac converter that feeds the following induction machine running at rated conditions: 2.2 kW, 415 V, 50 Hz, 4.7 A, 1,435 rpm; $R_s = 8.0\ \Omega$, $R_r = 8.3\ \Omega$, $X_s = X_r' = 14.11\ \Omega$.

4. A three-phase wound-type induction motor is rated 440 V, 25 A, 50 Hz, 1,440 rpm. It uses a SCR-controlled, three-phase semi-converter at the rotor side for rotor resistance control. For a firing angle of 60°, the time delay is found to be 6 ms. Calculate the speed of the machine.

5. Calculate the peak current for a three-phase induction motor driving a fan/pump load. The machine specifications are: 2.2 kW, 415 V, 50 Hz, 1,435 rpm, and 10 A.

6. Estimate the terminal voltage to be applied for a drive that delivers a constant torque of 5 Nm at 1,200 rpm. The drive is employing an induction motor with the following rating and parameters: 415 V, 50 Hz, 7.5 A, 1,440 rpm, $R_s = 3.46\ \Omega$, $R_r = 0.68$, $X_s = X_r' = 5.9\ \Omega$.

7. A three-phase, 4-kW, four-pole, 50-Hz, 440-V, 1,440-rpm induction motor has the following parameters: stator resistance 1.68 Ω, rotor resistance 2.3 Ω, and rotor and stator reactance is 5.8 Ω at 50 Hz. It is initially driving a fan-type load at rated conditions. Find the voltage to be applied across the motor terminals to drive the fan at 1,300 rpm.

8. A star-connected, three-phase induction motor, rated 2.2 kW, 415 V, 50 Hz, and 1,435 rpm with parameters $R_s = 2.745\ \Omega$, $R_r = 5\ \Omega$, $X_s = X_r' = 11.225\ \Omega$ is initially driving a fan-type load at rated conditions. Compare the power factor of the drive at rated conditions and 1,200 rpm.

9. A three-phase wound-rotor induction motor is connected directly to a three-phase supply of constant voltage and frequency. The rotor circuit consists of a chopper followed by a rectifier. A fixed resistance of 100 Ω is connected in the chopper output. Draw the approximate torque-slip characteristic as the duty ratio varies

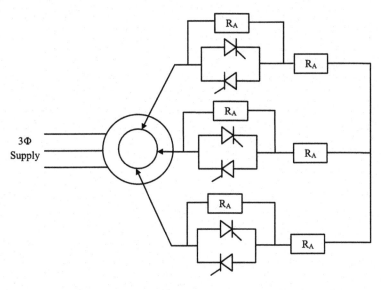

FIGURE 10.18
Circuit for Unsolved Problem 10.

from 100% to 0%. Machine parameters are 3 Ω, four-pole, 50 Hz, 415 V, 1,4350 rpm, $R_s = 8\ \Omega$, $R_r = 8.3\ \Omega$, $X_s = X_r = 14.2\ \Omega$.

10. A three-phase wound-rotor induction motor is connected directly to a three-phase supply of constant voltage and frequency. Pairs of inverse parallel SCRs are incorporated onto the secondary windings using the connection as shown in Fig. 10.18. In values referred to as primary turns, $R_A = 100\ \Omega$. Sketch the shifting of the torque slip characteristic when α is increased from zero toward 180°.

Answers

1. With transformer: 100.07° to 151.04°

   Without transformer: 92.86° to 138.59°

2. Slip = 0.166, efficiency = 83%

3. $I_s = 1.10\angle -3.7°$, minimum angle = 3.7°

4. Slip = 0.55, Speed = 666.6 rpm

5. $S_{FL} = 0.043$; $\dfrac{I_{s\,max}}{I_{sFL}} = \dfrac{2}{3\sqrt{3}\ (1 - s_{FL})\sqrt{S_{FL}}} = 1.93$; $I_{s\,max} = 19.3$

6. Slip = 0.2, $V_s = 143.05$ V

7. $T_{FL} = 26.52$ Nm, $S_{FL} = 0.04$, K = 28.77, T_L at 1,300 rpm = 21.626, slip at 1,300 rpm = 0.133, voltage = 223.90 V

8. Power factor at rated conditions = 0.98 lagging; at 1,200 rpm, power factor = 0.77 lagging

11

dc/ac Converters

11.1 Introduction

Popularly known as inverters, dc/ac converters deliver power from a dc grid or battery and produce variable-frequency, variable-amplitude ac output. Depending on the circuit topology, inverter output can be single-phase or three-phase, and they can possess either square wave, pulse-width–modulated wave, or stepped wave. Inverters are used extensively with variable-speed ac motors and in induction heating, uninterruptible power supply devices (UPS), active filters, and various applications in power systems. This chapter elaborates different inverter circuits and operations.

11.2 Basic Concept of Inverter Circuit-Voltage Source Inverters

Broadly speaking, there are two types of inverters: voltage-source inverters (VSIs) and current-source inverters (CSIs). An inverter is a power electronic circuit that converts a fixed dc voltage into variable-amplitude, variable-frequency voltage at its output. In a VSI, the input dc voltage is held constant at a specified value regardless of variations of output voltage and or output current. The schematic of a VSI is shown in Fig. 11.1. Various circuit configurations are explained below.

11.3 Bridge Inverters

There are two types of bridge inverters, namely half-bridge inverters and full-bridge inverters, which are shown in Fig. 11.2(a) and Fig. 11.3(a), respectively. With a half-bridge inverter, the existing supply is divided into two equal halves and applied to the load through switches S_1 and S_2. The gating signal to the switches, together with the output voltage waveforms, are shown in Fig. 11.2(b). Here, switch S_1 is turned ON for the first half-cycle of the output voltage, leading to $+V_{dc}/2$ across the load. For the negative half-cycle, S_1 is turned OFF and S_2 is turned ON for the remaining half-cycle interval. The output voltage is given in Fig. 11.2(c). The rms value of the output voltage remains $V_{dc}/2$; the output frequency is decided by the triggering frequency of switches, which enables variable-frequency operation.

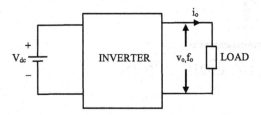

FIGURE 11.1
Block schematic of voltage source inverter.

A full-bridge inverter consists of four switches but requires only a single dc source of amplitude V_{dc}. The gating signals to the switches and the output voltage waveform are shown in Fig. 11.3(b). Here S_1 and S_3 conduct together during the positive half-cycle, and S_2 and S_4 are gated during the negative half-cycle of output voltage. The rms value of the output voltage is V_{dc}, and the output frequency can be changed by varying the triggering rate of the switches. Referring to the output voltage waveforms, it is evident that the rms value of the output voltage is V_{dc} and the rms value of the load current with resistive

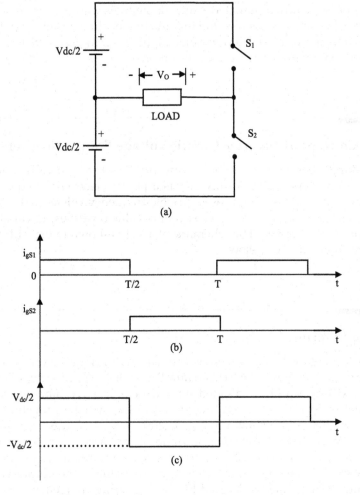

FIGURE 11.2
Single-phase, half-bridge inverter. (a) Circuit. (b) Gating signals. (c) Output voltage.

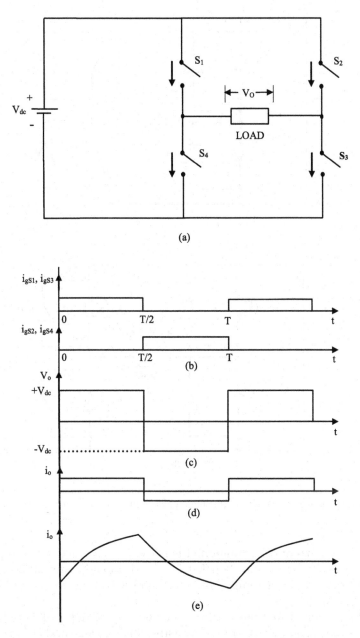

FIGURE 11.3
Single-phase, full-bridge inverter. (a) Circuit. (b) Gating signals. (c) Output voltage. (d) Load current with resistive load. (e) Load current with inductive load.

load is $\frac{V_{dc}}{R}$. Hence, the output power is $\frac{V_{dc}^2}{R}$. Furthermore, the peak reverse voltage across each switching element is $2V_{dc}$.

The switches shown in the inverter circuits can be realized using silicon-controlled rectifiers (SCRs), insulated-gate bipolar transistors (IGBTs), etc. If SCRs are employed as the switching elements, forced commutation, namely current commutation, is used to turn

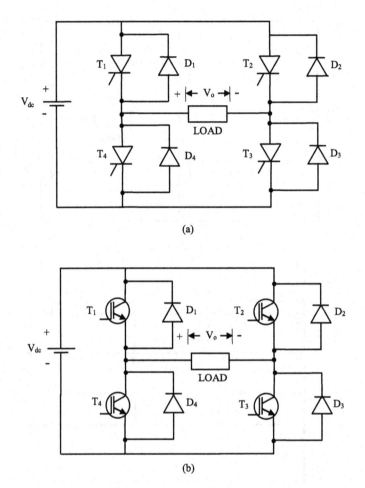

FIGURE 11.4
Single-phase, full-bridge inverter using SCR. (b) Single-phase, full-bridge inverter using IGBT.

the SCR OFF. With current commutation, it is mandatory to have one anti-parallel diode across each SCR. With gate-controlled IGBTs, an additional commutation circuit is not required. Full bridge inverters employing SCRs and IGBTs are shown in Fig. 11.4(a) and Fig. 11.4(b), respectively.

It should be mentioned that the nature of current in VSI depends largely on the type of load, such as R, R-L, and R-L-C. With a resistive load, voltage and current wave shapes are identical, whereas with R-L and R-L-C, load current wave shape is identified through mathematical analysis. In the case of an inductive load, the load current is similar to the one given in Fig. 11.3(e).

EXAMPLE 11.1

A single-phase full bridge inverter is supplied from a 24-V dc source and feeds power to a 10-Ω resistor. Compute the power developed and the peak reverse voltage across each switch.

SOLUTION:

$$\text{rms value of output voltage} = V_{dc} = 24 \text{ V}$$

$$\text{Output power} = \frac{V_{dc}^2}{R} = \frac{24^2}{10} = 57.6 \text{ W}$$

The peak reverse voltage across each switching element is $V_{dc} = 24$ V.

11.4 Three-Phase Inverter Circuits

The circuit representation of a three phase bridge type voltage source inverter is shown in Fig. 11.5. For simplicity, it is assumed that SCRs are employed as the switching elements. There are six SCRs paired into top and bottom groups; T_1, T_3 and T_5 belong to the upper group, while T_4, T_6 and T_2 belong to lower group. Each SCR is provided with anti-parallel diode connection so as to facilitate current commutation of each SCR. The three phase output from the inverter is available as R, Y and B. SCRs are triggered and commutated in a particular sequence. Generally, two types of patterns are used; 180 degree mode of operation and 120 degree mode of conduction. The two schemes are discussed below.

11.4.1 180° Mode of Conduction

In this type of inverter operation, each SCR is allowed to conduct one half-cycle period of the output voltage waveform. The SCRs are triggered in the order T_1, T_2, T_3, T_4, T_5, T_6, and a delay equivalent to $\pi/3$ radians is introduced between successive conducting SCRs. Accordingly, the gating signals for the six SCRs are shown in Fig. 11.6. For simplicity, each SCR is assumed to conduct for the duration of the respective gating pulse.

The calculation of output line and phase voltages is carried out assuming a purely resistive load. In general, a star-connected balanced load is assumed. There are six distinct

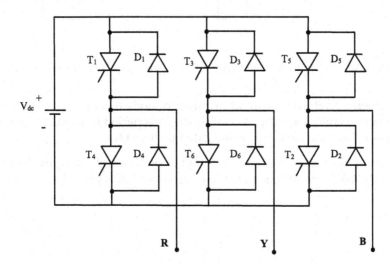

FIGURE 11.5
Three-phase full bridge inverter.

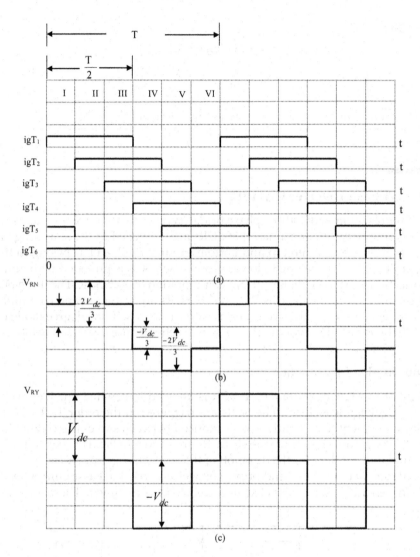

FIGURE 11.6
Three-phase full bridge inverter in 180° conduction mode. (a) Gating signals. (b) Phase voltage. (c) Line voltage.

switching patterns, referred to as mode 1 through mode 6, and during each mode three devices conduct together. Each thyristor conducts for a period of π radians, and each mode exists for a period of $\frac{\pi}{3}$ radians. This is indicated in Fig. 11.6.

MODE I: During this mode, thyristors 1, 5, and 6 conduct and the equivalent circuit is as shown in Fig. 11.7(a). From the equivalent circuit, current, i, is given by

$$i = \frac{V_{dc}}{1.5r}$$

$$V_{RN} = i \cdot \frac{r}{2} = \frac{V_{dc}}{1.5r} \cdot \frac{r}{2} = \frac{V_{dc}}{3}$$

$$V_{RY} = V_{dc}$$

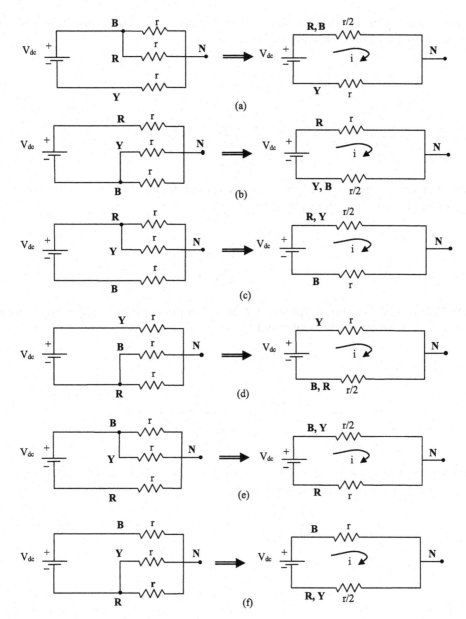

FIGURE 11.7
Equivalent circuits for (a) Mode I, (b) Mode II, (c) Mode III, (d) Mode IV, (e) Mode V, and (f) Mode VI.

MODE II: During this mode, thyristors 1, 2, and 6 conduct and the equivalent circuit is as shown in Fig. 11.7(b). From the equivalent circuit,

$$i = \frac{V_{dc}}{1.5r}$$

$$V_{RN} = i \cdot r = \frac{V_{dc}}{1.5r} \cdot r = \frac{2V_{dc}}{3}$$

$$V_{RY} = V_{dc}$$

MODE III: During this mode, thyristors 1, 2, and 3 conduct and the equivalent circuit is as shown in Fig. 11.7(c). From the equivalent circuit,

$$i = \frac{V_{dc}}{1.5r}$$

$$V_{RN} = i \cdot \frac{r}{2} = \frac{V_{dc}}{1.5r} \cdot \frac{r}{2} = \frac{V_{dc}}{3}$$

$$V_{RY} = 0$$

MODE IV: During this mode, thyristors 2, 3, and 4 conduct and the equivalent circuit is as shown in Fig. 11.7(d). From the equivalent circuit,

$$V_{RY} = -V_{dc}$$

$$V_{RN} = -i \cdot \frac{r}{2} = -\frac{V_{dc}}{1.5r} \cdot \frac{r}{2} = -\frac{V_{dc}}{3}$$

MODE V: During this mode, thyristors 3, 4, and 5 conduct and the equivalent circuit is as shown in Fig. 11.7(e). From the equivalent circuit,

$$i = \frac{V_{dc}}{1.5r}$$

$$V_{RY} = -V_{dc}$$

$$V_{RN} = -i \cdot r = -\frac{V_{dc}}{1.5r} \cdot r = -\frac{2V_{dc}}{3}$$

MODE VI: During this mode, thyristors 4, 5, and 6 conduct and the equivalent circuit is as shown Fig. 11.7(f). From the equivalent circuit,

$$i = \frac{V_{dc}}{1.5r}$$

$$V_{RY} = 0$$

$$V_{RN} = -i \cdot \frac{r}{2} = -\frac{V_{dc}}{1.5r} \cdot \frac{r}{2} = -\frac{V_{dc}}{3}$$

The gating pulses, line, and phase voltage waveforms of a three-phase full bridge inverter in 180° mode is shown in Fig. 11.6.

11.4.2 120° Mode of Operation

This mode is exactly the same as the previous mode, with the major difference being that each device is conducting only for $2\pi/3$ radians. As a result, only two devices conduct at any instant. The gating signals are depicted in Fig. 11.8. A star-connected resistive load is assumed, and modes I–VI are explained here.

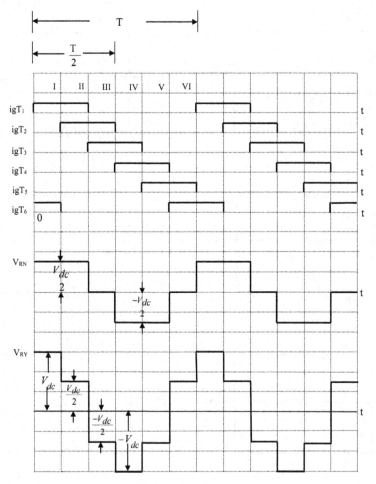

FIGURE 11.8
Three-phase full bridge inverter in 120° conduction mode. (a) Gating signals. (b) Phase voltage. (c) Line voltage.

MODE I: During this mode, thyristors 1 and 6 conduct and the equivalent circuit is as shown in Fig. 11.9(a). From the equivalent circuit,

$$V_{RY} = V_{dc}$$

$$V_{RN} = \frac{V_{dc}}{2}$$

MODE II: During this mode, thyristors 1 and 2 conduct and the equivalent circuit is as shown Fig. 11.9(b). From the equivalent circuit,

$$\overline{V_{RY}} = \overline{V_{RN}} + \overline{V_{NY}}$$

$$\overline{V_{NY}} = 0$$

$$V_{RY} = V_{RN} = \frac{V_{dc}}{2}$$

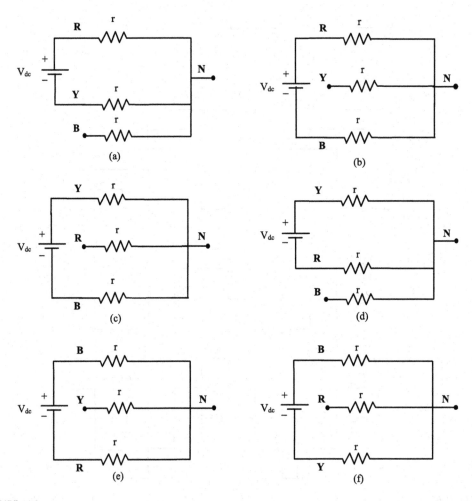

FIGURE 11.9
Equivalent circuits for (a) Mode I, (b) Mode II, (c) Mode III, (d) Mode IV, (e) Mode V, and (f) Mode VI.

MODE III: During this mode, thyristors 2 and 3 conduct and the equivalent circuit is as shown in Fig. 11.9(c). From the equivalent circuit,

$$\overline{V_{RY}} = \overline{V_{RN}} + \overline{V_{NY}}$$

$$\overline{V_{RN}} = 0$$

$$\overline{V_{RY}} = \overline{V_{NY}} = -\frac{V_{dc}}{2}$$

MODE IV: During this mode, thyristors 3 and 4 conduct and the equivalent circuit is as shown in Fig. 11.9(d). From the equivalent circuit,

$$\overline{V_{RN}} = -\frac{V_{dc}}{2}$$

$$\overline{V_{RY}} = -V_{dc}$$

MODE V: During this mode, thyristors 4 and 5 conduct and the equivalent circuit is as shown in Fig. 11.9(e). From the equivalent circuit,

$$\overline{V_{RN}} = \overline{V_{RY}} = -\frac{V_{dc}}{2}$$

MODE VI: During this mode, thyristors 5 and 6 conduct and the equivalent circuit is as shown in Fig. 11.9(f). From the equivalent circuit,

$$\overline{V_{RY}} = \overline{V_{RN}} + \overline{V_{NY}}$$

$$\overline{V_{RN}} = 0$$

$$\overline{V_{RY}} = \frac{V_{dc}}{2}$$

The gating pulses, line, and phase voltage waveforms of a three-phase full bridge inverter in 120° mode is shown in Fig. 11.8.

11.5 Voltage Control in Inverters

The frequency of output voltage of an inverter can be conveniently controlled by adjusting the SCR triggering frequency. In this section, various methods to achieve variable voltage operation are explained.

There are three commonly used methods of inverter voltage control:

- Voltage control at the inverter input
- Voltage control at the inverter output
- Voltage control within the inverter

11.5.1 Voltage Control at the Input dc Terminals

The magnitude of input dc voltage is varied so as to change inverter output voltage. This is feasible with the schemes given in Fig. 11.10. The chief disadvantage of this method is the requirement of additional power converters, which increases cost, space, and volume. Hence, this method is seldom employed.

11.5.2 Adjustment of Voltage Delivered by the Inverter

One way in which the amplitude of an inverter's output voltage can be controlled is by inserting an ac/ac converter between the inverter output and load; however, this is not a viable way due to increased harmonic content and the requirement of an additional power converter. If there are two inverters, it is possible to add the voltages delivered by the two circuits using transformers. The schematic is shown in Fig. 11.11(a). The output voltages of inverters are marked as V_{o1} and V_{o2}, and they are coupled to the load through the transformer secondaries. Here, it is mandatory that the two inverters operate at the same frequency but with a finite phase shift. This can be implemented by delaying the firing pulses to the second inverter. The phasor diagram representing the output voltages is shown in Fig. 11.11(c). It is evident that, for the delay angle $\theta_d = 0$, the inverter output

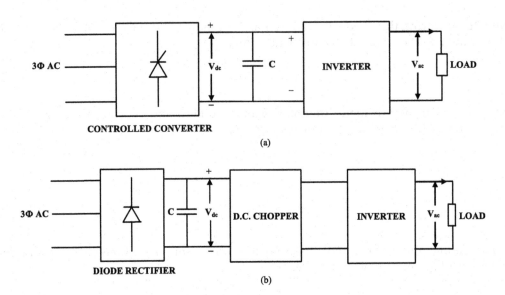

FIGURE 11.10
Voltage control at the input side of the inverter (a) employing a controlled converter, and (b) using dc choppers.

voltage is maximum, whereas with $\theta_d = \pi$ it is zero. Thus, by phase shifting the firing pulses to the second inverter, the output voltage can be smoothly controlled. However, this scheme is generally employed only for large power control.

11.5.3 Voltage Control within the Inverter

This is the most commonly used method in recent years. This method is also called pulse-width modulation. Here, instead of a single pulse, a large number of pulses of varying duration are developed at each cycle of the output voltage. This is done by turning the conducting devices ON and OFF several times per cycle.

Consider the single-phase bridge-type inverter in Fig. 11.3(a). Assume power switches S_1 and S_3 are turned ON such that the output voltage pulse width is d and this waveform is symmetric about $\frac{\pi}{2}$. Similarly, S_2 and S_4 are switched ON for the same duration during the negative half-cycle of output voltage such that the waveform is symmetric with respect to $\frac{3\pi}{2}$. The output voltage waveform is shown in Fig. 11.12. The rms value of output voltage can be derived as

$$V_{rms}^2 = \frac{1}{\pi} \int_{\left(\frac{\pi}{2}-\frac{d}{2}\right)}^{\left(\frac{\pi}{2}+\frac{d}{2}\right)} V_{dc}^2 \times d\omega_s t$$

$$V_{rms}^2 = V_{dc}^2 \left[\left(\frac{\pi}{2} + \frac{d}{2} \right) - \left(\frac{\pi}{2} - \frac{d}{2} \right) \right] \tag{11.1}$$

$$V_{rms} = V_{dc} \sqrt{\frac{d}{\pi}}$$

For $d = \pi$, $V_{o(rms)} = V_{dc}$, and this corresponds to a waveform without modulation as in Fig. 11.13(b). It is evident that this scheme enables the adjustment of output voltage by changing the value of d. It is worth mentioning that inverter output voltage can be split into

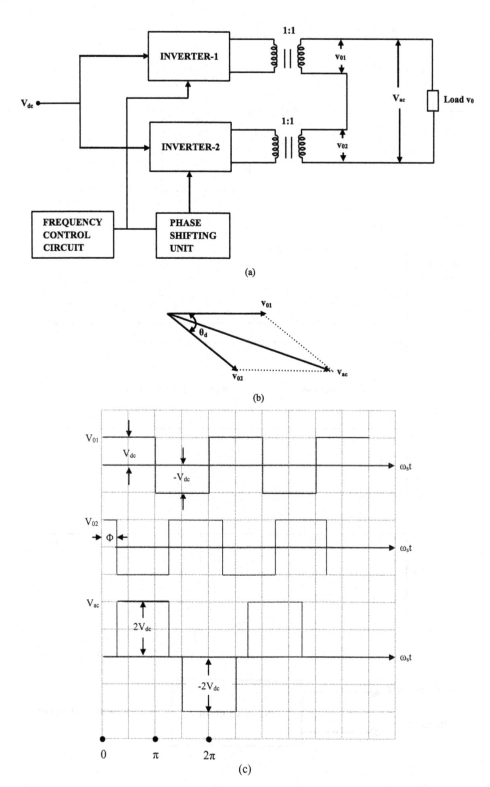

FIGURE 11.11
Voltage control at the output of inverter. (a) Circuit diagram. (b) Phasor diagram. (c) Output waveform.

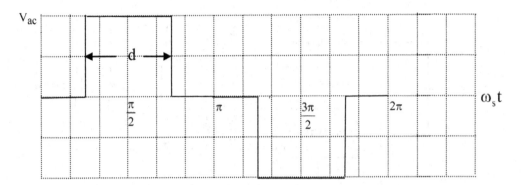

FIGURE 11.12
Single pulse-width modulation.

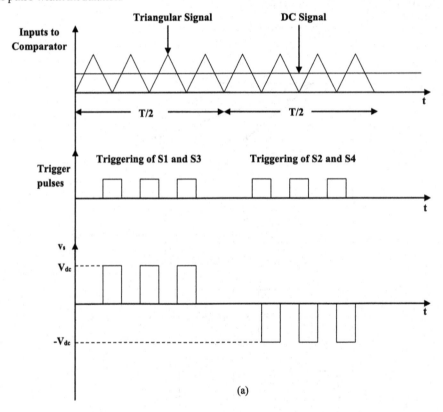

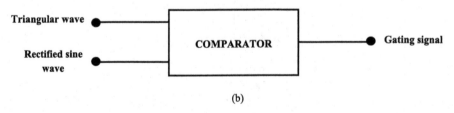

FIGURE 11.13
Multiple pulse-width modulation. (a) Waveforms. (b) Implementation.

two components: one is the fundamental component, which is sinusoidal in wave shape and has the same frequency as that of the inverter output voltage; and the other, known as harmonics, consists of combined sinusoidal and co-sinusoidal waveforms of multiple frequencies. The fundamental component alone is useful, and harmonics are undesired. The Fourier series method can be used to compute fundamental and harmonic components. The Fourier components of a single-pulse inverter can be derived as given below:

$$v_o(\omega_s t) = a_o + \sum_{n=1}^{n=\infty} (a_n \cos n\omega_s t + b_n \sin n\omega_s t)$$

$$a_o = \frac{2}{T} \int_0^T v_o(\omega_s t) \times dt$$

$$a_o = \frac{2}{2\pi} \int_0^{2\pi} v_o(\omega_s t) \times d\omega_s t$$

$$a_o = \frac{2}{2\pi} \left[\int_{\left(\frac{\pi}{2}-\frac{d}{2}\right)}^{\left(\frac{\pi}{2}+\frac{d}{2}\right)} V_{dc} \times d\omega_s t + \int_{\left(\frac{3\pi}{2}-\frac{d}{2}\right)}^{\left(\frac{3\pi}{2}+\frac{d}{2}\right)} -V_{dc} \times d\omega_s t \right] = 0$$

$$a_n = \frac{2}{T} \int_0^T v_o(\omega_s t) \times \cos(n\omega_s t) dt$$

$$a_n = \frac{2}{2\pi} \int_0^{2\pi} v_o(\omega_s t) \times \cos(n\omega_s t) d\omega t$$

$$a_n = \frac{1}{\pi} \left[\int_{\left(\frac{\pi}{2}-\frac{d}{2}\right)}^{\left(\frac{\pi}{2}+\frac{d}{2}\right)} V_{dc} \cos(n\omega_s t) \times d\omega_s t + \int_{\left(\frac{3\pi}{2}-\frac{d}{2}\right)}^{\left(\frac{3\pi}{2}+\frac{d}{2}\right)} -V_{dc} \cos(n\omega_s t) \times d\omega_s t \right] = 0$$

$$b_n = \frac{1}{T} \int_0^T v_o(\omega_s t) \times \sin(n\omega_s t) dt$$

$$b_n = \frac{2}{2\pi} \int_0^{2\pi} v_o(\omega_s t) \times \sin(n\omega_s t) d\omega_s t$$

$$b_n = \frac{1}{\pi} \left[\int_{\left(\frac{\pi}{2}-\frac{d}{2}\right)}^{\left(\frac{\pi}{2}+\frac{d}{2}\right)} V_{dc} \sin(n\omega_s t) \times d\omega_s t + \int_{\left(\frac{3\pi}{2}-\frac{d}{2}\right)}^{\left(\frac{3\pi}{2}+\frac{d}{2}\right)} -V_{dc} \sin(n\omega_s t) \times d\omega_s t \right]$$ (11.2)

$$b_n = \frac{4V_{dc}}{n\pi} \sin\left(n\frac{\pi}{2}\right) \sin\left(n\frac{d}{2}\right)$$

$$b_n = \frac{4V_{dc}}{n\pi} \sin\left(n\frac{d}{2}\right) \text{ for } n = 1,3,5\dots$$

$$b_n = 0 \text{ for } n = 2,4,6\dots$$

The above equations show that the maximum value of the fundamental component occurs at $d = \frac{\pi}{2}$ and is

$$V_{o1} = \frac{4V_{dc}}{\pi}$$ (11.3)

EXAMPLE 11.2

A single-phase, single-pulse, bridge-type inverter is supplied from 300-V dc bus bar and feeds 50 Ω resistance. If the pulse width of the output voltage is 60°, compute the power consumed.

SOLUTION:

The rms value of output voltage is

$$V_{dc}\sqrt{\frac{d}{\pi}}$$

$$d = 60° = \frac{\pi}{3}$$

$$\therefore V_{o(rms)} = 300 \times \sqrt{\frac{\pi/3}{\pi}} = 173.2 \text{ V}$$

$$\text{Power consumed} = \frac{V_{o(rms)}^2}{R} = \frac{173.2^2}{50} = 600 \text{ W}$$

EXAMPLE 11.3

A single-phase full bridge inverter supplied from 220-V ac rms has an output voltage of width 90° symmetrical to $\frac{\pi}{2}$. Compute the fundamental component and the third and fifth harmonic voltages.

SOLUTION:

The expression for output voltage is

$$v_o = \sum_{n=1,3,5...}^{\infty} \left(\frac{4V_{dc}}{n\pi} \sin nd \right) \sin(n\omega_s t)$$

Putting $d = \frac{\pi}{2}$ (i.e., 90°) and $n = 1$, the fundamental component is

$$V_{o(1)} = \frac{4V_{dc}}{\pi} \sin\left(\frac{\pi}{2}\right) \sin(\omega_s t)$$

$$= \frac{4 \times 220}{\pi} \times 1 \times \sin(\omega_s t)$$

$$= 280.11 \sin(\omega_s t)$$

Similarly, compute the third and fifth harmonics as:

$$V_{o(3)} = \frac{4V_{dc}}{3\pi} \sin\left(\frac{3\pi}{2}\right) \sin(3\omega_s t)$$

$$= -93.37 \sin(3\omega_s t)$$

$$V_{o(5)} = \frac{4V_{dc}}{5\pi} \sin\left(\frac{5\pi}{2}\right) \sin(5\omega_s t)$$

$$= 56.02 \sin(5\omega_s t)$$

There exists third, fifth, seventh, etc., harmonics along with the fundamental component. We can show that, as d decreases, the fundamental component reduces and harmonics increase. Thus voltage control using single-pulse width modulation is generally not recommended because it introduces more harmonics. This discussion points to the fact that voltage control in inverters demands the dual objective of adjusting the fundamental component to the desired value and minimizing the harmonic voltages. To achieve these dual goals, several voltage pulses per half-cycle of output voltage are required. This can be carried out in three ways.

11.5.3.1 Multiple Pulse-Width Modulation

Assume that a triangular waveform is compared with a variable amplitude dc voltage as shown in Fig. 11.13. The comparator output is used to gate the switching devices in the inverter, leading to multiple voltage pulses per half-cycle of output voltage from the inverter.

11.5.3.2 Sinusoidal Pulse-Width Modulation

Here, a rectified sine wave and a triangular wave are fed to a comparator, and the output is used to turn ON the switching devices in the inverter. This is shown in Fig. 11.14. The width of the voltage waveform is not uniform and increases towards $\frac{\pi}{2}$ and vice versa. Multiple pulse-width modulation has the advantage of reducing harmonics.

11.5.3.3 Selective Harmonic Elimination

In this method, the devices are triggered in such a way that the output voltage pulses are positioned at suitable instances, thus leading to the elimination of selective harmonics such as third, fifth, seventh, etc. This is shown in Fig. 11.15.

In Fig. 11.15, there are three voltage pulses per half-cycle and three switching instances (α_1, α_2, and α_3, respectively). This wave possesses half-wave as well as quarter-wave symmetries, and this property enables the elimination of all even harmonics as well as cosine

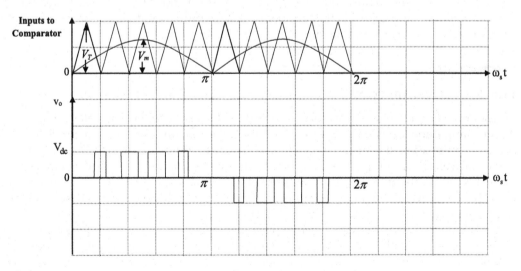

FIGURE 11.14
Sinusoidal pulse-width modulation.

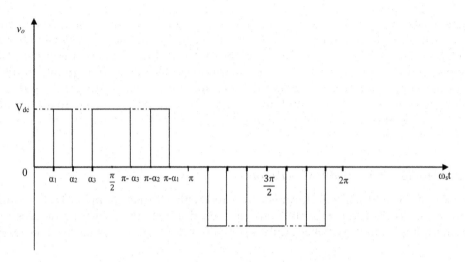

FIGURE 11.15
Selective harmonic elimination.

waveforms. Referring to the Fourier expansion explained in section 11.5.3, the coefficients are given below:

$$a_o = 0$$

$$a_n = 0$$

$$b_n = \frac{4V_{dc}}{n\pi}\left[\cos(n\alpha_1) - \cos(n\alpha_2) + \cos(n\alpha_3)\right] \quad \text{for } n = 1, 3, 5\ldots \tag{11.4}$$

$$b_n = 0 \quad \text{for } n = 2, 4, 6\ldots$$

In Fig. 11.15, there are three pulses per half-cycle and three unknowns: α_1, α_2, and α_3. The unknowns α_1, α_2, and α_3 are computed either iteratively or using optimization techniques to make the fundamental component equal to the required amplitude and thus eliminate lower-order harmonics. Thus, with α_1, α_2, and α_3, the third and fifth harmonic voltages can be eliminated. In general, if k is the number of voltage pulses existing per half-cycle, then $(k - 1)$ harmonics can be eliminated.

11.6 Multilevel Inverter

Consider the output voltage waveform of inverters discussed so far. The voltage has three distinct amplitude levels: $+V_{dc}$, 0, and $-V_{dc}$. Pulse-width modulation is often employed for harmonic mitigation. It is also possible to modify the inverter circuit such that the output voltage waveform has multiple amplitudes. Such inverters are named multilevel inverters. These inverters inherit high-quality output voltage waveforms at low switching frequencies. There are several topologies of multilevel inverters, but this section discusses only the five-level inverter.

Fig. 11.16(a) shows two bridge inverters cascaded together. There are eight power electronic switches, and the closed/opened status of these switches produces the required output voltage. The output voltage, V_o, is always $V_{01} + V_{02}$. Table 11.1 shows one possible combination, which results in a stepped output voltage waveform as given Fig. 11.16(b). It is evident that the output has five distinct amplitude levels, namely 0, V_{dc}, $2V_{dc}$, $-V_{dc}$, and $-2V_{dc}$.

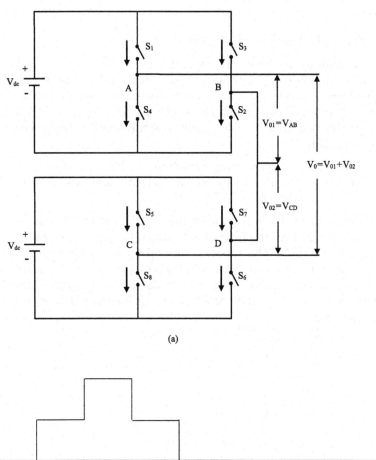

FIGURE 11.16
A cascaded bridge multilevel inverter. (a) Circuit diagram. (b) Output waveform.

TABLE 11.1

Multilevel Inverter Switching Patterns and Output Voltage

Switch Combinations	V_{01}	V_{02}	Output Voltage(V_{AD})
$S_1S_3 - S_5S_7$	0	0	0
$S_1S_2 - S_5S_7$	$+V_{dc}$	0	$+V_{dc}$
$S_1S_2 - S_5S_6$	$+V_{dc}$	$+V_{dc}$	$+2V_{dc}$
$S_3S_4 - S_6S_8$	$-V_{dc}$	0	$-V_{dc}$
$S_3S_4 - S_7S_8$	$-V_{dc}$	$-V_{dc}$	$-2V_{dc}$

11.7 McMurray Half-Bridge Inverter

The complete circuit of a McMurray half-bridge inverter is shown in Fig. 11.17. In this circuit, T_1 and T_2 are main SCRs; D_1 and D_2 are added to facilitate the turning OFF of T_1 and T_2, respectively. T_{A1} and T_{A2} are auxiliary SCRs used to turn OFF T_1 and T_2, and the LC circuit forms the commutation circuit. Diodes D_3 and D_4, together with resistance R, help clamp the capacitor voltage at V_{dc}. The entire operation of the bridge circuit is described through different stages.

STAGE 1: Here, T_1 alone is conducting, which makes the output voltage $V_0 = V_{dc}/2$. The capacitor is charged to V_{dc} with left side positive and the right side negative. The equivalent circuit for this stage is shown in Fig. 11.18(a).

STAGE 2: At the end of the half-cycle of the output voltage, T_1 has to be turned OFF and T_2 to be turned ON. For simplicity, it is assumed that commutation of T_1 starts at $t = 0$. At $t = 0$, T_{A1} is turned ON. The capacitor now discharges through C^+-L-T_1-T_{A1}-C^-. The equivalent circuit for this mode is shown in Fig. 11.18(b). The load current passes through V_1^+-T_1-load-V_1^-, whereas the capacitor current, i_c, also passes through T_1 but in the opposite direction. The waveform of i_c can be obtained by solving the following equation.

$$L\frac{di}{dt} + \frac{1}{C}\int i_c\, dt + V_{dc} = 0$$

$$L\frac{d^2i_c}{dt^2} + \frac{i_c}{C} = 0$$

$$\frac{d^2i_c}{dt^2} + \frac{i_c}{LC} = 0 \tag{11.5}$$

$$D^2 = -\frac{1}{LC}$$

$$D = \pm\sqrt{\frac{-1}{LC}} = \pm\frac{j}{\sqrt{LC}} = \pm j\omega_{LC}, \quad \text{where } \omega_{LC} = \frac{1}{\sqrt{LC}}$$

$$i_c = K_1\cos\omega_{LC}t + K_2\sin\omega_{LC}t$$

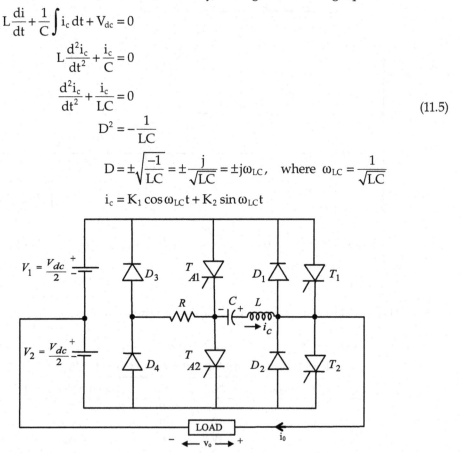

FIGURE 11.17
McMurray half-bridge inverter.

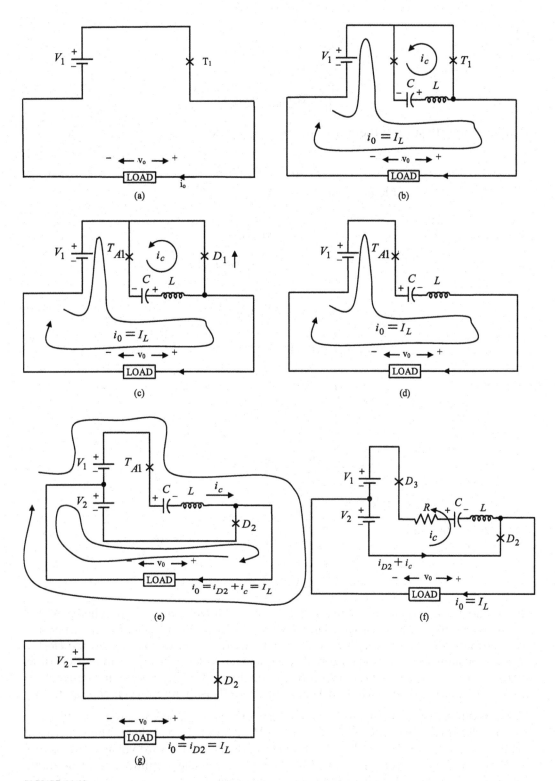

FIGURE 11.18
Equivalent circuits for (a) Stage I, (b) Stage II, (c) Stage III, (d) Stage IV, (e) Stage V, (f) Stage VI, and (g) Stage VII.

At t = 0, $i_c = 0$

$$K_1 = 0$$

$$i_c = K_2 \sin \omega_{LC} t$$

$$\frac{di_c}{dt} = K_2 \omega_{LC} \cos \omega_{LC} t$$

Multiplying both sides of the above equation by L yields

$$L \frac{di_c}{dt} = L K_2 \omega_{LC} \cos \omega_{LC} t$$

$$L K_2 \omega_{LC} \cos \omega_{LC} t \big|_{t'=0} = V_{dc}$$

$$K_2 = \frac{V_{dc}}{\omega_{LC} L}$$

$$i_c = \frac{V_{dc}}{\omega_{LC} L} \sin \omega_{LC} t$$

$$i_c = V_{dc} \sqrt{\frac{C}{L}} \sin \omega_{LC} t \tag{11.6}$$

$$V_c = \frac{1}{C} \int i_c \, dt + K_1$$

$$= \frac{1}{C} \int (-i_c) \, dt + K_1$$

$$= \frac{1}{C} \int - \left[V_{dc} \sqrt{\frac{C}{L}} \sin \omega_{LC} t \right] dt + K_1$$

$$= \frac{1}{C} V_{dc} \sqrt{\frac{C}{L}} \frac{\cos \omega_{LC} t}{\omega_{LC}} + K_1$$

$$V_c = V_{dc} \cos \omega_{LC} t + K_1$$

At t = 0, $V_c = V_{dc}$

$$V_{dc} = V_{dc} + K_1$$

$$K_1 = 0$$

$$V_c = V_{dc} \cos \omega_{LC} t \tag{11.7}$$

It is evident from Equations (11.6) and (11.7) that i_c changes sinusoidally, whereas v_c has a cosine wave shape. These variables together with i_0 are shown in Fig. 11.19. Because the total commutation time is very small (i.e., a few microseconds) compared to the electrical time constant of load, i_0 is assumed to be constant at i_L during commutation time. Referring to Fig. 11.19, at $\theta = \omega_{LC} t_1$, i_c becomes equal to i_L. Thus, T_1 carries equal currents in reverse directions and net current in T_1 becomes zero. T_1 is therefore open-circuited.

STAGE 3: When the net current in T_1 becomes zero, the excess i_c over i_L now passes through diode D_1. The load current $i_0 = i_L$ passes from V_1 through T_{A1}-C-L-load. The equivalent circuit is shown in Fig. 11.18(c). When diode D_1 conducts, the forward voltage drop across the device applies a reverse-bias across T_1, enabling reliable commutation. This mode continues for a period corresponding to $\omega_{LC} t_2$, where $\omega_{LC} t_2 = \pi - 2 \, \omega_{LC} t_2$. At $\omega_{LC} t_2$, i_c once again becomes equal to i_L and diode D_1 turns OFF.

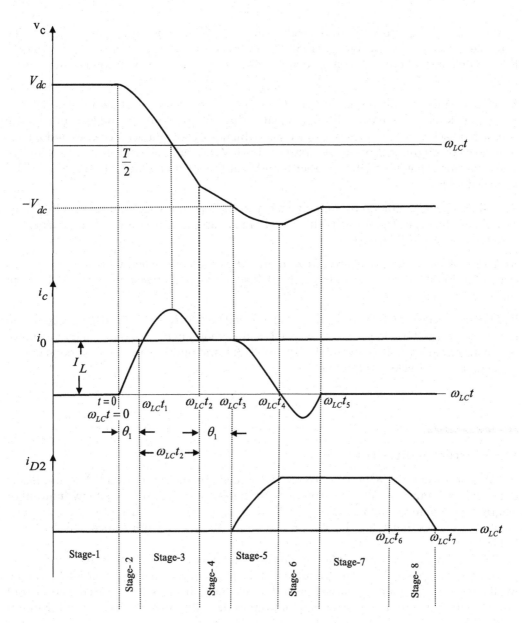

FIGURE 11.19
Waveforms during the commutation process.

STAGE 4: Now the load current i_L flows from V_1^+ through T_{A1}-C-L-load. Because i_c is constant, C now charges linearly. The capacitor voltage is given by

$$v_c = \frac{1}{C} \int i_c \, dt \tag{11.8}$$

$$v_c = \frac{1}{C} \int i_L \, dt = \frac{i_L}{C} t \tag{11.9}$$

At $\omega_{LC}t_3$, C charges to V_{dc} with the right plate positive and the other plate negative. The equivalent circuit is shown in Fig. 11.18(d). Now consider the loop comprising V_2-V_1-T_{A1}-C-L-D_2. With constant current, voltage drop across L is zero, which leads to forward-biasing of D_2.

STAGE 5: In this mode, there are two circuits carrying current as shown in Fig. 11.18(e). Here, v_0 is forcing current i_{D2} through V_2 such that energy stored in the load inductance is now fed back to V_2. Because $i_0 = i_{D2} + i_c = i_L$, which is a constant, i_{D2} increases and i_c falls. As i_c flows through the capacitor, C charges above V_{dc}. At $\omega_{LC}t_4$, i_c reaches zero and $i_{D2} = i_L$. The capacitor voltage is slightly higher than V_{dc} at this instant. Once i_c becomes zero, T_{A1} is turned OFF.

STAGE 6: The regeneration of trapped energy in the load inductor continues in this stage as well, as shown in Fig. 11.18(f). The overcharged capacitor discharges to a voltage, V_{dc}, through C-R-D_3-V_1-V_2-D_2 back to C.

STAGE 7: At $\omega_{LC}t_5$, i_c again becomes zero and D_3 is thereby turned OFF. The equivalent circuit for this stage is shown in Fig. 11.18(g). Regeneration of trapped energy continues in this stage as well.

STAGE 8: As the inductive energy in the load is fed back further and further, i_{D2} begins to fall from $\omega_{LC}t_6$ and becomes zero at $\omega_{LC}t_7$, when the whole energy from the load inductance is transferred to V_2. At this instant, D_2 is turned OFF and now T_2 can be triggered to obtain the negative half-cycle of output voltage.

11.8 Current-Source Inverter

A CSI is an inverter in which a current source feeds the inverter circuit. The amplitude of the current source is always maintained at the set value, regardless of any internal or external disturbances such as load change. In practice, an ideal current source is implemented by connecting a dc voltage source in series with a large inductance. This is shown in Fig. 11.20.

The dc voltage is varied to obtain the desired i_{dc}; the purpose of the inductance L is to produce ripple-free current. The frequency of the load current in a CSI is varied by changing the triggering frequency of the power devices in the inverter. In a CSI-fed load, output power control is achieved by changing the amplitude of the current. Hence, a CSI works in a closed loop in practice. A schematic of this is shown in Fig. 11.21.

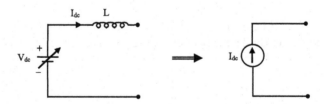

FIGURE 11.20
A practical current source.

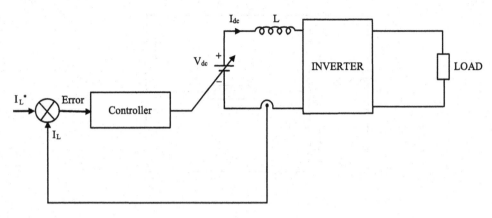

FIGURE 11.21
Closed-loop current control in CSI.

11.9 Single-Phase Current-Source Inverter

To explain the principle of operation of a CSI, the circuit configuration of a single-phase CSI employing ideal switches is shown in Fig. 11.22(a), with relevant waveforms in Fig. 11.22(b). The switches can be realized using switching devices such as SCRs, IGBTS, etc.

Here, switches S_1 and S_3 are gated for the first half-cycle period, leading to a load current of i_{dc}. At T/2, S_1 and S_3 are commutated and then S_2 and S_4 are triggered. This results in load current i_{dc} in the opposite direction. The resulting load current is alternating in nature. The amplitude of this current equals the value of the dc source current, and the frequency depends on the triggering frequency of the switches. The waveform of the load voltage depends on the composition of the load.

11.10 Capacitor-Commutated SCR-Based Single-Phase CSI

When the switching devices used are SCRs, forced commutation is used to turn OFF the respective SCRs. A capacitor is used at the output to perform forced commutation. The circuit of capacitor commutated CSI feeding a pure resistive load is shown in Fig. 11.23.

Initially, the capacitor is charged to a voltage of $-V_1$ with the right plate positive and the other plate negative; as seen from Fig. 11.23, $i_L R = V_c$ or $i_L = V_c/R$.

At $t = 0$, when T_1 and T_3 are fired, input current i_{dc} passes through the parallel combination of Capacitor, C and load resistance, R, and capacitor C charges to a voltage of $+V_1$ at $t = T/2$. The solution for V_c or i_L can be determined as follows:

$$\text{For } 0 \le t \le \frac{T}{2},$$

$$i_L R = V_c \tag{11.10}$$

$$i_L R = \frac{1}{C} \int i_c \, dt - V_1$$

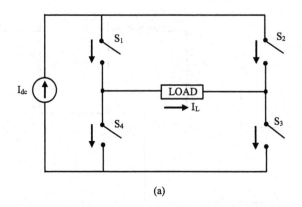

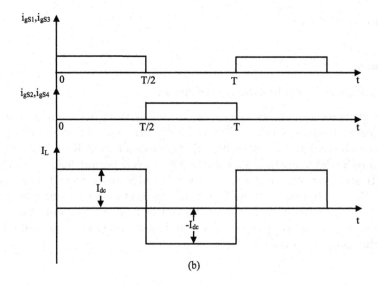

FIGURE 11.22
(a) Circuit of a single-phase CSI using ideal switches. (b) Waveforms of a single-phase CSI.

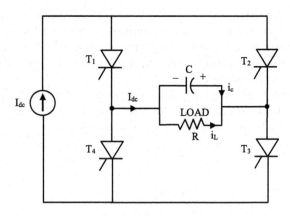

FIGURE 11.23
Capacitor-commutated CSI feeding a resistive load.

Differentiating both sides,

$$R\frac{di_L}{dt} = \frac{i_c}{C} \tag{11.11}$$

If the input current is held constant at i_{dc}, then

$$i_{dc} = i_c + i_L \text{ or } i_c = i_{dc} - i_L$$

Equation (11.11) becomes

$$R\frac{di_L}{dt} = \frac{i_{dc} - i_L}{C}$$

$$RC\frac{di_L}{dt} + i_L = i_{dc}$$

$$i_L = K_1 e^{\frac{-t}{RC}} + i_{dc} \tag{11.12}$$

At $t = 0$, $i_L = -I_1$

$$-I_1 = K_1 + i_{dc}$$
$$K_1 = -(I_1 + i_{dc})$$
$$\therefore \quad i_L = -(I_1 + i_{dc})e^{\frac{-t}{RC}} + i_{dc} \tag{11.13}$$

Rearranging the above equation yields

$$i_L = -I_1 e^{\frac{-t}{RC}} + i_{dc}\left(1 - e^{\frac{-t}{RC}}\right)$$

At $t = \dfrac{T}{2}$, $i_L = I_1$

$$\therefore I_1 = i_{dc}\left(1 - e^{\frac{-T}{2RC}}\right) - I_1 e^{\frac{-T}{2RC}}$$

$$I_1\left(1 + e^{\frac{-T}{2RC}}\right) = i_{dc}\left(1 - e^{\frac{-T}{2RC}}\right)$$

$$I_1 = i_{dc}\frac{\left(1 - e^{\frac{-T}{2RC}}\right)}{\left(1 + e^{\frac{-T}{2RC}}\right)}$$

Assume $I_1 \approx i_{dc}$:

$$\therefore \quad i_L \approx i_{dc}\left(1 - e^{\frac{-t}{RC}}\right) - i_{dc}e^{\frac{-t}{RC}}$$

$$i_L \approx i_{dc}\left(1 - 2e^{\frac{-t}{RC}}\right) \tag{11.14}$$

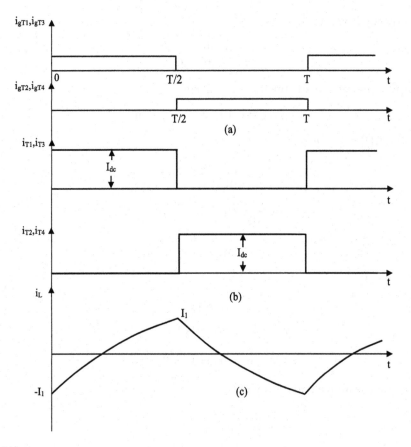

FIGURE 11.24
Output waveforms of a capacitor-commutated CSI. (a) Gating signals. (b) SCR currents. (c) Load current.

The variation of this circuit's parameters are shown in Fig. 11.24. Note that the capacitor voltage variations are identical.

Questions

1. What is a dc/ac converter?
2. Explain the operation of a three-phase inverter in 180° and 120° mode.
3. Discuss the different schemes to control the voltage of an inverter.
4. Explain the McMurray half-bridge inverter.
5. Explain the current-source inverter.
6. Elaborate the different types of current-source inverter.

Unsolved Problems

1. A single-phase bridge inverter supplied from a 48-V battery feeds a 22-Ω resistor. Find the output power and peak reverse voltage across each switching element.

2. A single-phase full-bridge inverter supplied from a 200-V ac rms has an output voltage of 120° symmetrical to $\frac{\pi}{2}$. Compute the fundamental, third, and fifth harmonic voltages.

3. A single-phase single-pulse bridge inverter is supplied from a 220-V dc bus bar and feeds 50 Ω resistance. If the pulse width of the output voltage is 90°, compute the power consumed.

Answers

1. 104.7 W, 48 V
2. $220.53\sin(\omega_s t)$, 0, $-44.10\sin(5\omega_s t)$
3. 484 W

12

Speed Control of Induction Motors and Synchronous Motors

12.1 Introduction

There are two types of electric motors run from an ac power supply: induction motors (asynchronous motors) and synchronous motors. An induction motor possesses a shunt-type speed-torque characteristic when supplied at rated voltage and frequency. A synchronous motor has a constant speed characteristic at all loads. When a variable-frequency inverter feeds ac motors, variable speed–variable torque operation can be conveniently achieved. This chapter focuses on fundamental principles of inverter fed variable speed ac motors. Operation of both voltage-source inverters and current-source inverters to supply power to ac motors is described with necessary characteristic curves. A qualitative treatment on vector control is included.

12.2 Inverter-Fed Induction Motor Drives

One of the best power converters suitable for induction motor speed control is a variable-frequency inverter. The output frequency of the inverter can be conveniently changed to achieve variable-speed operation. In other words, the synchronous speed, N_s, the number of poles, P, for which the stator is wound, and the frequency of stator voltage, f_1, are related by the following expression:

$$N_s = \frac{120 f_1}{P} \tag{12.1}$$

Equation (12.1) clearly indicates that the synchronous speed, N_s, is directly proportional to frequency, f_1. When f_1 changes, the N_s varies, and hence the rotor speed is adjusted to new value. The simple schematic of an inverter-fed induction motor drive is given in Fig. 12.1. The induction motor's efficiency as given in Equation (10.4) is

$$\text{Efficiency} \approx 1 - S \tag{12.2}$$

When the frequency changes, the synchronous speed changes, and hence the rotor speed also changes. Thus, at lower speeds the slip is low because synchronous speed is also very low, leading to higher efficiency of the motor. This is a major difference from the motor supplied from a variable-voltage converter.

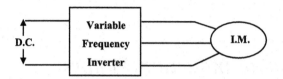

FIGURE 12.1
Inverter-fed induction motor drive.

There are two types of inverters, namely voltage-source inverters (VSIs) and current-source inverters (CSIs). It may be noted that, whether it is a VSI or a CSI, the fundamental principle of speed control is governed by Equation (12.1), and enhanced efficiency is obtained through Equation (12.2).

There are two schemes for speed control of inverter-fed induction motor speed control, namely scalar control and vector control. The scalar control can be achieved either employing VSI or CSI.

12.3 Scalar Control of Induction Motor Using Voltage-Source Inverters (VSI)

Consider a three-phase induction motor having the following specifications: 400 V, 50 Hz, four-pole. At rated voltage and frequency, the torque speed curve is sketched as shown in Fig. 12.2. Let this motor be now supplied at 40 Hz. The new synchronous speed is $\frac{120 \times 40}{4} = 1,200$ rpm. The torque-speed curve for 40 Hz is indicated in Fig. 12.2. In a similar way, a family of curves can be obtained for different stator frequencies. It may be noted that each curve has a specific synchronous speed dictated by the stator frequency. Each curve is shifted depending on frequency.

In the case of a three-phase induction motor, the stator-induced emf, E, is similar to emf induced in a transformer, and therefore it is related to air gap flux and stator frequency, f_1, by the following expression:

$$\frac{E_1}{f_1} \propto \Phi_m \qquad (12.3)$$

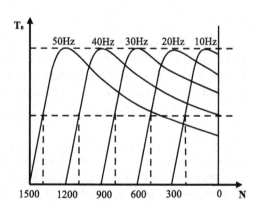

FIGURE 12.2
Torque-speed curves for an induction motor fed by a variable-frequency inverter.

Neglecting the stator drop, $E_1 \approx V_s$ (i.e., stator-applied voltage). Therefore, we can write

$$\frac{V_s}{f_1} \propto \Phi_m \tag{12.4}$$

From the above equation it is clear that, whenever stator frequency of an induction motor changes, it must be accompanied by a change in stator voltage such that the $\frac{V}{f}$ ratio is always maintained at the rated values. For example, for the induction motor specified earlier, $V_{phase} = \frac{400}{\sqrt{3}} = 231$, and the rated frequency is 50 Hz. For any frequency variation, the $\frac{V}{f}$ ratio must be proportional to $231/50 = 4.62$. Therefore, if a 40-Hz voltage is applied at the stator, the stator voltage must be reduced to 184.8 V. Thus, for any frequency change, the stator voltage must be changed to retain the air gap flux at its rated value. It is important to mention that the $\frac{V}{f}$ ratio on the motor stator is maintained at a constant such that the air gap flux is retained at the rated value regardless of frequency variation. A constant air gap flux will ensure that the machine can supply rated torque at any speed. Thus the $\frac{V}{f}$ ratio control results in large speed variations up to rated load torque.

Now consider a constant load torque of T_L that is being driven by the variable-frequency inverter–fed induction motor drive. For different stator frequencies, Fig. 12.2 clearly illustrates that a very large change in speed can be obtained. This demonstrates that a variable-frequency inverter is an ideal power converter for speed control of induction motor drives.

The basic schematic of a variable-voltage, variable-frequency (VVVF)-driven induction motor drive is shown in Fig. 12.3. Here, the three-phase ac is first rectified to dc using a thyristor converter, and made ripple free using a large filter capacitor, C. This dc voltage is supplied to the VSI, which in turn produces an output voltage having variable frequency and variable voltage.

The voltage adjustment at the motor terminal can be carried out in two ways, either by adjusting the firing angle of the silicon-controlled rectifier (SCR) voltage controller or employing pulse-width modulation (PWM) to the VSI. In the event we are employing the PWM method, then the SCR voltage controller can be replaced with three-phase diode rectifier.

A simple open-loop control of a VSI-fed induction motor drive is sketched in Fig. 12.4. In this diagram, the input command is the required synchronous speed, ω_s^*. The fundamental frequency, f_1, is then computed using $\omega_s = \frac{2\pi}{P} f_1$, and this frequency information is supplied to the inverter. For the given motor, the $\frac{V}{f}$ ratio is computed, and the required

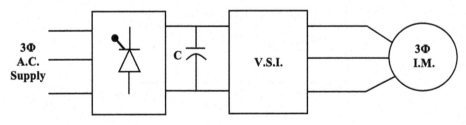

FIGURE 12.3
Variable-voltage, variable-frequency drive.

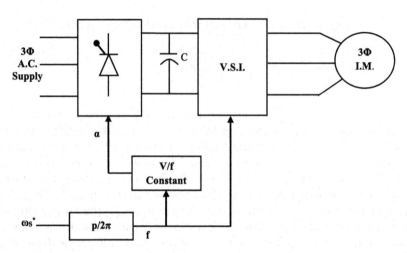

FIGURE 12.4
Open-loop control of VSI-fed induction motor drive.

α for the particular f_1 is also computed, so that the $\frac{V}{f}$ ratio is maintained constant on the machine. For illustration, the variation of rotor speed of the motor for two typical frequencies, say 50 Hz and 20 Hz, with load torque variation is plotted in Fig. 12.5.

Consider Fig. 12.5 where the stator frequency is 50 Hz and for the constant load torque T_L, let the speed $(N_r) = 1,450$ rpm.

Now slip is

$$S = \frac{N_s - N}{N_s} = \frac{1,500 - 1,450}{1,500} = 0.03$$

The approximate efficiency is

$$\text{Efficiency} = 1 - S = 1 - 0.03 = 0.97 = 97\%$$

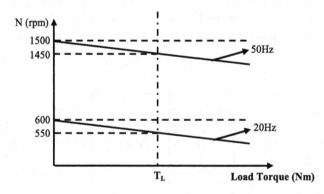

FIGURE 12.5
Drooping characteristics of an induction motor at different frequencies.

If the same load is now driven at 20 Hz, such that $N_r = 550$ rpm, then

$$S = \frac{N_s - N}{N_s} = \frac{600 - 550}{600} = 0.0833$$

$$\text{Efficiency} = 1 - S = 1 - 0.0833 = 0.9166 = 91.66\%$$

The above numerical example illustrates that a variable-frequency inverter ensures high efficiency even at very low speeds.

It can be seen from Fig. 12.5 that, for a load torque of T_L, two different speeds namely, 1,450 rpm and 550 rpm can be obtained with suitable commands for synchronous speed, ω_s^*. However, being in open-loop, precise speed control cannot be obtained. In other words, we get a speed close to ω_s^*. It is worth mentioning that the speed curves in Fig. 12.5 are always parallel. In other words, at any operating point, the slip speed, ω_{sl}, defined below remains the same for all frequencies, provided $\frac{V}{f}$ is constant.

$$\omega_{sl} = \omega_s - \omega_r \tag{12.5}$$

Further,

$$\frac{\omega_{sl}}{\omega_s} = s$$

$$\omega_{sl} = s\omega_s \tag{12.6}$$

The closed-loop scheme employing constant slip speed is given in Fig. 12.6.

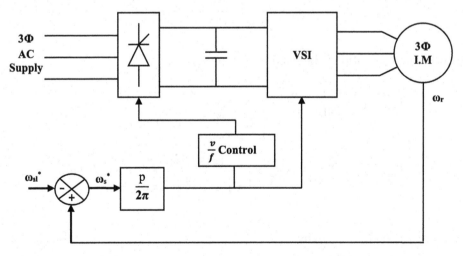

FIGURE 12.6
Closed-loop control of a VSI-fed induction motor drive.

EXAMPLE 12.1

Consider a 3Φ, Y-connected, 3.73-kW, 415-V, 7.5-A, 1,400-rpm, 50-Hz induction motor drive that is controlled by a VSI at constant $\frac{V}{f}$ ratio. Calculate the approximate speed for a frequency of 30 Hz at full load.

SOLUTION:

At rated conditions, slip speed $\omega_{sl} = \omega_s - \omega_r = 1,500 - 1,400 = 100$ rpm.

For 30 Hz, synchronous speed $= \dfrac{120f}{P} = \dfrac{120 \times 30}{4} = 900$ rpm.

The slip speed remains same.
At 30 Hz, speed $= 900 - 100 = 800$ rpm.

EXAMPLE 12.2

A 3Φ, Y-connected, 3.73-kW, 415-V, 7.5-A, 1,400-rpm, 50-Hz induction motor with the parameters of $R_s = 3.46\ \Omega$, $X_s = X_r' = 2.95\ \Omega$, $R_r = 0.68\ \Omega$, $X_m = 468.08\ \Omega$ is controlled by a VSI at a constant $\frac{V}{f}$ ratio. Calculate the maximum torque for a frequency of 30 Hz and compare it with rated maximum torque.

At rated conditions of 415 V and 50 Hz, the slip corresponding to maximum torque is given by Equation (9.29):

$$s_{max} = \frac{R_r}{\sqrt{R_s^2 + (X_{Ls} + X_{Lr})^2}}$$

Slip for rated maximum torque is given as

$$s_{max(50\ Hz)} = \frac{0.68}{\sqrt{3.46^2 + (2.95 + 2.95)^2}} = 0.1$$

Hence, at rated conditions, the maximum torque is

$$T_{max(50\ Hz)} = \frac{pm_1}{2\pi f_1} \times \frac{R_r}{s_{max}} \times \frac{V_s^2}{\left(R_s + \dfrac{R_r}{s_{max}}\right)^2 + \omega_s^2 (L_s + L_r)^2}$$

$$T_{max(50\ Hz)} = \frac{2 \times 3}{2\pi \times 50} \times \frac{0.68}{0.1} \times \frac{\left(415/\sqrt{3}\right)^2}{\left(3.46 + \dfrac{0.68}{0.1}\right)^2 + 5.9^2} = 53.225\ \text{N-m}$$

Parameters at 30 Hz is given by,

$$X_{s(30Hz)} = X_{r(30Hz)}' = 2.95 \times \frac{30}{50} = 1.77\ \Omega$$

From Equation (9.29),

$$S_{max(30\ Hz)} = \frac{0.68}{\sqrt{3.46^2 + (1.77 + 1.77)^2}} = 0.13$$

Because the $\frac{V}{f}$ ratio is constant, $V_{s(new)} = V_{s(old)} \times \frac{f_2}{f_1} = \frac{415}{\sqrt{3}} \times \frac{30}{50} = 143.76$ V.

$$T_{max(30\ Hz)} = \frac{2 \times 3}{2\pi \times 30} \times \frac{0.68}{0.13} \times \frac{(143.7)^2}{\left(3.46 + \frac{0.68}{0.13}\right)^2 + 3.54^2} = 43.483 \text{ N-m}$$

Hence,

$$\frac{T_{max(50\ Hz)}}{T_{max(30\ Hz)}} = 1.224$$

From the above results, it is clear that the maximum torque is different at 50 Hz and at 30 Hz. For $\frac{V}{f}$ control, the assumption of constant maximum torque is applicable near rated frequency alone. Maximum torque will slightly reduce at lower frequencies.

12.4 Scalar Control of Induction Motor Using Current-Source Inverter

A current source inverter (CSI) can be used to control speed or torque in an induction motor drive. A simple schematic of a CSI-fed induction motor is shown in Fig. 12.7. Here the input current is labeled as I_{dc} and is regulated to maintain its set value. Hence, a CSI-fed induction motor always works in a closed-loop mode.

The electromagnetic torque of an induction motor is given by the following equation:

$$T_e = \frac{p_1 m_1}{2\pi f_1} I_r^2 \frac{R_r}{S} \tag{12.7}$$

This equation shows that the torque is directly proportional to the square of the rotor current. In a CSI-fed induction motor, the amplitude of input current, I_{dc}, can be adjusted to any set value, and thus the motor stator current and rotor current can be instantaneously adjusted to any required value. Hence, the dynamic response of a CSI-fed induction motor

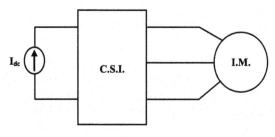

FIGURE 12.7
A CSI-fed induction motor drive.

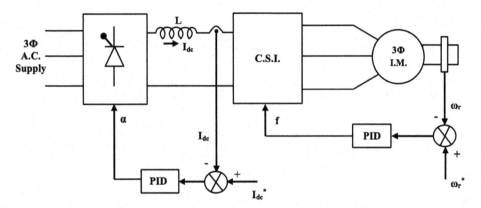

FIGURE 12.8
Closed-loop control of a CSI-fed induction motor drive.

is much faster than that obtained with a VSI. Therefore, even though a CSI-fed induction motor requires closed-loop operation, this scheme is recommended where rapid transient torque response is demanded. The scheme in Fig. 12.7 can be modified to regulate stator current and frequency adjustment as shown in Fig. 12.8.

In the block diagram in Fig. 12.8, there are two controlled loops, one of which corresponds to the motor stator current adjustment, which is carried out by varying the firing angle of the ac/dc converter as an adjustable current source. The frequency variation is performed on the inverter by comparing the actual speed with the reference value. In other words, torque and speed controls are independently carried out on the motor.

To identify the torque-speed curve of the variable-current, variable-frequency (VCVF) inverter drive, consider a 400-V, 50-Hz, 1,400-rpm, four-pole induction motor drive. The synchronous speed corresponding to rated speed operation is 1,500 rpm. Assume that this motor is supplied at the rated current and rated frequency from a CSI. The torque-speed characteristic curve will be exactly same as if it were supplied from a three-phase bus bar at rated frequency and rated current. This curve is labeled as A in Fig. 12.9. To find the torque-speed variation at another current and at the 50-Hz frequency, consider the following torque equation:

$$T_e = \frac{p_1 m_1}{2\pi f_1} I_r^2 \frac{R_r}{S}$$

For a given f_1 and S,

$$T_e \, \alpha \, I_r^2$$
$$T_e \, \alpha \, I_s^2 \tag{12.8}$$
$$T_e \, \alpha \, I_{dc}^2$$

Thus, it is evident that, by increasing the motor stator current, the torque curve will be stepped up as given in curve C, which corresponds to 200% of the rated current. Reducing the motor stator current will pull down the torque similar to curves D, E, and F. Another set of similar characteristic curves for different values of stator current can be obtained for 40 Hz as well. Thus, a CSI-fed induction motor drive enables accurate speed and torque control independently.

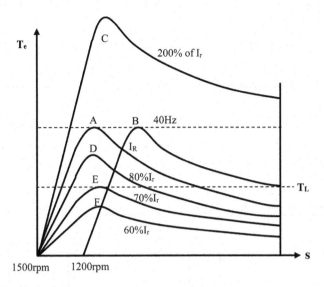

FIGURE 12.9
Torque-speed curves for a CSI-fed induction motor drive.

12.5 Three-Phase CSI Feeding a Three-Phase Induction Motor Load

The basic structure of a three-phase current source–fed induction motor drive is shown in Fig. 12.10. The switching devices are assumed to be mechanical switches for ease of simplicity. There are six switches and labeled in the usual way. These switches are controlled by controlling the gating signal of each switch. If the devices happen to be power metal-oxide semiconductor field-effect transistors (MOSFET) or insulated-gate bipolar transistors (IGBTs), then the gating signals are voltage pulses; in case of SCRs, the current signals are used to turn ON the device. For simplicity, it is assumed that each switch conducts as long as they receive the gating signal. The inverter is feeding power to a three-phase induction motor that is assumed to star-connected. The stator currents are marked as i_R, i_Y, i_B.

The switches are gated in a sequential manner. Each switch conducts for a period of $2\pi/3$ radians, and the interval between successive turning-ON of two switches is $\pi/3$.

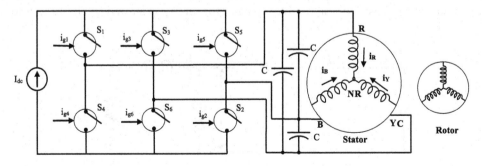

FIGURE 12.10
A three-phase, CSI-fed induction motor drive.

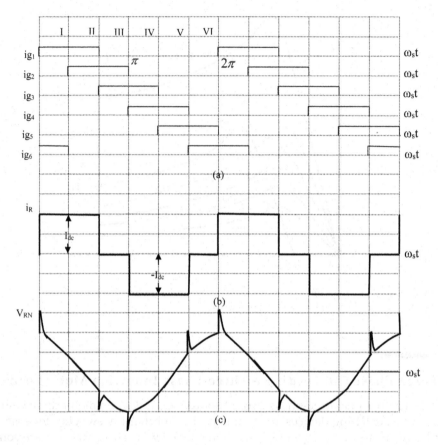

FIGURE 12.11
(a) Gate signals. (b) Motor current. (c) Motor terminal voltage.

Accordingly, the gating signals are indicated in Fig. 12.11. The interval 0–T can be divided into six equal parts corresponding to $\pi/3$ radians. The motor currents are sketched by following the ON/OFF position of each switch. The variation of i_R is shown in Fig. 12.11. The other two motor currents i_Y and i_B will resemble i_R but with phase shifts of $2\pi/3$ and $4\pi/3$ radians, respectively.

The determination of motor terminal voltage, V_{RN}, requires the per-phase equivalent circuit of a three-phase induction motor drive. This circuit is shown in Fig. 12.12 with the

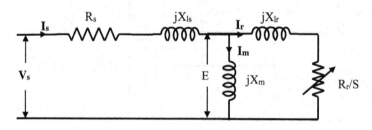

FIGURE 12.12
Equivalent circuit of three-phase induction motor.

usual notations. The square wave shape of i_R is split up into fundamental and harmonic components using Fourier series analysis. The fundamental component of the current passes through the magnetic circuit as well as the rotor path of the motor. For the harmonic sinusoidal currents, the magnetization reactance, X_m, offers a very large impedance, and thus only negligible current passes through a magnetic circuit. It can be concluded that the magnetic current in a CSI-fed induction motor is purely sinusoidal, leading to sinusoidal variation of motor voltage V_{RN}, which is plotted in Fig. 12.11. However, the motor current i_R undergoes instantaneous sudden changes at $\omega t = 0$, $2\pi/3$, etc. The sharp change in I_{dc} will lead to large induced spikes on the motor windings and will be superimposed on V_{RN}. To suppress the amplitude of voltage spikes of V_{RN}, capacitor banks are generally provided at the machine terminals as shown in Fig. 12.10.

12.6 Thyristorized CSI and the Commutation Process

Let us now consider a three-phase CSI using SCRs. It is mandatory that a commutation network should exist for the turning OFF of SCRs. If the capacitors are employed at machine terminals for suppressing voltage spikes and can be used for commutation of SCRs, then the scheme will be economical. Therefore, the capacitors are not directly connected across the machine terminals but are relocated for the dual purposes of commutation and voltage-spike reduction.

The SCR-based CSI inverter is represented in Fig. 12.13(a). There are six thyristors labeled as T_1, T_2, T_3, T_4, T_5, T_6; six diodes D_1, D_2, D_3, D_4, D_5, D_6; and six capacitors, which are connected to facilitate commutation of SCRs. The commutation process employed in three-phase CSI is called auto-sequential commutation because, when T_3 is turned ON, T_1 is turned OFF. When T_5 is turned ON, T_3 is commutated; turning T_1 ON causes T_5 to be switched OFF. A similar process exists for T_6, T_4, and T_2. It should be mentioned that no auxiliary thyristor is employed for commutation.

SCR T_1 is commutated at $2\pi/3$ radians. The commutation of T_1 is elaborated now, and the turning OFF process of T_3 and T_5 is similar. Prior to $\omega t = 2\pi/3$, the equivalent circuit is shown in Fig. 12.13(b). The capacitor, C_1, is charged to a finite voltage with the polarity as indicated in Fig. 12.13(b) due to previous commutation. At $\omega t = 2\pi/3$, T_3 is fired. When T_3 is fired, the capacitor voltage comes across T_1, reverse-biases it, and T_1 is turned OFF instantaneously. Because the motor current i_R cannot fall to zero instantaneously due to the inductive nature of load, i_R decays slowly. Because of the polarity of C_1, D_3 will not conduct until V_{c1} becomes negative. Therefore, i_R continues the path through

$$T_3 \begin{bmatrix} C_1 & \\ & \\ C_3{-}C_5 \end{bmatrix} D_1 - R \text{ phase} - B \text{ phase}$$

The equivalent circuit is shown in Fig. 12.13(c).

Thus C_3 and C_5 get charged to the polarities as shown in Fig. 12.13(d), while C_1 discharges first and then charges to the opposite polarity.

As time advances, the voltage across C_1 becomes zero and starts charging to reverse direction. At this instant, D_3 gets forward-biased and i_Y starts increasing as i_R starts receding. The equivalent circuit is shown in Fig. 12.13(e). Now T_3 carries two currents: one is i_R passing through

$$T_3 \begin{bmatrix} C_1 & \rule{1cm}{0.4pt} & D_1 - R\ phase \\ C_3 - C_5 \end{bmatrix}$$

and the other is $T_3 - D_3 - Y_{phase}$.

As the voltage across C_1, C_3, and C_5 increases, i_R falls to zero and only the path $T_3 - D_3 - Y_{phase}$ exists. At this time, C_3 is charged to a finite voltage with a polarity suitable to turn OFF T_3, when T_5 is triggered. The equivalent circuit at the end of commutation is shown in Fig. 12.13(e).

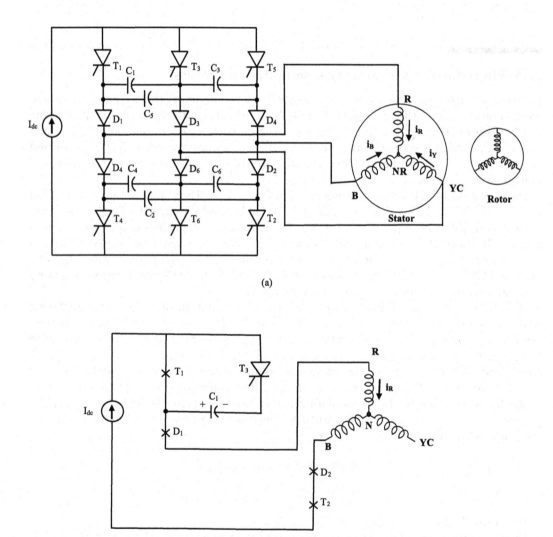

(a)

(b)

FIGURE 12.13

(a) Thyristorized CSI circuit. (b)–(e) Its equivalent circuits.

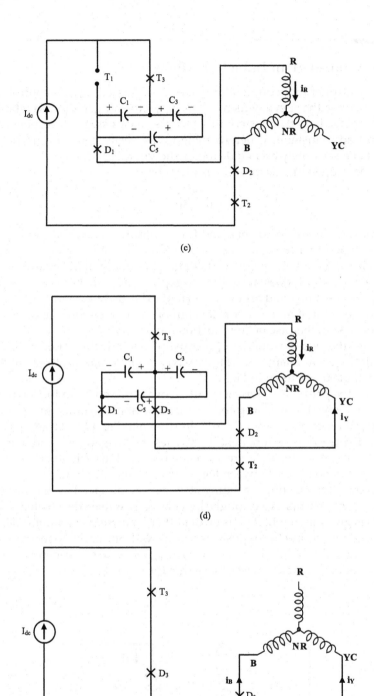

FIGURE 12.13 (*CONTINUED*)
(a) Thyristorized CSI circuit. (b)–(e) Its equivalent circuits.

12.7 Vector Control of an Induction Motor

So far, scalar control of induction motor drives is discussed. In this control, frequency is increased to increase the speed, which brings down the air gap flux. To retain the air-gap flux, the voltage needs to be boosted; this process causes sluggish response. Sometimes, this can even lead to instability. Before going to the vector control concept, consider the simple circuit of a separately excited dc motor shown in Fig. 12.14.

The torque developed in a dc machine is given by

$$T_e = k_f I_a I_f$$

where k_f is constant, I_f is field current, and I_a is armature current. Neglecting field saturation, it can be stated that the control variables I_f and I_a are completely decoupled in the sense that an increase in I_f increases the field flux but does not affect armature current I_a. Similarly, an increase in I_a does not affect air-gap flux Φ_m; furthermore, construction of a dc motor requires field flux and armature current to be always orthogonal to each other in space and independent of motor speed. It is important to mention that this type of decoupling does not exist in the case of an induction motor.

Now consider the two-axis model of a three-phase induction motor fixed to a stator, which was discussed in chapter 9. The two-axis model on an arbitrary reference frame rotating at an angle θ_a is shown in Fig. 12.15.

Now consider the per-phase equivalent circuit of a three-phase induction motor, neglecting core loss component as well as rotor leakage reactance as shown in Fig. 12.16.

Taking E as reference, the phasor diagram is drawn in Fig. 12.17. From this diagram, it is evident that I_r is the torque component, I_m decides the air-gap flux, and both phasors are rotating at $\omega_s = 2\pi f_1$. Also, I_r and I_m are always orthogonal to each other.

One-to-one correspondence between Fig. 12.15 and Fig. 2.17 shows that Fig. 12.18 resembles the arbitrary reference frame rotating at synchronous speed ω_s, with voltage phasors replaced by current phasors. Accordingly, I_r in Fig. 12.17 will be designated as $I_{qs(syn)}$ and I_m as $I_{ds(syn)}$. This is given in Fig. 12.18. It is evident that an increase of amplitude of $I_{qs(syn)}$ will lead to increased torque and the opposite is true as well. Similarly, an increase in $I_{ds(syn)}$ will enhance the air-gap flux, while a reduction of $I_{ds(syn)}$ will weaken air-gap flux. Furthermore, any change of stator frequency must ensure that $I_{qs(syn)}$ and $I_{ds(syn)}$ are always orthogonal to each other.

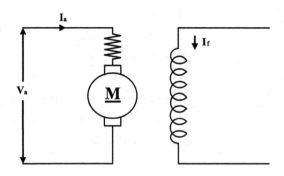

FIGURE 12.14
Circuit of a separately excited dc motor.

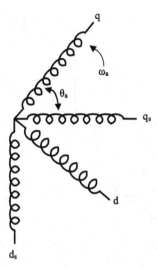

FIGURE 12.15
Arbitrary rotating reference frame.

The transformation equations are given below:

$$\left.\begin{array}{l} I_{qs} = I_{as} \\ I_{ds} = \dfrac{1}{\sqrt{3}}\left(I_{cs} - I_{bs}\right) \end{array}\right\} \tag{12.9}$$

Further

$$\left.\begin{array}{l} v_{qs} = v_{as} \\ v_{ds} = \dfrac{1}{\sqrt{3}}\left(v_{cs} - v_{bs}\right) \end{array}\right\} \tag{12.10}$$

Equations (12.9) and (12.10) correspond to a three-phase to two-phase transformation, i.e., a three-phase to two-phase transformation fixed to stator. The transformation to synchronously rotating reference frame is presented as

$$\begin{bmatrix} I_{qs} \\ I_{ds} \end{bmatrix} = \begin{bmatrix} \cos\theta_s & \sin\theta_s \\ -\sin\theta_s & \cos\theta_s \end{bmatrix} \begin{bmatrix} I_{qs(syn)} \\ I_{ds(syn)} \end{bmatrix} \tag{12.11}$$

Assume that the required air-gap flux is Φ_m^* and that the actual flux is Φ_m. The error is multiplied by gain, G, so as to obtain the required flux component of current $I_{ds(syn)}^*$. Similarly, let reference speed be ω_r^* and actual speed be ω_r. The difference when multiplied by gain G_2 gives $I_{qs(syn)}^*$ as shown in Fig. 12.19(a)–(b). Complete vector-control

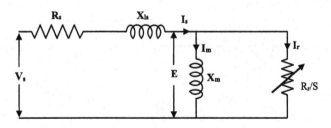

FIGURE 12.16
Equivalent circuit of an induction motor, neglecting rotor leakage reactance.

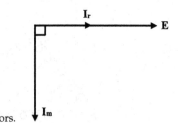

FIGURE 12.17
Rotor and magnetization currents as rotating orthogonal vectors.

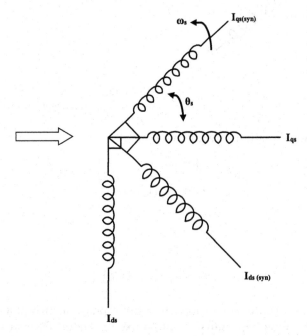

FIGURE 12.18
Mapping of induction motor currents to the synchronously rotating vectors.

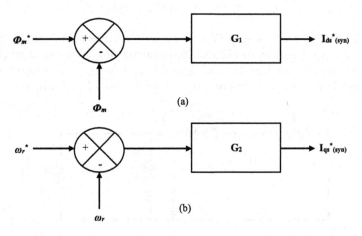

FIGURE 12.19
Generation of reference currents.

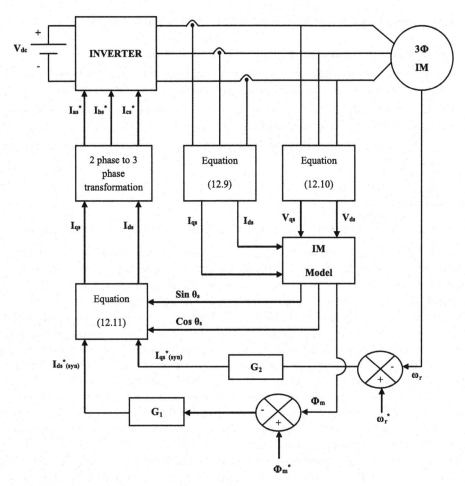

FIGURE 12.20
Implementation of vector control.

implementation of an inverter-supplied induction motor drive is given in Fig. 12.20. This figure shows that motor voltages and currents are sensed and supplied to the induction motor model. The model computes $\sin\theta_s$, $\cos\theta_s$ and air-gap flux Φ_m. The computed value of the air-gap flux is compared with the flux demand, and with the help of gain G_1, the required value of $I^*_{ds(syn)}$ is obtained. Similarly, the speed error is processed by gain G_2, and $I^*_{qs(syn)}$ is calculated. Equation (12.11) is then used to estimate I_{ds} and I_{qs}, which are then transformed into three-phase coordinates to obtain the required stator currents.

12.8 Speed Control of Synchronous Motor Drives

A synchronous motor has a three-phase stator winding and electromagnets on the rotor. The three-phase supply produces a rotating magnetic field, and if the electromagnets are made to align with this speed, the rotor also rotates at synchronous speed. Then this motor has two inputs, one to the stator and the other to the rotor. Therefore, a synchronous motor can be perceived as a beautiful combination of an ac motor and a dc motor.

The major advantages of a synchronous motor include the following:

- It always runs at a constant speed, unlike an induction motor.
- A synchronous motor has greater efficiency compared to an induction motor. This compensates for the higher initial cost and the running costs of the motor.
- A field can be suitably excited so as to give unity power factor at the stator side, which is impossible in an induction motor.
- A synchronous motor has less inertia than an induction motor, and hence dynamic response is faster.

12.9 Equivalent Circuit of Synchronous Motor and Basic Equations

Consider the per-phase equivalent circuit of a three-phase synchronous motor. The stator of the motor is same as that of a three-phase induction motor, and therefore the per-phase equivalent circuit consists of stator resistance, R_s, stator leakage reactance, L_{ls}, core loss component, R_m, and magnetizing inductance, L_m. The rotor of a synchronous motor consists of electromagnets wound with thick coils. These coils will have negligible resistance and inductance. The rotor is supplied from the voltage source V_f, which produces a field current, I_f, as seen from the stator side, and the rotor current is I_f' at the stator-supply frequency. Accordingly, the per-phase equivalent circuit is shown in Fig. 12.21(a). For the purpose of simplicity, the core loss equivalent, R_m, is neglected, and by applying Thevenin's theorem the equivalent circuit of the synchronous motor reduces to the one given in Fig. 12.21(b). This figure also shows that $L_s = L_{ls} + L_{lm}$. If the emf, e_f, is the excitation emf due to the field current I_f, the sum of the leakage reactance, $\omega_s L_{ls}$, and magnetizing reactance, $\omega_s L_m$, is called synchronous reactants and is denoted as X_s. In other words, $X_s = \omega_s(L_{ls} + L_{lm})$. The total impedance of the motor is denoted as Z_s and is known as synchronous impedance and is equal to $R_s + jX_s$. For most computations, the stator resistance is neglected and only synchronous reactance is considered; the reduced equivalent circuit is given in Fig. 12.21(c). The excitation emf, e_f, lags from the source voltage, V_f, by angle δ, and δ is called the torque angle or load angle of the motor.

The input power to the motor is $P_i = 3V_s I_s \cos\Phi$, where Φ is the phase angle between Vs and I_s. Because stator core and copper losses are neglected, this is the same as the mechanical output power, P_{mech}. That is,

$$P_{mech} = 3V_s I_s \cos\Phi \tag{12.12}$$

Taking V_s as the reference, $E_f \angle -\delta$, referring to Fig. 12.21,

$$I_s = \frac{V_s\angle 0 - E\angle -\delta}{jX_s} = \frac{V}{X_s}\angle -\frac{\pi}{2} - \frac{E}{X_s}\angle -\left(\frac{\pi}{2}+\delta\right)$$

$$I_s \cos\Phi = \frac{V_s}{X_s}\cos\left(\frac{\pi}{2}\right) - \frac{E_f}{X_s}\cos\left(\frac{\pi}{2}+\delta\right) = \frac{E_f}{X_s}\sin\delta \tag{12.13}$$

Substituting Equation (12.12) gives

$$P_{mech} = \frac{3V_s E_f}{X_s}\sin\delta \tag{12.14}$$

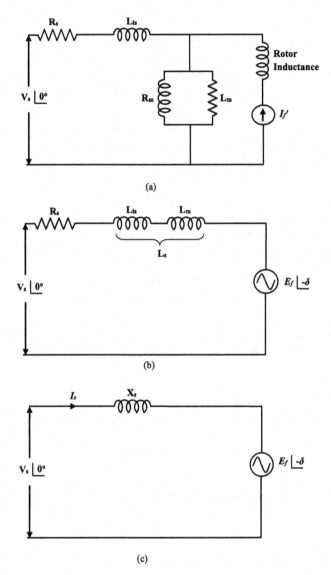

FIGURE 12.21
Synchronous motor equivalent. (a) Actual motor. (b) Core loss neglected. (c) Stator resistance neglected.

EXAMPLE 12.3

A 5-kW, three-phase, 415-V, 7-A, 1,500-rpm, 0.8(lag) power factor star-connected synchronous motor has a rated field current of 1.2 A at 220 V. The synchronous reactance is 11.23 Ω, and stator resistance is negligible.

  a. Find the stator emf at rated torque and field magnetomotive force.

  b. Find the stator emf at 60% of rated torque and at rated excitation.

  c. Find the required value of field current for unity power factor.

  d. Find the input power at 0.9(lag) power factor at a field current of 1 A.

SOLUTION:

(a) Input power $= 3V_sI_s \cos\Phi$. The motor speed remains at synchronous speed; hence, at rated torque, the output power is rated value, i.e., 5 kW. Substituting,

$$3V_sI_s \cos\phi = 5\times10^3$$

$$3\frac{415}{\sqrt{3}}\times I_s \times 0.8 = 5\times10^3$$

$$\therefore |I_s| = 8.69\ A$$

Thus,

$$I_s = 8.69\angle-36.86°\ A$$

Now referring to the equivalent circuit in Fig. 12.21(c),

$$V_s = I_sX_s + E_f$$
$$\therefore E_f = V_s - I_sX_s$$
$$= 197\angle-23.32\ V$$

(b) 60% of rated output $= 0.6\times5{,}000 = 3{,}000$ W.
Because the field current remains at rated value, $|E_f| = 197$ V. Now,

$$P_{mech} = \frac{3V_sE_f}{X_s}\sin\delta$$

$$3{,}000 = \frac{3x\dfrac{415}{\sqrt{3}}\times197}{11.23}\sin\delta$$

$$\therefore \delta = 13.76°$$

$$\text{Stator current, } I_s = \frac{V_s\angle0° - E_f\angle-13.76}{j11.23}$$

$$\text{i.e., } I_s = 5.9\angle-45.48°$$

(c) Find the required value of field current for unity power factor.
At unity power factor,

$$3V_sI_s = 5\ kW$$

$$3\times\frac{415}{\sqrt{3}}\times I_s = 5\times10^3$$

$$\therefore |I_s| = 6.956\ A$$

$$I_s = 6.956\angle0°$$

$$E_f = V_s - I_sX_s$$

$$= \frac{415}{\sqrt{3}} - (6.956\angle0°)(11.23\angle90°)$$

$$= 252\angle-18°$$

$$\text{now, } |E_f|\alpha\ I_f$$

$$\frac{E_{f\,rated}}{E_f} = \frac{I_{f\,rated}}{I_f}$$

$$\therefore I_f = \frac{E_f}{E_{f\,rated}} \times I_{f\,rated} = \frac{252}{197} \times 1.2 = 1.53 \text{ A}$$

(d)
$$E_f = \frac{I_f}{I_{f\,rated}} \times E_{f\,rated} = \frac{1}{1.2} \times 197 = 164.16 \text{ V}$$

$$|V_s - I_s X_s| = E_f$$

$$\left| \frac{415}{\sqrt{3}} - I_s(11.23\angle 90°) \right| = E_f$$

$$\sqrt{\left(\frac{415}{\sqrt{3}}\right)^2 - (11.23 I_s)^2} = 164.16$$

$$I_s = 15.54 \text{ A}$$

Input power $= 3V_s I_s \cos\phi$

$$= 3 \times \frac{415}{\sqrt{3}} \times 15.54 \times 0.9$$

$$= 10,053 \text{ W}$$

12.10 Scalar Control of Synchronous Motor

The block diagram shown in Fig. 12.22 gives the scalar control scheme for a synchronous motor drive. It is in the open-loop mode, and the reference signal required is synchronous speed, ω_s^*. The required frequency is generated by a suitable firing scheme and inverter. For the $\dfrac{V}{f}$ ratio to be constant, as in the case of an induction motor drive, the converter

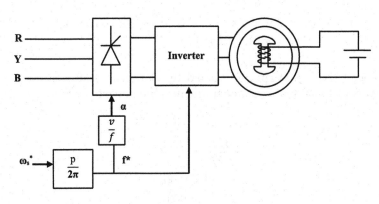

FIGURE 12.22
Scalar control of a synchronous motor.

firing angle α is adjusted so that any change in the stator frequency is accompanied by a change in the stator voltage.

In the case of an induction motor drive, a change in stator frequency leads to a change in the speed of the rotating magnetic field, which then leads to variations in rotor speed. In the case of a synchronous motor, if the frequency variation is large, then the variation has to be carried out linearly and smoothly so that the rotor field is always "locked up" with the rotating magnetic field.

EXAMPLE 12.4

A synchronous motor has the following ratings: three-phase, 415 V, 1,500 rpm, 5 kW, 0.8 (lag) power factor, field excitation 1.2 A at 220 V. The motor is supplied from a VSI maintaining $\frac{V}{f}$ constant.

a. If the motor is running at 900 rpm at rated current at 0.85 power factor, compute the output power developed and the field current.

b. At 70% of rated torque, 1,200 rpm, and 100% field excitation, calculate the stator current and power factor.

SOLUTION:

Under rated conditions,

$$V_s = \frac{415}{\sqrt{3}} = 239.6 \text{ V}$$

$$3 V_s I_s \cos\phi = 5 \times 10^3$$

$$3 \frac{415}{\sqrt{3}} \times I_s \times 0.8 = 5 \times 10^3$$

$$\therefore |I_s| = 8.18 \text{ A}$$

$$\text{Hence, } I_s = 8.18\angle -31.78° \text{ A}$$

$$X_s = 11.23 \ \Omega$$

$$V_s = I_s X_s + E_f$$

$$\therefore E_f = V_s - I_s X_s = 239.6\angle 0° - 8.18\angle -31.78° \times 11.23\angle 90°$$

$$= 206.55\angle -22.21 \text{ V}$$

(a) Motor speed (N_1 = 900 rpm) is varied by changing stator frequency. Because motor speed is proportional to stator frequency, the new stator frequency, f_1, is found as follows:

$$\frac{f_{rated}}{f_1} = \frac{N_{rated}}{N_1}$$

$$\frac{50}{f_1} = \frac{1500}{900}$$

$$f_1 = 30 \text{ Hz}$$

Because synchronous reactance is proportional to frequency, the new stator frequency of 30 Hz affects the synchronous reactance as follows:

$$X_{s(30\ Hz)} = \frac{30}{50} \times 11.23 = 6.738\ \Omega$$

Because $\frac{V}{f}$ is maintained constant, the stator voltage is adjusted to the new frequency of 30 Hz:

$$V_{s(30\ Hz)} = \frac{30}{50} \times 239.6 = 143.76\ V$$

Now, E_f at the new frequency is given by

$$E_{f(30\ Hz)} = V_{s(30\ Hz)} - I_s \times X_{s(30\ Hz)}$$
$$= 143.76\angle 0° - (8.18\angle - 31.78) \times (6.738\angle 90)$$
$$= 123.93\angle - 22.21°$$

Because field-induced emf is proportional to field current,

$$I_{f(30\ Hz)} = \frac{123.93}{206.55} \times 1.2 = 0.72\ A$$

$$P_{mech} = \frac{3 V_s E_f}{X_s} \sin \delta$$

$$P_{mech} = \frac{3 \times 143.76 \times 123.93}{X_s} \sin 22.21 = 2.998\ kW$$

(b) Field current is at rated value.

$$\therefore E_f = 206.55 \angle - 22.21°$$

Change of speed to 1,200 rpm is caused by a change of inverter frequency. The new frequency, f_2, is given by

$$f_2 = \frac{1,200}{1,500} \times 50 = 40\ Hz$$

Synchronous reactance at 40 Hz is

$$X_{s(40\ Hz)} = \frac{40}{50} \times 11.23 = 8.984\ \Omega$$

Because $\frac{V}{f}$ is constant, the changed value of V_s is

$$V_{s(40\ Hz)} = \frac{40}{50} \times 239.6 = 191.68\ V$$

$$\text{Now } I_{s(40\ Hz)} = \frac{V_{s(40\ Hz)} - E_{f(rated)}}{X_{s(40\ Hz)}}$$
$$= 8.69\angle - 0.22°$$

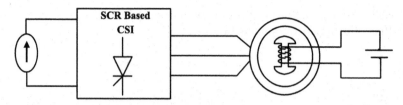

FIGURE 12.23
Load-commutated, CSI-fed, synchronous motor drive.

12.11 Load-Commutated CSI-Fed Synchronous Motor Drive

A CSI employs either SCRs or self-commutating devices such as IGBTs or power MOSFETs. Although the SCR-based CSI requires additional capacitors for the commutation of SCRs, the SCR-based CSI is preferred over the IGBT-based CSI. This has been detailed in section 12.5.

When a synchronous motor is excited at a suitable level, it can operate at leading power factor. Then the synchronous motor can be viewed as a capacitor-connected synchronous motor. Thus the SCR-based CSI does not require external capacitors to commutate the SCRs. In other words, the capacitors in a SCR-based CSI can be dispensed with in an asynchronous motor operating at leading power factor. Thus the advantages of SCRs, such as lower cost, reliability, and so on, can be integrated with a capacitor-less CSI. The commutation of SCRs and subsequent operation of a capacitor-less CSI feeding a synchronous motor at leading power factor is generally known as a load-commutated CSI-fed synchronous motor drive. It is shown in Fig. 12.23.

12.12 Permanent Magnet Synchronous Motor (PMSM) Drive

The thick coil wound around the electromagnets of a synchronous motor reduces the robustness of the rotor. To overcome this drawback, the rotor windings can be removed so that the rotor consists of permanent magnets. Such a motor is called a permanent magnet synchronous motor (PMSM) drive. The principle of operation is the same as that of the synchronous motor discussed earlier.

To ensure that the rotor field is always aligned with the rotating stator field, it is necessary to sense the position of the rotor poles and to adjust as needed; that is, to advance or delay the firing pulses to the inverter. Thus, if the stator is supplied at a fixed frequency, the advancement or the delay of the stator field can be achieved by suitably positioning the gating pulses to the inverter. A closed-loop scheme incorporating this aspect is merged with scalar control in Fig. 12.24. In other words, instead of controlling the inverter frequency independently, the frequency and phase of the inverter output voltage are adjusted by an absolute position sensor placed on the motor shaft. The speed control of a synchronous motor employing this scheme is called a self-controlled synchronous motor drive.

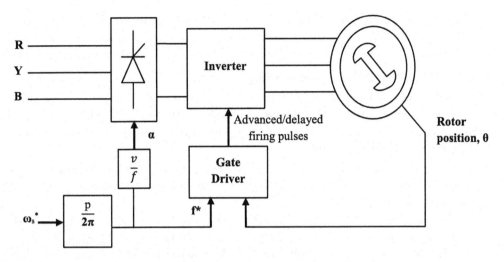

FIGURE 12.24
A permanent magnet synchronous motor (PMSM) drive.

Questions

1. Explain the inverter-fed induction motor drives.
2. Describe the scalar control with voltage-source inverters.
3. Explain the CSI-fed induction motor drive.
4. Elaborate the vector control of an induction motor.
5. Derive the equivalent circuit of a synchronous motor.
6. Discuss the scalar control of a synchronous motor.
7. What is a load-commutated CSI-fed synchronous motor drive?
8. Explain the permanent magnet synchronous motor drive.

Unsolved Problems

1. A 6.9-kW, three-phase, 415-V, 12-A, 1,200-rpm, 0.8 (lagging) power factor, star-connected, synchronous motor has a rated field current of 1 A at 200 V. The synchronous reactance is 15.6 Ω, and the stator resistance is negligible.

    a. Find the stator emf at rated torque and field magnetomotive force.
    b. Find the stator emf at 60% of rated torque and at rated excitation.
    c. Find the required value of field current for unity power factor.
    d. Find the input power at 0.8 (lagging) power factor at a field current of 0.8 A.

2. A synchronous motor has the following ratings: three-phase, 415 V, 50 Hz, 1,500 rpm, 6.9 kW, 0.8 (lagging) power factor; and field excitation is 1 A at 200 V. The motor is supplied from a VSI holding $\frac{V}{f}$ constant. The synchronous reactance is 14.25 Ω, and stator resistance is negligible.

   a. If the motor is running at 1,000 rpm at rated current at 0.85 power factor, compute the output power developed and the field current.

   b. At 80% of rated torque, 1,400 rpm, and 100% field excitation, calculate the stator current and power factor.

3. A 3Φ, Y-connected, 5-kW, 415-V, 10-A, 1,200-rpm, 50-Hz induction motor drive is controlled by a VSI at a constant $\frac{V}{f}$ ratio. Calculate the approximate speed for a frequency of 40 Hz at full load.

4. A 3Φ, Y-connected, 3.73-kW, 415-V, 8-A, 1,400-rpm, 50-Hz induction motor with the following parameters: $R_s = 4.2$ Ω, $X_s = X'_r = 3.30$ Ω, $R_r = 0.8$ Ω, $X_m = 490.08$ Ω, is controlled by a VSI at constant $\frac{V}{f}$ ratio. Calculate the maximum torque for a frequency of 30 Hz and compare it with the rated maximum torque.

Answers

1. (a) $E_f = 196.57$; (b) $\delta = 27.2°$, $I_s = 7.1\angle -35.785°$; (c) $I_s = 9.59\angle 0°$, $E_f = 284$, $I_f = 1.44$ A; (d) $E_{f(0.5A)} = 157.26$ V, $I_s = 11.58\angle -36.86°$, Power = 6.65 kW

2. $I_{s(rated)} = 11.477 \angle -36.86$, $E_{f(rated)} = 192.88\angle -43.196$, $f_{(1,000 \text{ rpm})} = 33.33$ Hz, $X_{s(33.33 \text{ Hz})} = 9.5$ Ω, $V_{s(33.33Hz)} = 159.71$ V, $E_f = 236.04\angle -23.10$, $I_f = 1.22$ A, $P_{mech} = 4.670$ kW, $f_{(1,400 \text{ rpm})} = 46.66$, $X_s = 13.29$, $V_{s(46.66 \text{ Hz})} = 223.59$, $I_s = 11.90\angle -32.3$, $P_f = 8.4$ (lagging)

3. $\omega_{sl} = 300$, $\omega_{s(40Hz)} = 960$, $\omega_r = 660$ rpm

4. $s_{max(50 \text{ Hz})} = 0.09$, $T_{max(50 \text{ Hz})} = 45.35$ N-m, $X_{s(30 \text{ Hz})} = X'_{r(30 \text{ Hz})} = 1.98$ Ω, $s_{max(30 \text{ Hz})} = 0.13$,

   $V_{s(new)} = 249$, $T_{max(30 \text{ Hz})} = 19.77$ N-m, $\dfrac{T_{max(50 \text{ Hz})}}{T_{max(30 \text{ Hz})}} = 2.29$

References

1. C. H. Ram Jethmalani, S. P. Simon, K. Sundareswaran, P. S. Rao Nayak, N. Prasad Padhy. Auxiliary Hybrid PSO-BPNN-Based Transmission System Loss Estimation in Generation Scheduling, *IEEE/Industrial Informatics*, Vol. No. 13, Issue 4, 2017.
2. S. K. Murugan, S. P. Simon, K. Sundareswaran, P. S. Rao Nayak, N. Prasad Padhy. An Empirical Fourier Transform-Based Power Transformer Differential Protection', *IEEE/Power Delivery*, Vol. No. 32, Issue 1, 2017.
3. K. Sundareswaran, V. Vigneshkumar, P. Sankar, S. P. Simon, P. S. Rao Nayak, and S. Palani. Development of an Improved P&O Algorithm Assisted Through a Colony of Foraging Ants for MPPT in PV System, *IEEE Transactions on Industrial Informatics*, Vol. No. 12, Issue 1, 2016.
4. S. K. Murugan, S. P. Simon, P. S. Rao Nayak, K. Sundareswaran, and N. Prasad Padhy. Power Transformer Protection Using Chirplet Transformer IET Generation', *Transmission & Distribution*, Vol. No. 10, Issue 10, 2016.
5. A. T. Thankappan, S. P. Simon, P. S. Rao Nayak, K. Sundareswaran, and N. Prasad Padhy. Pico-hydel Hybrid Power Generation System with an Open Well Energy Storage. *IET Generation, Transmission & Distribution*, Vol. No. 11, Issue 3, 2015.
6. K. Sundareswaran, P. Sankar, P. Srinivasa Rao Nayak, S. P. Simon, and S. Palani. Enhanced Energy Output from a PV System Under Partial Shaded Conditions Through Artificial Bee Colony *IEEE Transactions on Sustainable Energy*, Vol. No. 6, Issue 1, 2015.
7. K. Sundareswaran, V. Vignesh Kumar, and S. Palani. Application of a Combined Particle Swarm Optimization and Perturb and Observe Method for MPPT in PV Systems Under Partial Shading Conditions, *Renewable Energy*, Vol. No. 75, 2015.
8. K. Sundareswaran, V. Vigneshkumar, and S. Palani. Development of a Hybrid Genetic Algorithm/Perturb and Observe Algorithm for Maximum Power Point Tracking in Photovoltaic Systems Under Non-Uniform Insolation, *IET Renewable Power Generation*, Vol. No. 9, 2015.
9. K. A. Kumar, K. Sundareswaran, and P. R. Venkateswaran. Performance Study on a Grid Connected 20 kW Solar Photovoltaic Installation in an Industry in Tiruchirappalli (India), *Energy for Sustainable Development*, Vol. No. 23, 2014.
10. K. Sundareswaran and P. S. Nayak. Particle Swarm Optimization Based Feedback Controller Design for Induction Motor Soft-starting, *Australian Journal of Electrical & Electronics Engineering*, Vol. No. 11, Issue 1, 2014.
11. K. Sundareswaran and Nayak, P. S. R. Design of Feed Back Controller for Soft-starting Induction Motor Drive System Using Genetic Algorithm, *Internationl Journal of Industrial Electronics and Drives*, Vol. No. 1, Issue 2, 2014.
12. K. Sundareswaran, P. S. R. Nayak, and A. C. Sekhar. Praiseworthy Prize, Development of an Improved Particle Swarm Optimization (PSO) and its Application to Induction Motor Soft-Starting, *International Review of Automatic Control*, Vol. No. 7, Issue 2, 2014.
13. K. Sundareswaran, P. Sankar, and S. Palani. MPPT of PV Systems Under Partial Shaded Conditions Through a Colony of Flashing Fireflies, *IEEE Transactions on Energy Conversion*, Vol. 29, 2014.
14. R. Sheeba, M. Jayaraju, and K. Sundareswaran. Performance Enhancement of Power System Stabilizer Through Colony of Foraging Ants, *Electric Power Components and Systems*, Vol. No. 42, Issue 10, 2014.
15. K. Sundareswaran and P. Sankar. Development of Two Novel Power Electronic Circuits for Dynamic Braking of Motors, *Electric Power Components and Systems*, Vol. No. 41, Issue 12, 2013.
16. K. Sundareswaran, P. Sankar, and S. Palani. Application of Random Search Method for Maximum Power Point Tracking in Partially Shaded Photovoltaic Systems, *IET Renewable Power Generation*, Vol. No. 12, Issue 5, 2014.

17. K. Sundareswaran and P. S. Nayak. Ant Colony based Feedback Controller Design for Soft-Starter Fed Induction Motor Drive, *Applied Soft Computing*, Vol. No. 40, Issue 6, 2012.

18. K. Sundareswaran and V. Devi. Feedback controller Design for a boost converter through s Colony of Foraging Ants, *Electric Power Components and Systems*, Vol. No. 40, Issue 6, 2012.

19. K. Sundareswaran and P. Sankar. Buck-Boost Converter Design Using Queen-Bee Assisted GA, *International Journal of Power Electronics*, Vol. No. 4, Issue 5, 2012.

20. K. Sundereswaran, V. Devi, and N. A. Shrivastava, Design and development of a Feedback Controller Using Artificial Immune System, *Electric Power Components and Systems*, Vol. No. 39, Issue 10, 2011.

21. K. Sundereswaran and V. Devi. Application of a Modified Particle Swarm Optimization Technique for Output Voltage Regulation of Boost Converter, *Electric Power Components and Systems*, Vol. No. 39, Issue 3, 2011.

22. K. Sundareswaran, V. Devi, S. Sankar, P. S. Rao Nayak, and S. Peddapati, Feedback Controller Design for a Boost Converter Through Evolutionary Algorithms, *IET Power Electronics*, Vol. No. 7, Issue 4, 2011.

23. K. Sundareswaran, V. Devi, S. K. Nadeem, V. T. Sreedevi, and S. Palani. Buck-Boost Converter Feedback Controller Design via Evolutionary Search, *International Journal of Electronics*, Vol. No. 97, Issue 11, 2010.

24. K. Sundareswaran, N. Rajasekar, and S. Palani. Performance Evaluation of Energy Efficient Speed Control Techniques for Capacitor-Run Single-Phase Induction Motor Driving Domestic Fans, *Electric Power Components and Systems*, Vol. No. 39, Issue 4, 2010.

25. K. Sundareswaran and V. T. Sreedevi. Boost Converter Controller Design Using Queen Bee Assisted GA, *IEEE Transactions on Industrial Electronics*, Vol. No. 56, Issue 3, 2009.

26. K. Sundareswaran and V. T. Sreedevi. Inverter Harmonic Elimination using Honey Bee Intelligence, *Australian Journal of Electrical and Electronics Engineering*, Vol. No. 6, Issue 2, 2009.

27. K. Sundareswaran and V. T. Sreedevi. Design and Development of Feed-back Controller for a Boost Converter Using a Colony of Foraging Bees, *Electric Power Components and Systems*, Vol. No. 37, Issue 5, 2009.

28. K. Sundareswaran, V. Devi, S. Kaul, N. Rajasekar, and S. Palani. Optimal Feedback Controller Design for a Buck Converter Using Gaussian Type Pheromone Profile, *European Transactions on Electric Power*, Vol. No. 20, Issue 8, 2009.

29. K. Sundareswaran, K. Jayant, T. N. Shanvas. Inverter Harmonic Elimination Through a Colony of Continuously Exploring Ants, *IEEE Transactions on Industrial Electronics*, Vol. No. 54, Issue 5, 2007.

30. K. Sundareswaran. Line Current Harmonic Elimination and Voltage Control of PWM AC/DC Converter Using Hybrid Genetic Algorithm, *Electric Power Components and Systems*, Vol. No. 35, Issue 2, 2007.

31. K. Sundareswaran, N. Rajasekar, and V. T. Sreedevi. Performance Comparison of Capacitor Run Induction Motors Supplied from AC Voltage Regulator and SPWM AC Chopper, *IEEE Transactions on Industrial Electronics*, Vol. No. 53, Issue 3, 2006.

32. K. Sundareswaran and B. M. Jos. Development and Analysis of a Novel Soft-Starter/Energy Saver Topology for Delta Connected Induction Motors, *IEE Proceedings - Electric Power Applications*, Vol. No. 152, Issue 4, 2005.

33. K. Sundareswaran. Steady State Analysis and Simulation of PWM Inverter Fed Capacitor Run Induction Motor, *IETE Journal of Research*, Vol. No. 51, Issue 6, 2005.

34. K. Sundareswaran and A. Pavan Kumar. Performance Enhancement of PWM AC Chopper Using Random Search Method, *Journal of Institution of Engineers (India) Electrical Division*, Vol. No. 85, 2005.

35. K. Sundareswaran and A. Pavan Kumar. Voltage Harmonic Elimination in PWM AC Chopper Using Genetic Algorithm, *IEE Proceedings - Electric Power Applications*, Vol. No. 151, Issue 1, 2004.

36. K. Sundareswaran and P. S. Manujith. Analysis and Performance Evaluation of Triac-Voltage Controlled Capacitor Run Induction Motor, *Electric Power Components and Systems*, Vol. No. 32, Issue 9, 2004.

37. K. Sundareswaran and S. Razia Begum, Genetic Tuning of Power System Stabilizer, *European Transactions on Electrical Power*, Vol. No. 14, Issue 3, 2004.
38. K. Sundareswaran. A Simplified Model for Speed Control of AC Voltage Controller Fed Induction Motor Drives, *IETE Journal of Research*, Vol. No. 49, Issue 4, 2003.
39. K. Sundareswaran and P. S. Manujith. Analysis and Simulation of Phase Controlled Capacitor Run Induction Motors, *Journal of Institution of Engineers (India) Electrical Division*, Vol. No. 83, 2002.
40. K. Sundareswaran and M. Chandra. An Evolutionary Approach for Line Current Harmonic Reduction in AC/DC Converters, *IEEE Transactions on Industrial Electronics*, Vol. No. 49, Issue 3, 2002.
41. K. Sundareswaran and D. Laxminarayana. An Evolutionary Approach for Speed Controller Design of AC Voltage Controller Fed-Induction Motor Drive, *Electric Power Components and Systems*, Vol. No. 30, Issue 10, 2002.
42. K. Sundareswaran, V. Devi, S. Kaul, N. Rajasekar, and S. Palani. An Improved Energy Saving Scheme for Capacitor Run Induction Motor Drive, *IEEE Transactions on Industrial Electronics*, Vol. No. 48, Issue 1, 2001.
43. K. Sundareswaran and P. S. Manujith. Steady-State Analysis and Simulation of AC Chopper Fed Capacitor Run Induction Motors, *IETE Journal of Research*, Vol. No. 47, Issue 6, 2001.
44. K. Sundareswaran, J. Babu, P. Kaviarasu, and R. Ayinapudi. Dynamic Modeling and Simulation of Thyristor Converter Fed DC Motor Drive, *AMSE Journal (France) on Modeling, Measurement and Control*, Vol. No. 74, Issue 5, 2001.
45. K. Sundareswaran and S. Palani. Optimal Efficiency Control of Induction Motor Drive Using Neural Networks, *AMSE Journal (France) on Advances in Modeling and Analysis*, Vol. No. 41, Issue 1, 1999.
46. K. Sundareswaran and S. Palani. A Novel Technique for Sensor Less Speed Estimation of Variable Voltage Induction Motor Drive via Neural Networks, *AMSE Journal (France) on Advances in Modeling and Analysis*, Vol. No. 40, Issue 1, 1998.
47. K. Sundareswaran, V. Vigneshkumar, S. P. Simon, and P. S. Rao Nayak. Gravitational Search Algorithm Combined with P&O Method for MPPT in PV Systems, *Proc 2016 IEEE Annual India Conference (INDICON)*, Bangalore, India, 2016.
48. K. Sundareswaran, V. Vigneshkumar, S. P. Simon, and P. S. Rao Nayak. Cascaded Simulated Annealing/Perturb and Observe Method for MPPT in PV Systems, *Proc 2016 IEEE International Conference on Power Electronics, Drives and Energy Systems (PEDES)*, Trivandram, India, 2016.
49. K. Sundareswaran, K. Kiran, and P. S. Rao Nayak. Application of Particle Swarm Optimization for Output Voltage Regulation of Dual Input Buck-Boost Converter, *Proc Second International Conference on ICGCCEE-14*, Coimbatore, India, 2014.
50. K. Sundareswaran, K. Kiran, I. Gangadhar, P. Sankar, P. S. Nayak, and V. Vignesh. Output Voltage Control and Power Management of a Dual Input Buck–Boost Converter Employing P&O Algorithm, *Proc Third International Conference on Advances in Control and Optimization of Dynamical Systems*, Kanpur, India, 2014.
51. K. Sundareswaran, K. Kiran Kumar, H. Prasad, P. Sankar, P. S. Nayak, and V. Vignesh. Optimization of Dual Input Buck Converter Control through Genetic Algorithm, *Proc Third International Conference on Advances in Control and Optimization of Dynamical Systems*, Kanpur, India, 2014.
52. K. Sundareswaran, B. Hariharan, A. S. Daniel, P. P. Fawas, B. Subair, O. V. Asokan. Optimal Placement of Static VAr Compensator Using Particle Swarm Optimisation, *Proc 2010 International Conference on Power, Control and Embedded Systems (ICPCES)*, Allahabad, India, 2010.
53. K. Sundareswaran, P. Bharathram, M. Siddharth, G. Vaishnavi, N. A. Shrivastava, and H. Sharm. Voltage Profile Enhancement Through Optimal Placement of FACTS devices using Queen Bee Assisted GA, *Third International Conference on Power systems (ICPS)*, Kharagpur, India, 2009.
54. K. Sundareswaran and V. T. Sreedevi. Speed Regulator Design for a Thyristor-Voltage Controlled Induction Motor Drive Using Bee Colony Intelligence, *Proc 2009 TIMA Conference*, Chennai, India, 2009.

55. K. Sundareswaran, H. N. Shyam, S. Palani, and J. James. Induction Motor Parameter Identification Using Hybrid Genetic Algorithm, *Proc 2008 IEEE Region 10 Colloquium*, Kharagpur, India, 2008.

56. K. Sundareswaran, H. N. Shyam, S. Abraham, G. Varakumar, S. Kaul, and R. Sheeba. A Genetic Algorithm Based Approach Towards Induction Motor Starting with Minimum Torque Pulsations, *Proc 2008 IEEE Region 10 Colloquium*, Kharagpur, India, 2008.

57. K. Sundareswaran, K. V. S. Manoj Kumar Vadali, S. K. Nadeem, and H. N. Shyam. Robust Controller Identification for a Boost Type DC-DC Converter Using a Genetic Algorithm, *Proc 2008 IEEE Region 10 Colloquium*, Kharagpur, India, 2008.

58. K. Sundareswaran and V. T. Sreedevi. DC Motor Speed Controller Design through a Colony of Honey Bees, *Proc 2008 IEEE International Conference (TENCON)*, Hyderabad, India, 2008.

59. K. Sundareswaran and V. T. Sreedevi. Development of Novel Optimization Procedure Based on Honey Bee Foraging Behaviour, *Proc 2008 IEEE International Conference on Systems Man Cybernetics (SMC)*, Singapore, 2008.

60. K. Sundareswaran, N. Rengarajan, S. Palani, K. A. Hameed. Analysis on the Application of Soft Computing Methodologies for the Design of Power System Stabilizer, *Proc 2007 International Conference (TIMA)*, 2007.

61. K. Sundareswaran, B. M. Jos. Analysis. Simulation and Performance Comparison of AC Voltage Controlled Fed Three-Wire and Four-Wire Connected Induction Motor Drives, *Proc 2005 INDCON*, Chennai, India, 2005.

62. K. Sundareswaran, B. M. Jos. Comprehensive Study on Starting Performance of Thyristor Controlled Induction Motor Drives, *2005 Proc International Conference on Computer Applications in Electrical Engineering*, Roorkee, India, 2005.

63. K. Sundareswaran, N. Rajasekar. Harmonic Elimination in PWM AC Chopper Using a Hybrid GA, *2005 Proceedings International Conference on Computer Applications in Electrical Engineering*, Roorkee, India, 2005.

64. K. Sundareswaran, R. Madhan, C. V. N. Harish, and K. Jayant. Development of an Immigrant Genetic Algorithm for Enhanced Computational Efficiency and Its Application to Power Quality Improvement, *Proc 2005 IEE International Conference on Energy, Information Technology and Power Sector (PEITSICON)*, Kolkata, India, 2005.

65. K. Sundareswaran, C. Palaniappan, M. D. Anand, and R. Lingesham. Performance Evaluation of Genetic Algorithm & Conventional Optimization Techniques to the Design of Feedback Controller for Thyristorised Induction Motor Drives, *Proc 2004 International Conference on Intelligent Signal Processing and Robotics Indian Institute of Information Technology*, Allahabad, India, 2004.

66. K. Sundareswaran and S. Palani. Design of High Gain Controller for Part-Load Performance Optimization of AC Voltage Controller Fed Induction Motor Drive, *Proc 2001 International Conference on Energy, Automation and Information Technology*, Kharagpur, India, 2001.

67. K. Sundareswaran, High Performance AC Voltage Controller Fed Induction Motor Drive Using Fuzzy Logic Estimator, *Proc 2000 International Power Electronics Conference*, Tokyo, Japan, 2000.

68. K. Sundareswaran and M. Vasu. Genetic Tuning of PI Controller for Speed Control of DC Motor Drive, *Proc 2000 IEEE International Conference on Industrial Technology (ICIT)*, Goa, India, 2008.

69. K. Sundareswaran and S. Palani. Performance Enhancement of AC Voltage Controller Fed Induction Motor Drive Using Neural Networks, *Proc 2000 IEEE International Conference on Industrial Technology (ICIT)*, Goa, India, 2000.

70. K. Sundareswaran and S. Palani. Design of High Gain Controller for Part Load Performance Optimization of Variable Voltage Induction Motor Drive, *Proc 1999 IEEE International Conference on Power Electronics and Drive Systems*, Hong Kong, China, 1999.

71. K. Sundareswaran and S. Palani. Fuzzy Logic Approach for Energy Efficient Voltage Controlled Induction Motor Drive, *Proc 1999 IEEE International Conference on Power Electronics and Drive Systems*, Hong Kong, China, 1999.

72. K. Sundareswaran and S. Palani. Artificial Neural Network Based Voltage Controller for Energy Efficient Induction Motor Drive, *Proc 1998 IEEE Region 10 International Conference*, New Delhi, India, 1998.

73. K. Sundareswaran and S. Palani. Speed Identification of Voltage Controlled Induction Motor Drive Using Neural Networks, *Proc 1997 International Conference on Computer Applications in Electrical Engineering*, Roorkee, India, 1997.

74. K. Sundareswaran. Development of an Intelligent Optimization Algorithm Through a colony of Foraging Ants and its Application to DC Motor Speed Controller Design, *Proc 2008 National Conference on Electrical Engineering and Embedded Systems*, Chennai, India, 2008.

75. K. Sundareswaran, K. Jayant, C. V. N. Harish, and R. Madhan. Optimum Feedback Controller Design for a Non-Linear Plant Using the Ant Colony Metaphor, *Proc 2005 USC-SAP (DRS) National Conference on Computing and Mathematical Modeling*, Gandhigram, India, 2005.

76. K. Sundareswaran, S. Hemamalini, T. N. Shanavas, and R. Madhan. Parameter Estimation of Unknown Plants Using Particle Swarm Optimization, *Proc 2005 USC-SAP (DRS) National Conference on Computing and Mathematical Modeling*, Gandhigram, India, 2005.

77. K. Sundareswaran and N. Rajasekar. A Comprehensive Study of Capacitor-Run Induction Motor Speed Control with Integral Switching, *Proc 2004 National Power Systems Conference (NPSC)*, Chennai, India, 2004.

78. K. Sundareswaran, B. M. Jos, and V. T. Sreedevi. Dynamic Simulation of AC Voltage Controller Fed-Induction Motor Drive using Simulink, *Proc 2004 AICTE-Sponsored National Conference on Control, Communication and Information Systems*, Goa, India, 2004.

79. K. Sundareswaran. A Quantitative Study of Performance Characteristics of Variable Voltage Operation of Domestic Fans, *Proc 2003 National Conference on Instrumentation & Control*, Tiruchirappalli, India, 2003.

80. K. Sundareswaran and B. M. Jos. Transient Response Analysis of Delta Connected Induction Motor Starting Using Star-Delta Switch, *Proc 2004 National Conference on Power Conversion and Industrial Control (PCIC)*, San Francisco, California, 2004.

81. K. Sundareswaran, M. D. Anand, and C. Palaniappan. Modeling and Simulation of DC-Fed Braking of Induction Motor Drive, *Proc 2003 National Conference on Power Conversion and Industrial Control (PCIC)*, Houston, Texas, 2003.

82. K. Sundareswaran, M. D. Anand, and A. P. Ann. FPGA Based IC Design for Performance Enhancement of Delta Connected Induction Motor Drives, *Proc 2003 National Conference on Power Conversion and Industrial Control (PCIC)*, Houston, Texas, 2003.

83. K. Sundareswaran. Transient Simulation of Reversible DC Drive Using Power System Block Set, *Proc 2001 National Systems Conference*, 2001.

84. K. Sundareswaran, J. S. Babu, S. H. Kumar, and R. Ayinapudi. Application of Genetic Algorithm for Speed Control of DC Motor Drive, *Proc 2000 National Symposium on Intelligent Measurement and Control*, Chennai, India, 2000.

85. K. Sundareswaran and S. Palani. Energy Efficient Induction Motor Drive Using Neural Networks, *Proc 1999 National Renewable Energy Convention*, Indore, India, 1999.

86. K. Sundareswaran and S. Palani. Minimum-Time Maximum-Efficiency Control of Voltage Controlled Induction Motor Drive, *Proc 1998 National Systems Conference*, Calicut, India, 1998.

87. K. Sundareswaran and K. Madhukar. Digital Simulation and Experimental Results of Energy Efficient Voltage Controlled Single-Phase Induction Motor, *Proc 1997 National Conference on Electric Drives and Control for Transportation Systems*, Vidisha, India, 1997.

Index